Advances in Marine Gas Hydrate Exploration and Discovery—2nd Edition

Advances in Marine Gas Hydrate Exploration and Discovery—2nd Edition

Guest Editors

Wei Zhang
Pibo Su
Jiliang Wang
Qianyong Liang

MDPI

Basel • Beijing • Wuhan • Barcelona • Belgrade • Novi Sad • Cluj • Manchester

Guest Editors

Wei Zhang
Sanya Institute of South China Sea Geology
Guangzhou Marine Geological Survey
Sanya
China

Pibo Su
Sanya Institute of South China Sea Geology
Guangzhou Marine Geological Survey
Sanya
China

Jiliang Wang
Institute of Deep-sea Science and Engineering
Chinese Academy of Sciences
Sanya
China

Qianyong Liang
National Engineering Research Center of Gas Hydrate Exploration and Development
Guangzhou Marine Geological Survey
Guangzhou
China

Editorial Office

MDPI AG
Grosspeteranlage 5
4052 Basel, Switzerland

This is a reprint of the Special Issue, published open access by the journal *Journal of Marine Science and Engineering* (ISSN 2077-1312), freely accessible at: https://www.mdpi.com/journal/jmse/special_issues/VJEO7661R9.

For citation purposes, cite each article independently as indicated on the article page online and as indicated below:

Lastname, A.A.; Lastname, B.B. Article Title. *Journal Name* **Year**, *Volume Number*, Page Range.

ISBN 978-3-7258-8141-3 (Hbk)
ISBN 978-3-7258-8142-0 (PDF)
https://doi.org/10.3390/books978-3-7258-8142-0

Cover image courtesy of Zhang Wei

Contents

Journal of
Marine Science and Engineering

MDPI

Article

Numerical Study of the Gas Production Enhancement Effect of Boundary Sealing and Wellbore Heating for Class 1 Hydrate Reservoir Depressurization with Five-Spot Wells

ingli Wang [1,2], Zhibin Sha [1,2,*], Zhanzhao Li [1,2], Jianwen Wu [1,2] and Tinghui Wan [1,2,*]

1 Guangzhou Marine Geology Survey, China Geological Survey, Ministry of Natural Resources, Guangzhou 511458, China; 13650780173@163.com (Z.L.); m15889942847@163.com (J.W.)
2 National Engineering Research Center for Gas Hydrate Exploration and Development, Guangzhou 511458, China
* Correspondence: shazb2008@163.com (Z.S.); atomion@126.com (T.W.)

Abstract

Natural gas hydrates (NGHs) are a promising alternative energy source with huge global reserves, but they face significant challenges in commercial production and require more efficient exploitation methods. Based on field data from China's first offshore NGH pilot production, this study systematically investigates the enhancement effect of boundary sealing and wellbore heating on the development of Class 1 hydrate reservoirs with five-spot wells. Numerical simulation findings illustrate that when the sealing layer thickness is 1 m and the permeability is 0.001 mD, it can effectively expand the radial propagation of pressure, promote the gas output, and significantly reduce water production. When the heating power is 100 W/m, the highest energy efficiency ratio can be achieved, which can promote dissociation and inhibit the secondary hydrate generation. The combination of two technologies shows a synergistic effect, which increases the cumulative gas production and gas-to-water ratio to 197.4% and 224.3% of the base case, respectively, achieving the optimal balance between high recovery rate and economic efficiency, which provides key insights for the effective development of Class 1 hydrate reservoirs.

Keywords: natural gas hydrate; numerical simulation; boundary sealing; wellbore heating; five-spot wells

1. Introduction

Natural gas hydrates (NGHs) are a compact crystalline substance composed of water and gas molecules under low-temperature, high-pressure environments. It is found extensively in the deep-sea, deep lake sediments, and permafrost areas on land [1,2]. It is estimated that there are approximately 10^{15}–10^{18} m^3 of CH_4 in the global NGH resources. As a highly promising alternative energy source, the strategic value of NGHs in the energy sector is becoming increasingly prominent [1–4]. The recent offshore NGH test conducted by China and Japan has confirmed the recoverability of this emerging resource, but the trial production results are far from commercial production capacity [5–8]. How to significantly increase gas production is currently a research hotspot. It is generally recognized that the depressurization method and its derivative improvement schemes are the optimal path to promote the industrial exploitation of marine hydrates. The production enhancement technologies mainly include complex structured wells (horizontal wells, multilateral wells, well-net, etc.), new mining methods like depressurization assisted with thermal stimulation, and a range of reservoir stimulation methods, including hydraulic fracturing, near-wellbore stimulation, etc. The mechanisms of these technologies include improving the drainage area, enhancing the efficiency of hydrate dissociation, and optimizing reservoir flow conditions [9,10].

Boundary sealing, as one of the important directions for reservoir stimulation, has received increasing attention in recent years. The primary principle is to manually reduce the permeability of the sealing layers, thereby preventing water invasion and improving pressure propagation. Zhao et al. (2020) [11] proposed forming artificial impermeable barriers via gel injection into sealing layers to improve the recovery rate of the Shenhu NGH reservoir. Numerical simulations showed that the barriers prevented water invasion and enhanced the depressurization effect; the hydrate dissociation degree increased by 36.5%, gas production increased 137.2%, and the water production decreased by 74.5%. Li et al. (2021) [12] proposed a novel approach integrating hydraulic fracturing with sealing layers to boost gas production. Numerical simulation results demonstrate that when sealing lengths exceed the fracturing radial length, this approach boosts gas production by 93.2% and reduces water production by 62.9%. Lv et al. (2022) [13] analyzed the adaptability of artificial barriers to depressurization production in vertical wells and found that artificial barriers can reduce cumulative water production by 20.88% and improve the gas-to-water ratio. Wang et al. (2024) [14] studied the effect of boundary sealing on the development of NGHs through horizontal wells combined with thermal injection. Numerical simulations revealed that the boundary sealing inhibited water invasion, resulting in an increase of 22.0~30.1% in hydrate dissociation and 63.9~85.1% in gas production efficiency, respectively. Nie et al. (2024) [15] put forward an innovative approach integrating boundary sealing and hot water injection. Numerical results showed their combination achieved a synergistic effect; Compared with three-spot horizontal well patterns, the five-spot horizontal well pattern reduced the production cycle by 680 days and improved the gas-to-water ratio by 17%. Guo et al. (2024) [16] evaluated the yield-increasing effect of network fracturing under boundary sealing and demonstrated that boundary sealing can prevent water invasion and improve pressure propagation, resulting in an increase of 11.1 times in gas production and 13.3 times in gas-to-water ratio. Qin et al. clarified the effect of sealing layer properties and proposed optimal vertical well sealing layer parameters, including a sealing radius of 150 to 200 m, a sealing thickness of 0.5 m, and a permeability ratio of 5000 [17]. The above work proposes a combination mode of boundary sealing and other technologies, optimizes the key parameters of sealing, and promotes in-depth research on boundary sealing technology.

The above work proposes a combination mode of boundary sealing and other technologies, optimizes key parameters of sealing layers, and promotes in-depth research on boundary sealing technology. However, most of the well types involved are single vertical or single horizontal wells. There is a lack of research on the combination of well-net development modes, such as five-spot wells with boundary sealing and wellbore heating, their systematic integration and the quantitative analysis of their synergistic effect, which have great potential for large-scale exploitation of NGHs [18]. Therefore, this work established a 3D numerical model of five-spot wells, and the synergistic effect of boundary sealing and wellbore heating in the development of Class 1 hydrate reservoirs is systematically evaluated. It includes the influence of boundary sealing permeability and thickness, as well as heating power on pressure propagation. The results can provide new insights for the high-efficiency exploitation of Class 1 hydrate reservoirs.

https://doi.org/10.3390/jmse14020134

2. Methodology

2.1. Geological Background

The SHSC4 well is located in the Shenhu area (Figure 1) [7]. The reservoir in this sea area is characterized by abundant gas supply and suitable geological and thermodynamic conditions, with a water depth of approximately 1266 m, seabed temperature around 3.0 °C, and a geothermal gradient of 43.653 °C/km [19]. The NGH reservoir system in this sea area is classified as a Class 1 NGH reservoir [20]. It can be categorized into three distinct layers: the upper Gas Hydrate-Bearing Layer (GHBL), rich in hydrate and water, occurs at 201–236 m below seafloor (mbsf) with a thickness of 35 m; the middle Three-Phase Layer (TPL), containing hydrate, high-saturation free gas, and water, is found at 236–251 mbsf and is 15 m thick; and the lower Free Gas Layer (FGL), composed of low-saturation free gas and water, extends from 251 to 278 mbsf with a thickness of 27 m [7].

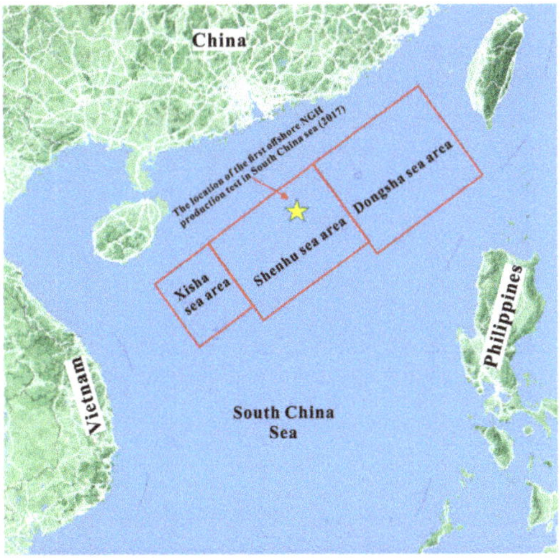

Figure 1. SHSC4 well site (adapted from Hao et al., 2018. Copyright 2022 American Chemical Society) [21].

2.2. Simulation Code

TOUGH + HYDRATE V1.0 combines both kinetic and equilibrium modeling approaches to characterize NGH formation and dissociation, supporting various dissociation strategies such as depressurization, thermal stimulation, salting-out effects, and inhibitor injection [22]. The simulation system incorporates four phases (gas, water, ice, and solid hydrate) and multiple components (water, methane, hydrate, and soluble inhibitors), with comprehensive phase behavior accounting for inter-phase mass transfer [22]. This code has been extensively validated against laboratory experiments and field tests [23,24]. The original version supports a maximum of 50,000 grids, limiting its applicability to large-scale field simulations. In order to improve computational efficiency, the parallel version pT + H V1.0 was used for production prediction [25]. The equilibrium model was adopted for simulating gas production under reservoir conditions [26]. The following assumptions were followed during the simulation process: Fluid migration in porous media is governed by Darcy's law; Mechanical diffusion of dissolved gases and inhibitors is neglected; Geomechanical responses and stress coupling effects are not incorporated; The aqueous phase is assumed present in the presence of salt; Thermophysical properties of the aqueous phase are unaffected by dissolved inhibitors; Inhibitors are treated as non-volatile components. The governing equations are as follows [22]:

https://doi.org/10.3390/jmse14020134

1. Mass Conservation Equation:

$$\frac{d}{dt}\int V_n M^{\kappa} dV = \int \Gamma_n F^{\kappa} \cdot nd\Gamma + \int V_n q^{\kappa} dV \quad (1)$$

where M^{κ} is mass accumulation, F^{κ} denotes mass flux, and q^{κ} represents source/sink terms.

2. Energy Conservation Equation:

$$\frac{d}{dt}\int V_n M^{\theta} dV = \int \Gamma_n F^{\theta} \cdot nd\Gamma + \int V_n q^{\theta} dV \quad (2)$$

where θ is heat component, is heat flux, M^{θ}, F^{θ}, and q^{θ} correspond to heat accumulation, flux, and source/sink ratio, respectively.

2.3. Model and Case Design

The schematic diagram of the model is shown in Figure 2a. The geological model spans 800 m in both the X and Y directions (Figure 2b), with a vertical extent (Z) of 137 m (Figure 2c). The domain was discretized using a structured grid system comprising 47 layers in the X-axis, 47 layers in the Y-axis, and 95 layers in the Z-axis, with a total of 209,855 grids. A local grid refinement strategy was employed, particularly in the vicinity near the wellbore, to enhance resolution in regions experiencing significant changes in physical properties during gas production. The refinement follows with varying grid sizes, including 1 m, 2 m, 4 m, 8 m, and 12 m, to accurately capture multiphase flow and heat transfer dynamics.

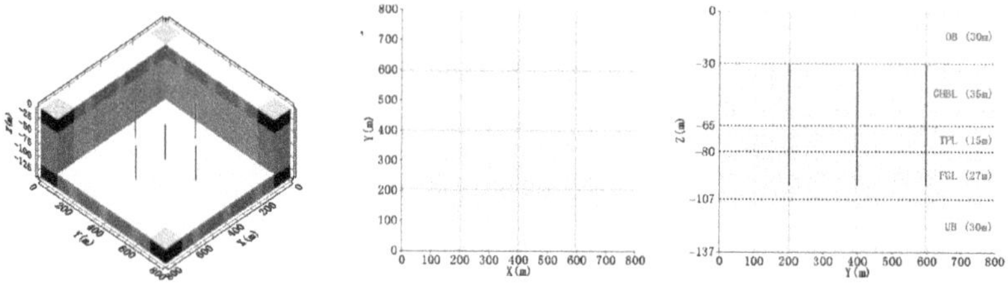

Figure 2. Schematic diagram of the model: (**a**) five-spot wells design. (**b**) X-Y plane mesh discretization. (**c**) Y-Z plane mesh discretization.

Eight simulation cases were established, as summarized in Table 1. Case 1 serves as the reference case without any sealing layers. Sealing layers are located at the top of the GHBL and the bottom of the FGL. Cases 2 to 4 are designed to investigate the effect of sealing permeability, which ranges from 0.0001 to 0.01 mD, with a thickness of 1 m. Case 5 was designed to investigate the influence of sealing thickness (permeability of 0.001 mD and thickness of 2 m). Cases 6 to 8 introduce thermal stimulation by applying wellbore heating at power levels of 100 W/m, 200 W/m, and 300 W/m, respectively, with a fixed sealing configuration (permeability of 0.001 mD and thickness of 1 m).

Table 1. Simulation cases.

Cases	Boundary Sealing	Sealing Permeability (mD)	Sealing Thickness (m)	Heating Power (W/m)	Pressure Difference (MPa)
Case 1	No	-	-	-	4.0
Case 2	Yes	0.0001	1	-	4.0
Case 3	Yes	0.001	1	-	4.0
Case 4	Yes	0.01	1	-	4.0
Case 5	Yes	0.001	2	-	4.0
Case 6	Yes	0.001	1	100	4.0
Case 7	Yes	0.001	1	200	4.0
Case 8	Yes	0.001	1	300	4.0

https://doi.org/10.3390/jmse14020134

2.4. Initial and Boundary Conditions

The model initialization begins with defining the pore water pressure. Considering the connectivity of the overlying layer, this pressure is given by the hydrostatic pressure formula [27–29]:

$$P_{pw} = P_{atm} + \rho_{sw} g(H + Z) \times 10^{-6} \tag{3}$$

Within the formula, each parameter is defined as follows: P_{pw}: Porewater pressure (MPa), P_{atm}: Normal atmospheric pressure (MPa), ρ_{sw}: Average density of seawater (kg/m^3), g: Gravitational acceleration (m/s^2), H: Water depth (m), Z: Sediment depth from the seabed (m). Subsequently, the simulator's self-balancing function was used to calculate a stable initial temperature and pressure profile consistent with the geothermal gradient (Figure 3) [30]. The model boundaries were set as the first type boundary [31]. The production wellbore was conceptualized as an internal boundary, and the well grids maintain a fixed pressure difference of 4.0 MPa.

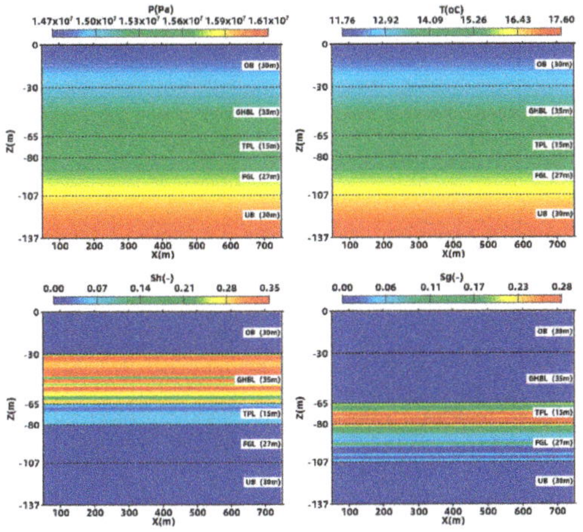

Figure 3. Model's initial conditions.

Key model parameters for multiphase flow were set based on previous research. The parameters for the *van Genuchten* model were set as: the maximum reference water saturation (S_{mxA}) was assigned a value of 1.0, the irreducible water saturation (S_{irA}) was fixed at 0.30, the pore size distribution index (λ) was designated as 0.45, and the initial capillary pressure (P_0) was set to 1×10^4 Pa. The parameters for the *Stone* model were set as: the irreducible water saturation (S_{irA}) remained consistent at 0.30, the irreducible gas saturation (S_{irG}) was defined as 0.03, the water-phase permeability reduction exponent (n_A) was assigned a value of 3.5, and the gas-phase permeability reduction exponent (n_G) was set to 2.5 [32–34].

Table 2 specifies the physical properties of the model. The formation thicknesses are configured as follows: overburden (OB) and underburden (UB) are 30 m each, GHBL is 35 m, TPL is 15 m, and FGL is 27 m. The permeability and porosity of each layer are directly adopted from field data: GHBL has 2.9 mD and a porosity of 0.35, TPL has 1.5 mD and a porosity of 0.33, and FGL has 7.4 mD and a porosity of 0.32. The OB and UB layers follow conventional assumptions with a permeability of 2.0 mD and porosity of 0.30 [7]. Sealing layers are located at the top of the GHBL and the bottom of the FGL. Other properties include a grain density of 2600 kg/m^3 and a geothermal gradient of 43.653 °C/km, and the gas composition is 100% methane.

https://doi.org/10.3390/jmse14020134

Table 2. Basic physical properties for the model.

Parameter Type	Parameters	Value and Unit
Formation thickness	OB/UB GHBL TPL FGL	30 m 35 m 15 m 27 m
Initial permeability	OB/UB GHBL TPL FGL	2.0 mD 2.9 mD 1.5 mD 7.4 mD
Porosity	OB/UB GHBL TPL FGL	0.30 0.35 0.33 0.32
Initial saturation	Hydrate saturation of GHBL and TPL Free gas saturation of FGL	cited from logging data cited from logging data
Multiphase flow	Capillary pressure model [32–34]	$P_{cap} = -P_0\left[(S^*)^{-1/\lambda} - 1\right]^{1-\lambda}$, $S^* = \frac{(S_A - S_{irA})}{(S_{mxA} - S_{irA})}$
	Relative permeability model [32–34]	$K_{rA} = [(S_A - S_{irA})/(1 - S_{irA})]^{nA}$, $K_{rG} = [(S_G - S_{irG})/(1 - S_{irA})]^{nG}$
	S_{mxA} (Maximum aqueous saturation)	1
	λ (Capillary pressure exponent)	0.45
	P_0 (Capillary pressure reference value)	10^4 Pa
	n_A (Aqueous relative permeability exponent)	3.5
	n_G (Gas relative permeability exponent)	2.5
	S_{irG} (Gas irreducible saturation)	0.03
	S_{irA} (Aqueous irreducible saturation)	0.30
Production well	Production pressure Wellbore radius	4.0 MPa 0.1 m
Others	Geothermal gradient Grain specific heat Dry thermal conductivity Wet thermal conductivity Grain density Salinity Gas composition Location of sealed burdens	43.653 °C/km 1000 $J \cdot kg^{-1} \cdot K^{-1}$ 1.0 $W \cdot m^{-1} \cdot K^{-1}$ 3.1 $W \cdot m^{-1} \cdot K^{-1}$ 2600 kg/m^3 3.5% 100% CH_4 Top of GHBL, Bottom of FGL

2.5. Model Validation

To validate the numerical model, a 70 m vertical well was implemented at the model's center, replicating the design of the production well from China's first NGH trial production. The well was perforated from −201 to −271 mbsf and operated under a constant pressure differential of 3 MPa [35]. As shown in Figure 4, the close agreement between the simulated 60-day cumulative gas production and the field data confirms the model's accuracy for subsequent work.

https://doi.org/10.3390/jmse14020134

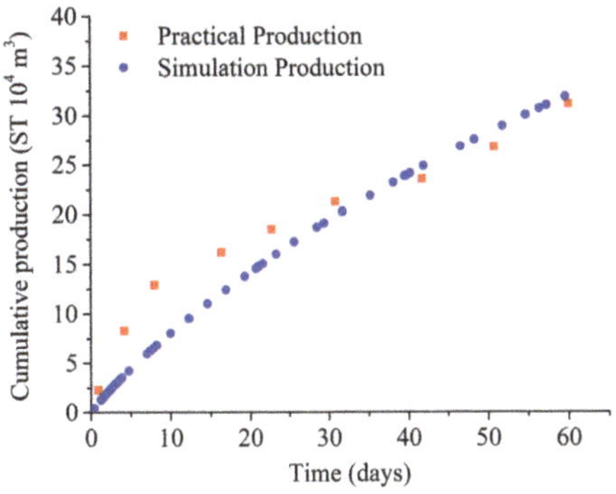

Figure 4. Gas production history fitting.

3. Results and Analysis

3.1. Gas and Water Production

Figure 5a depicts the trends of gas production rate (Q_g) and cumulative gas production (V_g) with or without boundary sealing and wellbore heating within 360 days. Under the conditions of a boundary sealing permeability of 0.001 mD, a thickness of 1 m, and heating power set to 100 W/m, the Q_g of Case 6 is significantly higher than that of Case 1 (base case). In comparison to Case 1, the V_g of Case 6 increased from 744.3 $\times$ 10^4 to 1469.4 $\times$ 10^4 ST m^3, an increase of 197.4%. This is because when both boundary sealing and wellbore heating are used simultaneously, the boundary sealing at the top of GHBL and the bottom of FGL makes the radial propagation of pressure more uniform, thereby promoting hydrate dissociation and gas production. And wellbore heating helps to promote the dissociation of hydrates around the wellbore, while effectively preventing and eliminating secondary hydrates, ensuring smooth channels for free gas.

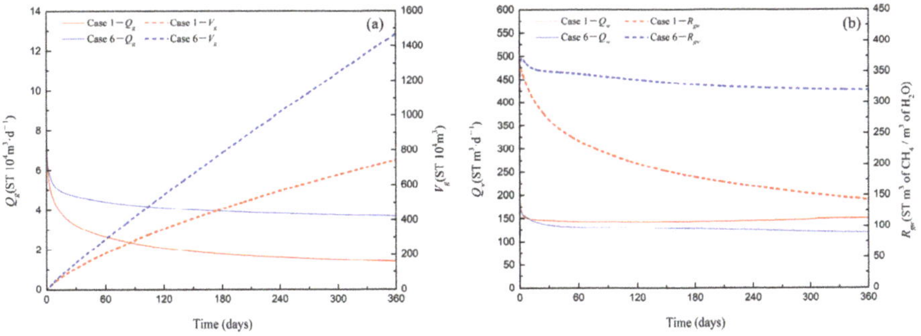

Figure 5. Production behavior with or without boundary sealing and heating over 360 days. (**a**) Evolution of gas production and cumulative gas. (**b**) Evolution of water production and gas-to-water ratio.

Figure 5b presents the variation in water production rate (Q_w) and gas-to-water ratio (R_{gw}) over 360 days. The trend of Q_w curves in the two cases is different from Q_g; the R_{gw} of the base case is only 142.6, while the case with boundary sealing and wellbore heating maintains a high R_{gw} during the whole production cycle. By the end of production, the Rgw reaches 319.9, which is 224.3% of the base case. The sealing layer serves to reduce water production by inhibiting water flow into the wellbore from both the top of the GHBL and the bottom of the FGL. Simulation results demonstrate that integrating boundary sealing and wellbore heating significantly improves gas output in five-spot wells.

https://doi.org/10.3390/jmse14020134

3.2. Spatial Distribution of Reservoir Physical Parameters

3.2.1. Pore Pressure

The spatial distribution of pore pressure within 360 days is shown in Figure 6. In Case 1, the pressure gradient of the reservoir around the wellbore located in the GHBL is greater compared to the TPL and FGL. Solid hydrates in the GHBL lower its permeability, while the gas expansion effect of free gas of the TPL and FGL limits pressure propagation. Moreover, due to the lack of boundary sealing, the rapid invasion of top and bottom water into the wellbore causes the pressure propagation range to remain almost unchanged throughout the entire production period. In case 6 with boundary sealing and wellbore heating, the presence of the sealing layer greatly reduces the invasion of water, and heating accelerates the hydrates' dissociation. The combination of the two methods maintains a more uniform distribution of pressure propagation in the reservoir. Furthermore, the area influenced by pressure around the wellbore expands over time.

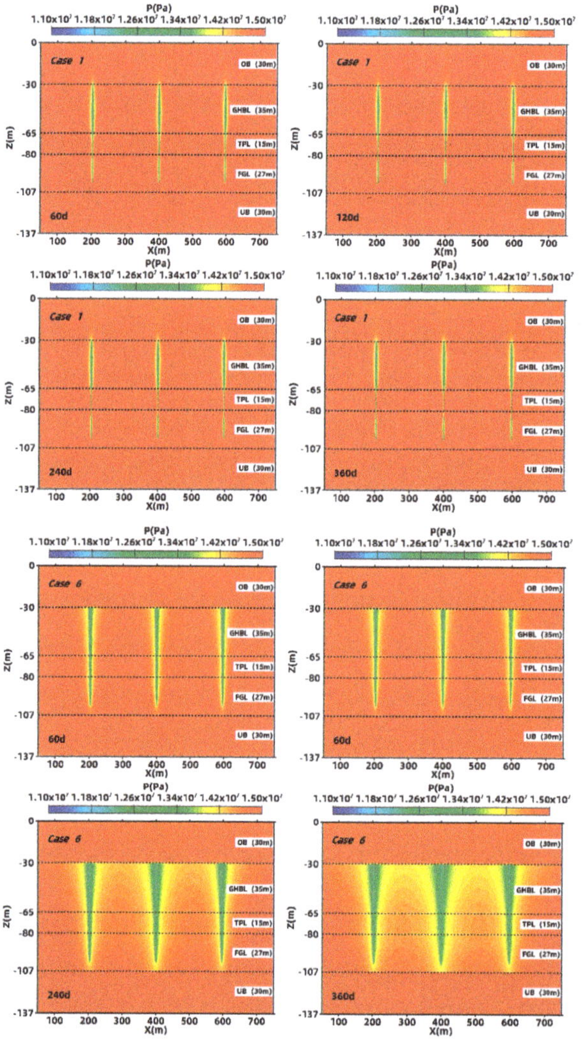

Figure 6. Pore pressure with or without boundary sealing and heating within 360 days.

https://doi.org/10.3390/jmse14020134

3.2.2. Reservoir Temperature

The spatial distribution of reservoir temperature within 360 days is shown in Figure 7. In Case 1, a decrease in reservoir temperature near the wellbore can be observed due to the dual effects of heat absorption of hydrate dissociation and the *Joule-Thomson* effect caused by a large amount of free gas entering the wellbore. As production proceeds, the high-temperature bottom water gradually moves towards the bottom of the wellbore. In Case 6, with boundary sealing and wellbore heating, wellbore heating caused an increase in reservoir temperature around the wellbore. In addition, the presence of a sealing layer effectively weakens the invasion of water.

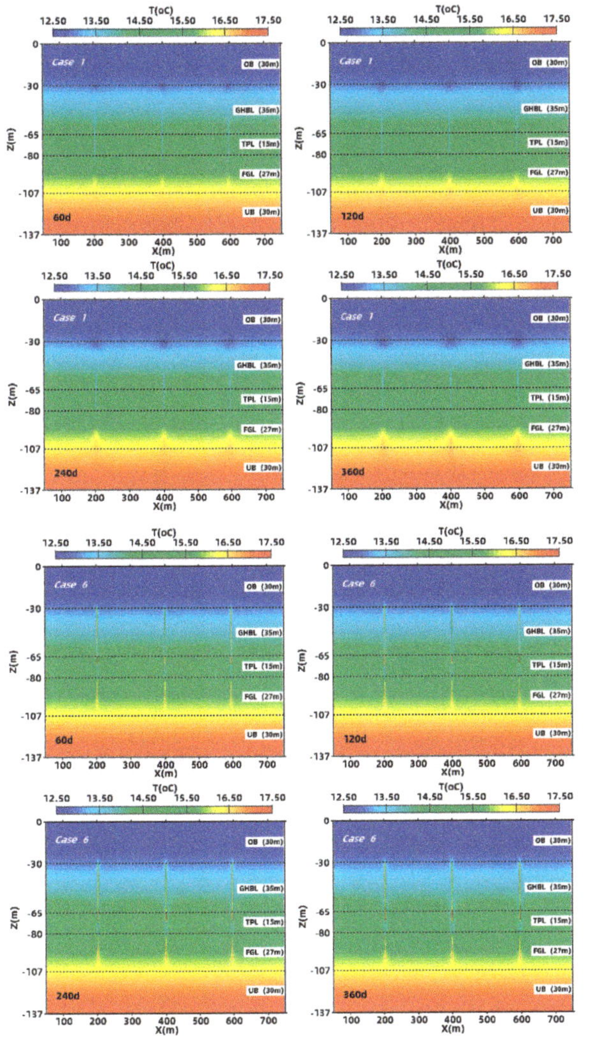

Figure 7. Reservoir temperature with or without boundary sealing and heating within 360 days.

3.2.3. Hydrate and Gas Saturation

The spatial distribution of saturation of hydrates and gas within 360 days is shown in Figures 8 and 9. In Case 1, the dissociation of hydrates is limited to the near-wellbore area, and the gas saturation around the wellbore is lower. The formation of secondary hydrates

https://doi.org/10.3390/jmse14020134

in the reservoir near the TPL's wellbore is also observed in the initial production period. As production proceeds, the quantity and saturation of secondary hydrates increase. In Case 6 with boundary sealing and wellbore heating, the hydrate dissociation front is wider, and the gas saturation is higher. Wellbore heating effectively avoids the generation of secondary hydrates. The combination of boundary sealing and wellbore heating can promote more effective pressure propagation in the radial direction of the reservoir, facilitate efficient dissociation of hydrates, and maximize recovery efficiency.

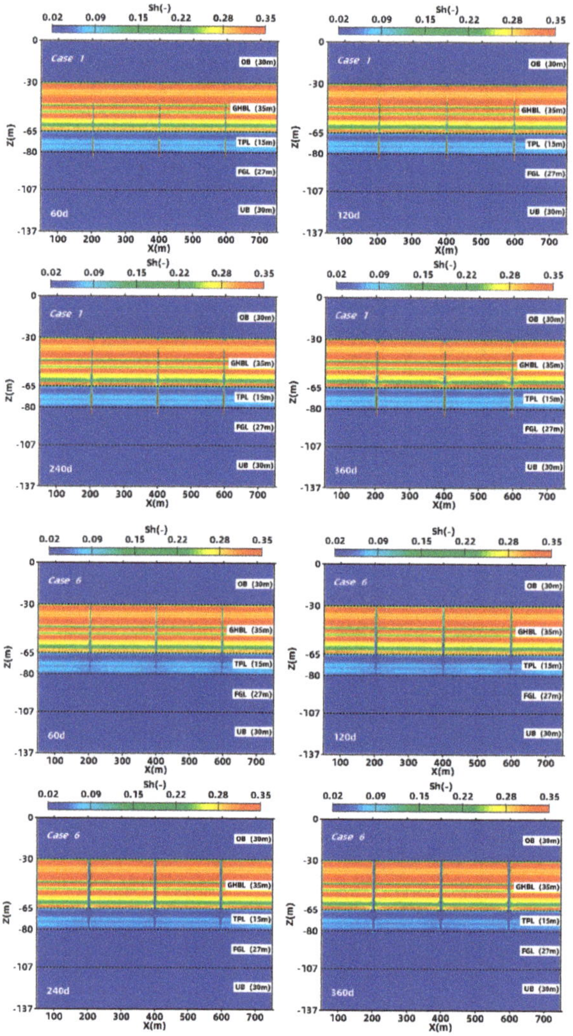

Figure 8. Hydrate saturation with or without boundary sealing and heating within 360 days.

https://doi.org/10.3390/jmse14020134

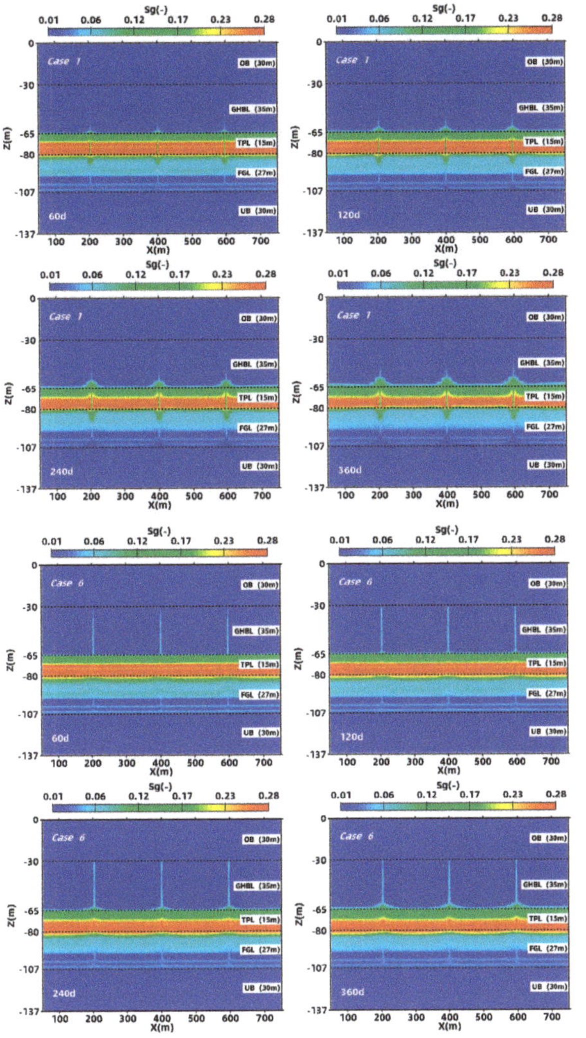

Figure 9. Gas saturation with or without boundary sealing and heating within 360 days.

4. Discussion

This work further investigated the effects of different sealing parameters, including permeability and thickness, and different heating powers on production capacity.

4.1. Effects of Boundary Sealing Parameters

Figure 10a shows the effects of different sealing parameters on gas production within 360 days. The thickness of the sealing layer in Cases 2 to 4 is 1 m, and the permeability is set to 0.0001, 0.001, and 0.01 mD, respectively. The thickness of the sealing layer in Case 5 is 2 m, and the permeability is set to 0.001 mD. The simulation results show that the V_g in cases 2 to 5 is 866.1, 863.9, 846.9, and 864.9 $\times$ 10^4 ST m^3, respectively, which is 116.3%, 116.1%, 113.7%, and 116.2% higher than Case 1. Figure 10b shows the effects of different sealing parameters on water production. Simulation results show that the R_{gw} of Cases 2 to 5 are 241.1, 237.9, 216.2, and 239.9, respectively, which are 169.1%, 166.8%, 151.6%, and 168.2% higher than Case 1. From the simulation results of Case 2 to Case 4, it can

https://doi.org/10.3390/jmse14020134

be concluded that as the permeability of the sealing layer increases, the V_g and R_{gw} both decrease. In terms of the above three sealing layer permeability settings, Case 3 performs the best. From the simulation results of Cases 3 and 5, it can be seen that as the thickness of the sealing layer increases, the V_g and R_{gw} both increase, but the magnitude of the change is small. In terms of the above two sealing layer thickness settings, Case 3 performs the best. This work is similar to previous research results in terms of the optimal sealing thickness and significant reduction in water production [11,17]. It is worth noting that compared with the research work of Lv et al. (2022), the multi-well system has achieved greater improvement in gas-to-water ratio [13].

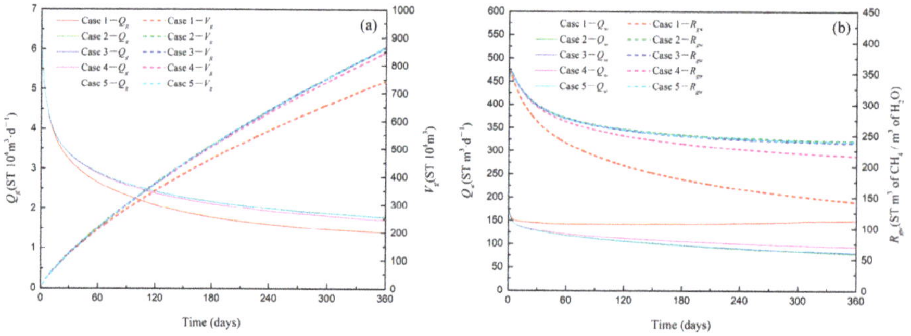

Figure 10. Production behavior with different boundary sealing parameters over 360 days. (**a**) Evolution of gas production and cumulative gas. (**b**) Evolution of water production and gas-to-water ratio.

4.2. Effects of Heating Power

Sealing layer parameter setting based on Case 3. Figure 11a presents the impacts of various heating powers on gas output within 360 days. The heating power of cases 6 to 8 is set to 100, 200, and 300 W/m, respectively. The simulation results show that the V_g in cases 6 to 8 is 1469.4, 1483.9, and 1492.2 $\times$ 10^4 ST m^3, respectively, which is 169.6%, 171.3%, and 172.2% higher than Case 2. Figure 11b shows the effects of different heating powers on water production. Simulation results show that the R_{gw} of cases 6 to 8 are 319.9, 318.8, and 317.6, respectively, which are 132.6%, 132.2%, and 131.7% higher than Case 2. From the simulation results of Case 6 to Case 8, it can be concluded that as the heating power increases, the V_g slightly increases and R_{gw} slightly decreases. In terms of the above three heating power settings, Case 6 performs the best.

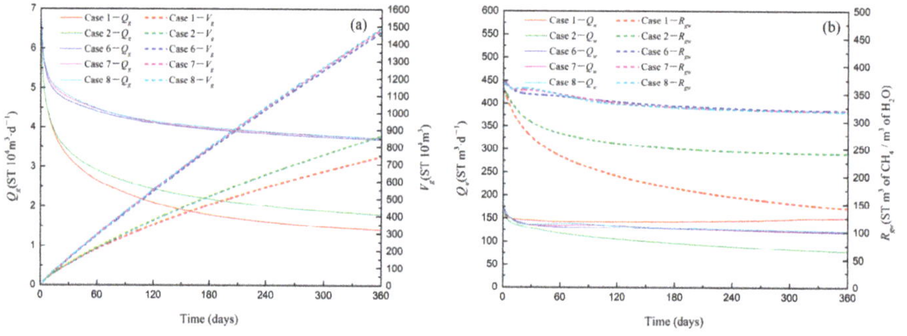

Figure 11. Production behavior with different heating powers over 360 days. (**a**) Evolution of gas production and cumulative gas. (**b**) Evolution of water production and gas-to-water ratio.

https://doi.org/10.3390/jmse14020134

In order to further explore the association between the increasing gas output and the extra heat demanded, a custom production efficiency index, $EROI_{custom}$, is introduced, which is defined as follows:

$$EROI_{custom} = E_O/E_I \tag{4}$$

E_O (Energy Output): Refers to the gas output increased throughout the production period multiplied by the natural gas calorific value (approximately 35.8 MJ/m^3); E_I (Energy Input): Refers to the total energy consumed by wellbore heating throughout the entire production period. The $EROI_{custom}$ curves corresponding to the increase in heating power from 100 to 300 W/m are shown in Figure 12. Among them, Case 6 corresponds to the optimal energy output-input ratio. Unlike the hot water injection method used by Nie et al. (2024), this work employs wellbore heating, which has a stronger inhibitory effect on the formation of secondary hydrates and is more favorable for energy output balance [15]. The custom production efficiency index quantifies this trade-off, which has been less emphasized in previous research.

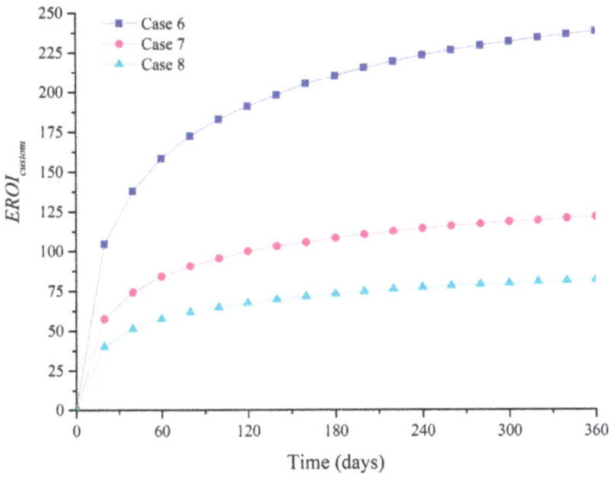

Figure 12. $EROI_{custom}$ curves with different heating power.

4.3. Implications and Future Recommendations

Although this work confirms that the combination of boundary sealing and wellbore heating has great potential for increasing production, its engineering application and conclusion interpretation need to consider multiple limitations. At the engineering level, building an effective seal layer (such as by injecting gel, polymer, cement, etc.) faces challenges such as accurate deployment of vertical and large horizontal distances, material stability under long-term low temperature/high pressure conditions, and substantial costs [36–38]. At the simulation level, the current model carries several simplifying assumptions. First, it does not consider the coupling effect of geomechanics, making it impossible to evaluate the potential deformation of the formation, permeability evolution, and their impact on wellbore stability, sand production, and sealing component integrity caused by depressurization. Second, the model operates with a fixed initial salinity and does not track the dynamic changes in produced water salinity. Consequently, it cannot account for the risk of salting out effect in the vicinity of the wellbore during production [39]. In addition, the energy efficiency indicators ($EROI_{custom}$) used are simplified screening tools and do not include the full cycle energy consumption and costs of sealing construction, fluid pumping, subsea facility construction, and operation. Therefore, the optimal parameters obtained in this work (such as sealing layer permeability of 0.001 mD and heating power of 100 W/m) are only applicable to the specific geological conditions and parameter ranges simulated. To promote the on-site application of this technology, future work needs to focus on the

https://doi.org/10.3390/jmse14020134

research and development of sealing materials and process optimization, consider numerical models coupled with geomechanics and salting out effects, and conduct comprehensive technical and economic evaluations covering the entire life cycle to support site scheme design and feasibility decision-making.

5. Conclusions

Based on China's initial NGH trial production field data, the enhancement effect of boundary sealing and wellbore heating on the gas production with five-spot wells was systematically analyzed. The research's findings can offer theoretical support for the effective development of Class 1 NGH reservoirs in different regions. The key conclusions are summarized below:

(1) The low-permeability sealing layer allows pressure to propagate more widely in the radial direction of the reservoir, effectively promoting the gas production. The presence of a sealing layer can effectively reduce the entry of top and bottom water into the wellbore, thereby significantly reducing water production and effectively improving the gas-to-water ratio. Under different reservoir and engineering conditions, there exists an optimal solution for the sealing parameters. In this work, it was found that a permeability of 0.001 mD and a thickness of 1 m are optimal.

(2) Heating the wellbore can not only accelerate hydrate dissociation but also prevent the secondary hydrate formation around the wellbore located in the TPL, ensuring a smooth channel for free gas and improving overall productivity. Under different reservoir and engineering conditions, there exists an optimal solution for wellbore heating power. In this work, a thermal power of 100 W/m is optimal, and higher heating power results in lower energy efficiency ratios.

(3) There is a synergistic effect between boundary sealing and wellbore heating. In this work, when using a boundary seal with a permeability of 0.001 mD and a thickness of 1 m, combined with a wellbore heating power of 100 W/m, the optimal balance between high gas recovery rate and economy can be achieved. Compared with the base case, the V_g and R_{gw} increase to 197.4% and 224.3%, respectively.

Author Contributions: J.W. (Jingli Wang): Conceptualization, Methodology, Software, Writing—Original Draft. Z.S.: Formal Analysis, Investigation, Funding Acquisition. Z.L.: Resources, Software. J.W. (Jianwen Wu): Data Curation, Visualization. T.W.: Writing—Review and Editing, Supervision, Project Administration. All authors have read and agreed to the published version of the manuscript.

Funding: This research was supported by the National Natural Science Foundation of China (42576254) and the Guangdong MEPP Fund (No.GDOE[2019]A39).

Data Availability Statement: The original contributions presented in this study are included in the article. Further inquiries can be directed to the corresponding authors.

Conflicts of Interest: The authors declare that they do not have any commercial or associative interests that could represent conflicts of interest in connection with the submitted work.

References

1. Sloan, E. Fundamental principles and applications of natural gas hydrates. *Nature* **2003**, *426*, 353–359. [CrossRef] [PubMed]
2. Boswell, R. Is gas hydrate energy within reach? *Science* **2009**, *325*, 957–958. [CrossRef] [PubMed]
3. Chong, Z.; Yang, S.; Babu, P.; Linga, P.; Li, X. Review of natural gas hydrates as an energy resource: Prospects and challenges. *Appl. Energy* **2016**, *162*, 1633–1652. [CrossRef]
4. Boswell, R.; Collett, T.S. Current perspectives on gas hydrate resources. *Energy Environ. Sci.* **2011**, *4*, 1206–1215. [CrossRef]
5. Yamamoto, K.; Terao, Y.; Fujii, T.; Ikawa, T.; Seki, M.; Matsuzawa, M.; Kanno, T. Operational Overview of the First Offshore Production Test of Methane Hydrates in the Eastern Nankai Trough. In Proceedings of the Offshore Technology Conference, Houston, TX, USA, 5–8 May 2014.

https://doi.org/10.3390/jmse14020134

6. Yamamoto, K.; Wang, X.; Tamaki, M.; Suzuki, K. The second offshore production of methane hydrate in the Nankai Trough and gas production behavior from a heterogeneous methane hydrate reservoir. *RSC Adv.* **2019**, *9*, 25987–26013. [CrossRef]
7. Li, J.; Ye, J.; Qin, X.; Qiu, H.; Wu, N.; Lu, H.; Xie, W.; Lu, J.; Peng, F.; Xu, Z.; et al. The first offshore natural gas hydrate production test in South China Sea. *China Geol.* **2018**, *1*, 5–16. [CrossRef]
8. Ye, J.; Qin, X.; Xie, W.; Lu, H.; Ma, B.; Qiu, H.; Liang, J.; Lu, J.; Kuang, Z.; Lu, C.; et al. The second natural gas hydrate production test in the South China Sea. *China Geol.* **2020**, *3*, 197–209. [CrossRef]
9. Wu, N.; Li, Y.; Wan, Y.; Sun, J.; Huang, L.; Mao, P. Prospect of marine natural gas hydrate stimulation theory and technology system. *Nat. Gas Ind. B* **2021**, *40*, 173–187. [CrossRef]
10. Huang, M.; Wu, L.; Ning, F.; Wang, J.; Dou, X.; Zhang, L.; Liu, T.; Jiang, G. Research progress in natural gas hydrate reservoir stimulation. *Nat. Gas Ind. B* **2023**, *10*, 114–129. [CrossRef]
11. Zhao, E.; Hou, J.; Liu, Y.; Ji, Y.; Liu, W.; Lu, N.; Bai, Y. Enhanced gas production by forming artificial impermeable barriers from unconfined hydrate deposits in Shenhu area of South China sea. *Energy* **2020**, *213*, 118826. [CrossRef]
12. Li, S.; Wu, D.; Wang, X.; Hao, Y. Enhanced gas production from marine hydrate reservoirs by hydraulic fracturing assisted with sealing burdens. *Energy* **2021**, *232*, 120889. [CrossRef]
13. Lv, T.; Pan, J.; Cai, J.; Li, R.; Shen, P. Adaptability of artificial barrier to depressurization production of marine stratified gas hydrate reservoir. *Nat. Gas Ind.* **2022**, *42*, 132–140.
14. Wang, Y.; Zeng, Y.; Zhong, X.; Pan, D.; Chen, C. Investigation into the Effect of Permeable Boundary Sealing on the Behavior of Hydrate Exploitation via Depressurization Combined with Heat Injection. *Energies* **2024**, *17*, 5172. [CrossRef]
15. Nie, S.; Liu, K.; Xu, K.; Zhong, X.; Tang, S.; Song, J.; Zhang, H.; Li, J.; Wang, Y. Numerical study on the stimulation effect of boundary sealing and hot water injection in marine challenging gas hydrate extraction. *Sci. Rep.* **2024**, *14*, 15280. [CrossRef]
16. Guo, W.; Zhong, X.; Chen, C.; Zhang, P.; Liu, Z.; Wang, Y.; Tu, G. Stimulation effect of network fracturing combined with sealing boundaries on the depressurization development of hydrate reservoir. *Energy* **2024**, *302*, 131752. [CrossRef]
17. Qin, F.; Sun, J.; Cao, X.; Mao, P.; Zhang, L.; Lei, G.; Jiang, G.; Ning, F. Numerical simulation on combined production of hydrate and free gas from silty clay reservoir in the South China Sea by depressurization: Formation sealing. *Appl. Energy* **2025**, *377*, 124343.
18. Gu, Y.; Li, S.; Song, Z.; Lu, H.; Xu, C.; Sun, J.; Wang, Y.; Li, X.; Linga, P.; Yin, Z. Analysis on a five-spot wells for enhancing energy recovery from silty natural gas hydrate deposits in the South China Sea. *Appl. Energy* **2024**, *376*, 124237. [CrossRef]
19. Zhang, W.; Liang, J.; Lu, J.; Wei, J.; Su, P.; Fang, Y.; Guo, Y.; Yang, S.; Zang, G. Accumulation features and mechanisms of high saturation natural gas hydrate in shenhu area, northern south china sea. *Pet. Explor. Dev.* **2017**, *44*, 708–719. [CrossRef]
20. Moridis, G.; Collett, T.; Pooladi-Darvish, M.; Hancock, S.; Santamarina, C.; Boswell, R.; Kneafsey, T.; Rutqvist, J.; Kowalsky, M.; Reagan, M.; et al. Challenges, uncertainties, and issues facing gas production from gas-hydrate deposits. *SPE Reserv. Eval. Eng.* **2011**, *14*, 76–112. [CrossRef]
21. Hao, Y.; Yang, F.; Wang, J.; Fan, M.; Li, S.; Yang, S.; Wang, C.; Xiao, X. Dynamic analysis of exploitation of different types of multilateral wells of a hydrate reservoir in the South China sea. *Energy Fuels* **2022**, *36*, 6083–6095. [CrossRef]
22. Moridis, G.; Kowalsky, M.; Pruess, K. *TOUGH+ Hydrate V1.0 User's Manual*; Report LBNL-0149E; Lawrence Berkeley National Laboratory: Berkeley, CA, USA, 2008.
23. Su, Z.; Moridis, G.; Zhang, K.; Wu, N. A huff-and-puff production of gas hydrate deposits in Shenhu area of South China Sea through a vertical well. *J. Pet. Sci. Eng.* **2012**, *86–87*, 54–61. [CrossRef]
24. Yin, Z.; Moridis, G.; Chong, Z.; Linga, P. Effectiveness of multi-stage cooling processes in improving the CH4-hydrate saturation uniformity in sandy laboratory samples. *Appl. Energy* **2019**, *250*, 729–747. [CrossRef]
25. Zhang, K.; Moridis, G.; Wu, Y.; Pruess, K. A domain decomposition approach for large-scale simulations of flow processes in hydrate-bearing geologic media. In Proceedings of the 6th International Conference on Gas Hydrates, Vancouver, BC, Canada, 6–10 July 2008.
26. Kowalsky, M.; Moridis, G. Comparison of kinetic and equilibrium reaction models in simulating gas hydrate behavior in porous media. *Energy Convers. Manag.* **2007**, *48*, 1850–1863. [CrossRef]
27. Yu, T.; Guan, G.; Wang, D.; Song, Y.; Abudula, A. Numerical investigation on the long-term gas production behavior at the 2017 Shenhu methane hydrate production-site. *Appl. Energy* **2021**, *285*, 116466. [CrossRef]
28. Sun, J.; Zhang, L.; Ning, F.; Lei, H.; Liu, T.; Hu, G.; Lu, H.; Lu, J.; Liu, C.; Jiang, G.; et al. Production potential and stability of hydrate-bearing sediments at the site GMGS3-W19 in the South China Sea: A preliminary feasibility study. *Mar. Pet. Geol.* **2017**, *86*, 447–473. [CrossRef]
29. Yuan, Y.; Xu, T.; Xin, X.; Xia, Y. Multiphase Flow Behavior of Layered Methane Hydrate Reservoir Induced by Gas Production. *Geofluids* **2017**, *2017*, 7851031. [CrossRef]
30. Sun, J.; Ning, F.; Li, S.; Zhang, K.; Liu, T.; Zhang, L.; Jiang, G.; Wu, N. Numerical simulation of gas production from hydrate-bearing sediments in the Shenhu area by depressurising: The effect of burden permeability. *J. Unconv. Oil Gas Resour.* **2015**, *12*, 23–33. [CrossRef]

31. Feng, Y.; Chen, L.; Suzuki, A.; Kogawa, T.; Okajima, J.; Komiya, A.; Maruyama, S. Enhancement of gas production from methane hydrate reservoirs by the combination of hydraulic fracturing and depressurization method. *Energy Convers. Manag.* **2019**, *184*, 194–204. [CrossRef]
32. Sun, Y.; Ma, X.; Guo, W.; Jia, R.; Li, B. Numerical simulation of the short- and long-term production behavior of the first offshore gas hydrate production test in the South China Sea. *J. Pet. Sci. Eng.* **2019**, *181*, 106196. [CrossRef]
33. Ma, X.; Sun, Y.; Liu, B.; Guo, W.; Jia, R.; Li, B.; Li, S. Numerical study of depressurization and hot water injection for gas hydrate production in China's first offshore test site. *J. Nat. Gas Sci. Eng.* **2020**, *83*, 103530. [CrossRef]
34. Cao, X.; Sun, J.; Qin, F.; Ning, F.; Mao, P.; Gu, Y.; Li, Y.; Zhang, H.; Yu, Y.; Wu, N. Numerical analysis on gas production performance by using a multilateral well system at the first offshore hydrate production test site in the Shenhu area. *Energy* **2023**, *270*, 126690. [CrossRef]
35. Qin, X.; Liang, Q.; Yang, L.; Qiu, H.; Xie, W.; Liang, J.; Lu, J.; Lu, C.; Lu, H.; Ma, B.; et al. The response of temperature and pressure of hydrate reservoirs in the first gas hydrate production test in South China Sea. *Appl. Energy* **2020**, *278*, 115649. [CrossRef]
36. Zhu, D.; Peng, S.; Zhao, S.; Wei, M.; Bai, B. Comprehensive review of sealant materials for leakage remediation technology in geological CO_2 capture and storage process. *Energy Fuels* **2021**, *35*, 4711–4742. [CrossRef]
37. Sun, Y.; Cao, B.; Zhong, J.; Kan, J.; Li, R.; Niu, J.; Chen, H.; Chen, G.; Wu, G.; Sun, C.; et al. Gas production from unsealed hydrate-bearing sediments after reservoir reformation in a large-scale simulator. *Fuel* **2022**, *308*, 121957. [CrossRef]
38. Sun, Z.; Li, N.; Jia, S.; Cui, J.; Yuan, Q.; Sun, C.; Chen, G. A novel method to enhance methane hydrate exploitation efficiency via forming impermeable overlying CO_2 hydrate cap. *Appl. Energy* **2019**, *240*, 842–850. [CrossRef]
39. Dumitrache, L.; Nistor, I.; Suditu, S.; Badea, A. Simulating Salt Precipitation in Dry Gas Reservoirs Using ECLIPSE Thermal CO2STORE. *Rev. Chim.* **2018**, *69*, 251–254. [CrossRef]

Journal of
Marine Science and Engineering

MDPI

Article

Influence Mechanism of Well Location and Near-Well Secondary Hydrates on Gas Production of Class 1S Hydrate Reservoirs

Xian Li [1,2,†], Chenlu Xu [1,2,†], Hongfeng Lu [1,2,*], Zihao Zhao [3], Jiawang Chen [4], Liwen Nan [4], Lu Yu [1,2], Jinwen Du [1,2], Changwen Xiao [1,2], Bo Liu [5] and Zhejun Pan [5,*]

1 Guangzhou Marine Geological Survey, China Geological Survey, Guangzhou 511458, China
2 National Engineering Research Center of Gas Hydrate Exploration and Development, Guangzhou 511458, China
3 Hildebrand Department of Petroleum and Geosciences Engineering, University of Texas at Austin, Austin, TX 78712, USA
4 Donghai Laboratory, Zhoushan 316021, China
5 State Key Laboratory of Continental Shale Oil, Northeast Petroleum University, Daqing 163318, China
* Correspondence: lhongfengcgs@126.com (H.L.); zhejun.pan@nepu.edu.cn (Z.P.)
† These authors contributed equally to this work.

Abstract

In recent years, a new type of natural gas hydrate reservoir (designated as Class 1S reservoir) has been discovered in the Qiongdongnan Basin. Within this hydrate reservoir, free gas and hydrate coexist within the same stratum. The Class 1S reservoir is comprised of three distinct zones: the gas accumulation zone, the three-phase zone, and the hydrate-bearing zone. It exhibits significant commercial development potential. This paper analyzes the formation mechanism and geological context of Class 1S hydrates. A geological model was established and numerical simulation methods were employed to evaluate its production capacity, elucidating the evolutionary patterns of hydrate saturation distribution at different well locations. The simulation results indicate that production wells should be prioritised in gas accumulation zones in order to achieve the highest cumulative gas production. Additional production wells may be considered in later stages to enhance recovery rates. Secondary hydrate formation significantly impacts production in Hydrate-bearing zone and three-phase zone. Measures such as wellbore heating can be employed to minimize secondary hydrate formation around the wellbore.

Keywords: natural gas hydrate; Qiongdongnan basin; secondary hydrate formation; numerical simulation

1. Introduction

Natural gas hydrates are defined as crystalline compounds that form through the interaction of natural gas and water, resembling ice [1,2]. It has been established that an increase in temperature or a decrease in pressure will result in the escape of methane gas, thereby causing the solid hydrate to disintegrate. Terrestrial permafrost regions and deep-sea environments are the main places where natural gas hydrates can be found [3]. The combustion of these materials produces only minimal amounts of carbon dioxide and water, resulting in significantly lower levels of pollution when compared with coal, oil and other fossil fuels [4]. In consideration of the substantial natural gas reserves, there is an emerging global consensus that natural gas hydrates may present a viable alternative energy source [5].

In the latter half of the 20th century, a significant number of hydrate drilling and test extraction projects were initiated on a global scale [6–8]. Nevertheless, the systematic understanding of hydrate deposits remains elusive, and no effective means for commercial exploitation of these deposits have been developed [9–11]. Moridis' classification of natural gas hydrate reservoirs, which was conducted between 2003 and 2007, resulted in the identification of three distinct categories [12,13]. Among these, Class I hydrate reservoirs are considered to offer the most promising commercial prospects [14]. Class I reservoirs manifest a discernible two-layer configuration, comprising an upper hydrate-bearing sedimentary layer and a lower gas–liquid two-phase fluid layer [15]. Moreover, it is noteworthy that the boundary of hydrate deposits within Class I reservoirs frequently coincides with the hydrate phase equilibrium boundary. Consequently, it can be deduced that the appropriate alteration of temperature or pressure conditions within Class I hydrate reservoirs has the capacity to induce hydrate decomposition [16–18]. This classification system, which has been widely adopted by subsequent researchers in the hydrate field, is valued for its practicality and scientific rigour. Building upon this framework, scholars have employed diverse strategies to assess gas production potential, significantly advancing research into the commercial development of hydrates [19–22].

In recent years, international academic research on natural gas hydrates has primarily focused on these three conventional types. However, a 2018 geological survey in the Qiongdongnan Sea area yielded new discoveries. The Guangzhou Marine Geological Survey (GMGS) has identified a novel hydrate reservoir. In comparison with conventional Type I hydrate reservoirs, this novel reservoir displays a distinctive spatial distribution pattern, characterised by the presence of gas accumulation zone at the centre and hydrate enrichment zone forming an annular distribution surrounding them. This hydrate deposit presents novel opportunities for the commercial development of hydrate reservoirs [23].

A hallmark of such hydrate reservoirs is the concurrent presence of highly saturated hydrates and free gas within the same stratum. The boundary of the hydrate stability zone is often observed to coincide with the boundary of the free gas phase. Therefore, a three-phase region exists between the hydrate stability zone and the free gas zone. In the context of natural gas production, this reservoir type is considered to be the most optimal. It is notable that minor variations in pressure and temperature can induce hydrate decomposition, which is a critical factor in the reservoir's functionality. This aligns with the characteristics of Class I deposits. For the purpose of subsequent reference, this particular category of hydrate reservoir is designated as a Class 1S reservoir. However, no systematic study has yet explored the development potential of Class 1S hydrate reservoir. Therefore, this study will comprehensively analyze the geological background, reservoir characteristics, and formation mechanism of Class 1S hydrate reservoir, and construct a geological model based on them to optimize the development strategy. In addition, this study will explore the impact of well deployment on gas production efficiency, and evaluate potential problems such as secondary hydrate generation during the development process to reveal its genesis mechanism. The findings of this research endeavour have the potential to provide a theoretically robust foundation and practical technological assistance for the effective development of hydrate reservoirs of this nature.

2. Geological Background

The Qiongdongnan Basin is located in the northwestern part of the South China Sea. It is naturally separated from the Yingjihai Basin to the west. To the east, it connects with the Pearl River Estuary Basin [24]. The basin is remarkable for its deep structure and sedimentary evolution, with the maximum water depth exceeding 3000 m, and the cumulative thickness of the Cenozoic sedimentary sequences exceeding 10,000 m [25]. The

Cenozoic stratigraphic sequence in the Qiongdongnan Basin is relatively well-preserved, with only the Paleocene Lingtou Formation lacking direct well exposure. The Yacheng Formation, formed during the late rift stage, constitutes the core hydrocarbon source rock layer in the basin [26]. The Lingshui Formation exhibits distinct petrological differentiation, with its base consisting of marine-terrestrial transitional deposits and its middle-upper sections dominated by marine sediments, forming the primary gas-bearing zone in the basin's deeper sections. Quaternary Ledong Formation: Predominantly clayey in lithology, interbedded with thin layers of silt and fine sand, rich in bioclastic material and unconsolidated. This formation constitutes the primary stratigraphic setting for hydrate reservoirs [27,28].

In recent years, systematic geological exploration has been conducted in the target area of the Qiongdongnan Basin, with over 20 wells (including core-recovery wells) drilled and a high-resolution three-dimensional seismic dataset acquired [29,30]. Through integrated interpretation of logging curves and seismic attributes, a deeper understanding of hydrate resources in the Qiongdongnan Basin (QDNB Block) has been gained [31,32].

The peripheral faults formed during the rapid rifting phase of the deepwater uplift and its surrounding areas in the QDNB basin provide rapid transport pathways for deep fluids. Gas plumes at the uplift's structural high points transport deep pyrothermal and biogenic gases to the shallow hydrate thermopressure stability zone, creating hydrate leakage pathways and pore-fracture seepage [33]. These pathways serve as secondary conduits for the diffusion and accumulation of shallow free gas. Simultaneously, gas chimneys with high temperature gradients (65–105 °C/km) significantly alter the regional geothermal gradient, forming annular low-temperature zones and central high-temperature areas. These temperature variations have been shown to influence the distribution of hydrates and gas, resulting in the presence of both free gas and hydrates within the same horizontal stratum. The distribution range of free gas shows a marked expansion trend with increasing depth. Thermodynamic field characteristics reveal that gas-rich zones correspond to high geothermal gradient zones, with the gradient gradually decreasing to 65 °C/km toward the hydrate stability zone, forming distinct temperature zoning. Based on well logging data and hydrate phase equilibrium data, a large-scale gas–water–hydrate three-phase coexistence zone is inferred to exist within this system.

Figure 1 shows the hydrate system in this region: the leftmost image depicts the hydrate distribution, with blue-green areas indicating hydrate-bearing zone; the middle image shows the gas distribution, where red areas represent gas accumulation zone; the rightmost image illustrates the combined distribution of hydrates and gas. This diagram reveals that the hydrates exhibit a ring-shaped distribution, with a high-saturation gas reservoir at its center. The hydrate distribution zone completely encloses the gas distribution zone, representing a significant spatial distribution difference compared to conventional hydrate reservoirs.

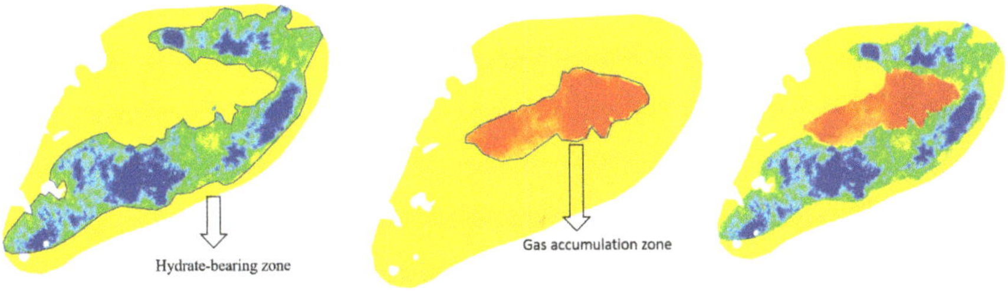

Figure 1. Hydrate and gas distribution of the gas hydrate reservoir.

This hydrate reservoir exhibits typical thin-bedded characteristics (thickness < 10 m), with lithology dominated by chalky fine-grained sediments. Notably, the upper reservoir is overlain by a suite of ultra-low-permeability deep-sea mudstone caps (permeability < 0.1 mD), which effectively block vertical gas migration from deeper layers, leading to significant accumulation of free gas.

Thus, the primary formation mechanisms for Class 1S hydrate reservoirs are: a robust deep gas supply system, a gas chimney-dominated transport system, and the hydrate storage system within stable domains coupled with the sealing effect of the caprock.

3. Model Construction

Since the original geological model has problems such as low computational efficiency and strong non-homogeneity, and cannot systematically summarize the characteristics and development effect of the new hydrate reservoir, this study extracts typical geological units to construct a simplified geologic model (Figure 2). The model consists of three areas: (1) Hydrate-bearing zone (highly saturated hydrate + water); (2) three-phase zone (hydrate + free gas + water); and (3) Gas accumulation zone. Based on the simplified geologic model, the corresponding computational numerical model will be established in the subsequent chapters, and multi-parameter analyses such as gas production will be carried out.

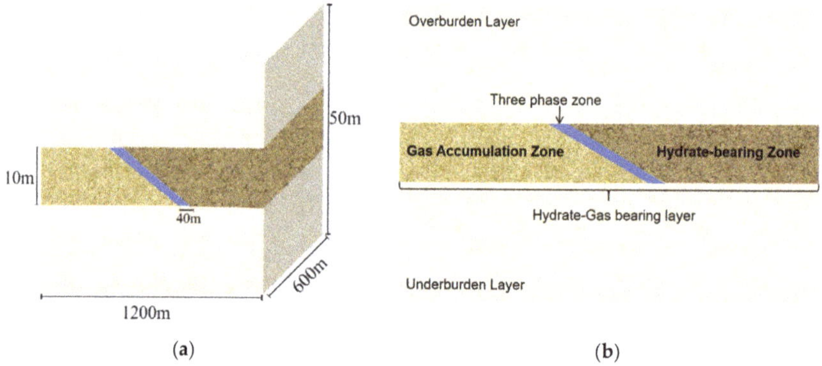

Figure 2. Three-dimensional schematic diagram and vertical cross-section diagram of the geologic model. (**a**) Three-dimensional schematic diagram of the geologic model. (**b**)vertical cross-section diagram of the geologic model.

3.1. Model Description

The simplified geological model described in this paper is shown in Figure 2. This model features a rectangular structure that extends in the X, Y and Z directions, with corresponding lengths of 1200 m, 600 m and 50 m. There are a total of 576,250 grid cells. The maximum grid spacing in the Z direction is 4 m, with a minimum spacing of 1 m. The maximum spacing in the X- and Y-directions is 5 m, with a minimum spacing of 1 m in each direction.

The distribution of temperature and pressure within the model is demonstrated in Figure 3. The model's temperature ranges from 13.7 to 19.2 °C, with the highest temperature of 19.2 °C occurring in the lower left corner. Temperatures decrease progressively toward the upper right corner. The model's pressure ranges from 18.9 MPa to 19.3 MPa, with pressure increasing with depth—a characteristic consistent with conventional hydrate reservoirs.

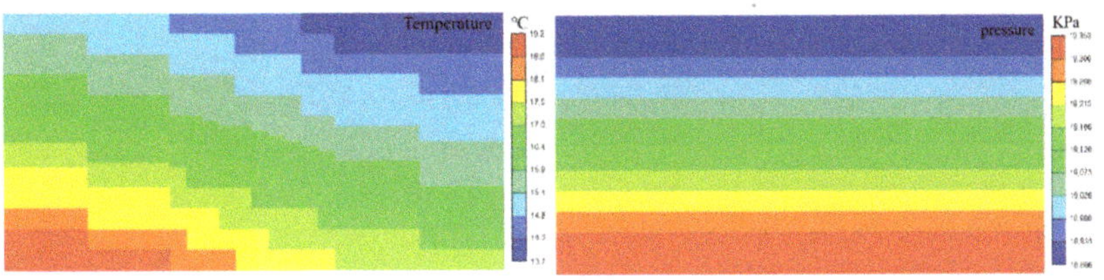

(**a**) (**b**)

Figure 3. Schematic diagram of temperature and pressure distribution in the geologic model. (**a**) Schematic diagram of temperature distribution. (**b**) Schematic diagram of pressure distribution.

Hydrates and free gas mainly accumulate in the Hydrate-Gas bearing layer (HGBL), which is 10 m thick, with a horizontal permeability of 20–50 mD (as shown in Figure 4), and a vertical permeability of 0.5 times the horizontal permeability. The porosity is between 0.4 and 0.5 (as shown in Figure 5). The left half of the model is characterised by a higher temperature and is predominantly gas. In contrast, the right half exhibits a lower temperature and is rich in highly saturated hydrates. The middle is a 40 m wide three-phase zone, thus forming a gas-three-phase zone-hydrate composite layer. In this study, the hydrate saturation level in the Hydrate-bearing Zone is set at 0.5. The gas saturation level in the gas accumulation zone is also set at 0.5. In the three-phase zone, both the hydrate and free gas saturation levels are recorded at 0.2 with specific parameters as shown in Table 1. This paper's model employs a closed boundary with no transfer of matter or heat. Relative Permeability and Capillary-Pressure Curves: refer to the first International Hydrate Simulator Comparison Project for Problem 5.

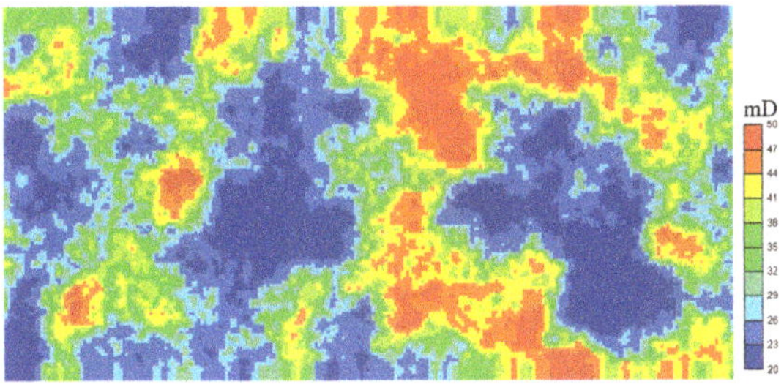

Figure 4. Schematic Diagram of Permeability Parameters in Hydrate Reservoirs.

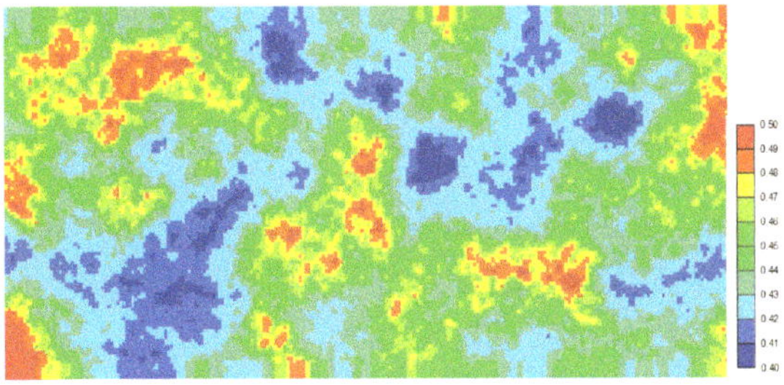

Figure 5. Schematic Diagram of Porosity Parameters in Hydrate Reservoirs.

Table 1. Physical parameter table of the model.

Parameters	Value
Rock thermal conductivity (W/m/K)	3.1
Rock density (kg/m^3)	2600
Perforation section/m	10
Overburden Layer Thickness/m	20
Underburden Layer Thickness/m	20
Gas composition	100%CH_4
HGBL Porosity	0.4–0.5
HGBL Permeability/mD	20–50
Well bottom hole pressure/MPa	10
Initial pressure-1860 m/MPa	18.9
HGBL Thickness/m	10
Initial water saturation (Hydrate-bearing Zone)	0.5
Initial hydrate saturation (Hydrate-bearing Zone)	0.5

3.2. Production Modes

The current gas hydrate extraction technology system mainly includes multiple technology paths such as (1) reservoir depressurization method, (2) thermodynamic excitation method, (3) chemical inhibitor injection method, and (4) CO_2-CH_4 replacement method [34–38]. Since the boundary of the hydrate-bearing zone in Class 1S nearly coincides with that of the free gas enrichment zone, appropriately reducing pressure can disrupt the phase equilibrium of hydrates [39–41]. Therefore, this paper employs the pressure reduction method as the base mining method [42,43].

In terms of well optimization, although horizontal wells are theoretically better able to adapt to the strong horizontal inhomogeneity of the reservoir, the vertical well development mode was finally chosen due to engineering constraints such as the thin thickness of the target layer (<10 m) and the high trajectory control accuracy required ($\pm$0.5 m) [44,45].

In order to systematically study the influence of vertical well spatial location on the exploitation dynamics, this study established five case scenarios (Case 1, Case 2, Case 3, Case 4, Case 5). Case 1 and Case 2 both have perforation sections entirely within the gas accumulation zone. Case 3 features the most complex perforation section, traversing three distinct zones: gas accumulation zone, hydrate-bearing zone and three-phase zone. The lengths of the perforation sections within these three zones are 5 m, 1 m, and 4 m, respectively. Case 4 and Case 5 share a total perforation section length of 10 m, with both sections entirely located within the hydrate-bearing zone. All five scenarios employ single-well extraction, differing only in well location (specific locations shown in Figure 6). The design of the simulation scheme follows (1) uniform wellbore flow pressure control (10 MPa); (2) constant production cycle (3650 days) (3) injection of the hole stratum through the entire Hydrate-Gas bearing layer, and the perforation section length is uniformly 10 m. In particular, the completion locations are selected to cover key fluid units such as: gas accumulation zone, three-phase zone and hydrate-bearing zone, in order to reveal the law of production capacity difference under different mining paths. This spatial discretization research method can provide a new theoretical basis for well location optimization in Class 1S hydrate reservoir.

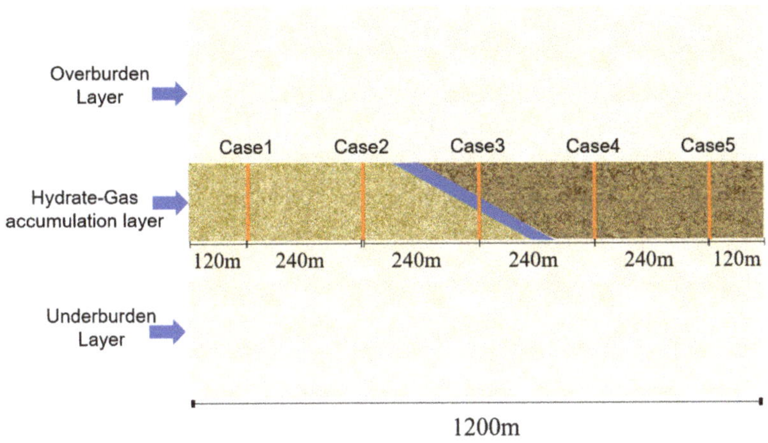

Figure 6. Schematic Diagram of Well Locations for Five Cases.

3.3. Model Verification

The present study employs the CMG-STARS (2021.10) software for numerical calculations [46–49]. The STARS simulator incorporates multiple advanced well management and control options. The software in question incorporates sophisticated fully implicit geochemical simulation capabilities and advanced reaction kinetics modelling. In 2008, the inaugural international gas hydrate code comparison evaluated five mainstream hydrate reservoir simulators. This finding serves to substantiate the robust performance of the CMG-STARS software in simulating hydrate reservoir exploitation [50–53].

This paper provides further validation of the reliability of the CMG-STARS model, as demonstrated through hydrate decomposition experiments [54]. The pressure-reduction hydrate decomposition experiment was conducted in the CHS experimental facility. A vertical well was used for gas and water production during pressure-reduction extraction. The experiment consisted of two phases: pressure reduction and constant pressure. This paper primarily focuses on fitting run 1, which involved a pressure reduction duration of 22 min. The specific experimental process and simulation parameters are referenced in Xiao's article. The fitting results are shown in Figure 7. A comparison of the simulation and experimental results shows that the cumulative gas production differs by 6%. Overall, however, the numerical simulation results are consistent with the experimental results, demonstrating the strong reliability of the CMG-STARS programme.

The size of the wellbore grid may significantly impact the results. To ensure simulation accuracy, five distinct wellbore grid sizes were set for comparison: 0.5 m, 1 m, 2.5 m and 5 m. Generally, smaller wellbore grids yield higher accuracy. Therefore, the 0.5 m grid size was designated as the control group. The cumulative gas production error for each alternative scheme was calculated relative to the control group. As shown in Table 2, when the grid size was 1 m, the error remained within 0.01 compared to the control group—a sufficiently small margin. Taking simulation speed into account, the 1 m grid size scheme was selected for subsequent simulations.

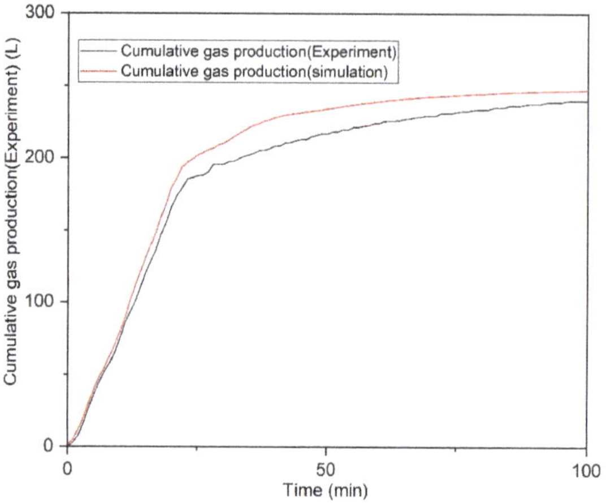

Figure 7. Comparison of Experimental Results with Simulation Results.

Table 2. Comparison of Cumulative Gas Production and Error Among Different Schemes.

	Case 1 ($\times 10^8$ m^3)	Error	Case 3 ($\times 10^7$ m^3)	Error
5 m	1.66976	0.034368	4.05816	0.195039
2.5 m	1.64801	0.020895	3.5422	0.0431
1 m	1.62977	0.009596	3.40666	0.003186
0.5 m	1.61428	-	3.39584	-

4. Results and Discussions

4.1. Production Characteristic

The gas production rates of Case 1 and Case 2 located in the gas accumulation zone are significantly higher than those of Case 3, Case 4, and Case 5 in Figure 8. Among these, Case 1 exhibits exceptionally high gas production during days 1–30, reaching approximately 83,000 m^3/d (The gas volume and hydrate volume in this paper are both expressed in standard conditions). Production gradually declines between days 31–120, then gradually increased again from days 121 to 600 to approximately 81,346 m^3/d, after which it began to decline steadily, reaching 17,130 m^3/d at day 3650. The gas production curve for Case 2 closely resembles that of Case 1, with the primary difference being that Case 2's production increased steadily from days 1 to 800, reaching a maximum daily production of 74,000 m^3/d. Production then gradually declined from 800 to 3650 days, reaching 13,860 m^3/d at 3650 days. Case 3 has a relatively low daily gas production capacity, approximately 10,000 m^3/d. Case 4 and Case 5 are primarily situated within the hydrate zone. Both cases exhibited low gas production rates, with Case 4 demonstrating higher output than Case 5.

As illustrated in Figure 9, the cumulative gas production over 3650 days for Case 1, Case 2, Case 3, Case 4, and Case 5 is 16.2932×10^7 m^3, 14.98×10^7 m^3, 3.41×10^7 m^3, 3.41×10^7 m^3, and 1.34×10^7 m^3, respectively. The cumulative gas production of Case 1 and Case 2 is significantly higher than that of the other three cases. The most unexpected result in this simulation is likely that of Case 3. Although half of the perforation section in Case 3 is located within the gas accumulation zone, both the daily gas production and cumulative gas production are relatively low. Compared to Case 4 and Case 5, the simulation results for Case 3 show no significant advantage.

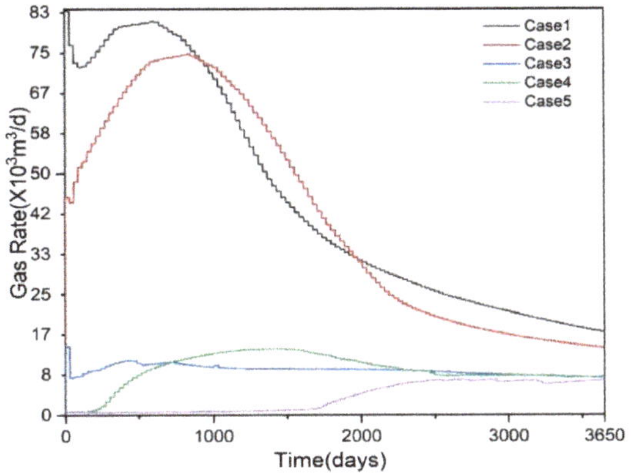

Figure 8. Gas rate of five cases.

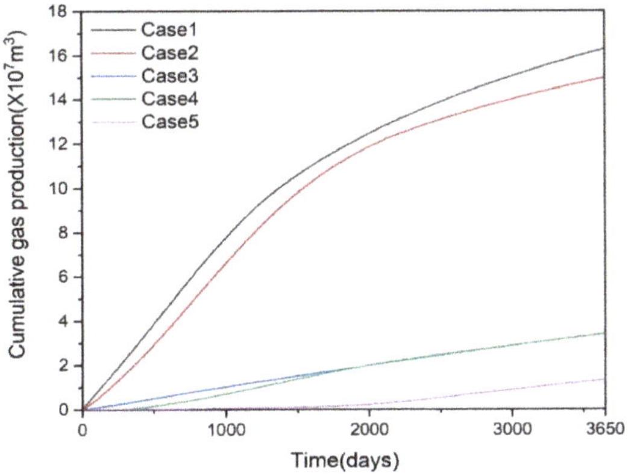

Figure 9. Cumulative gas production of five cases.

Figure 10 shows the variation patterns of hydrate volume across different scenarios. In Case 1 and Case 2, hydrate decomposition is more significant and exhibits a consistent trend. At 3650 days, Case 1 retains 2.0×10^5 m^3 of hydrate, while Case 2 retains 2.3×10^5 m^3. This is because the wells in Case 1 and Case 2 are primarily located within the gas accumulation zone, enabling rapid gas production that reduces reservoir pressure and subsequently triggers hydrate decomposition. Case 3 exhibited the least hydrate decomposition, with a decomposition volume of only 1.26×10^5 m^3, the lowest among the five scenarios. In Case 4, the hydrate decomposition rate remained relatively uniform between days 1 and 2700. From days 2700 to 3300, hydrate decomposition was negligible, with the hydrate volume stabilizing around 5.75×10^5 m^3. From days 3300 to 3650, hydrate decomposition continued, ultimately reducing the hydrate volume to 5.5×10^5 m^3. In Case 5, the hydrate decomposition rate was relatively uniform, with a final residual hydrate volume of 526,489 m^3.

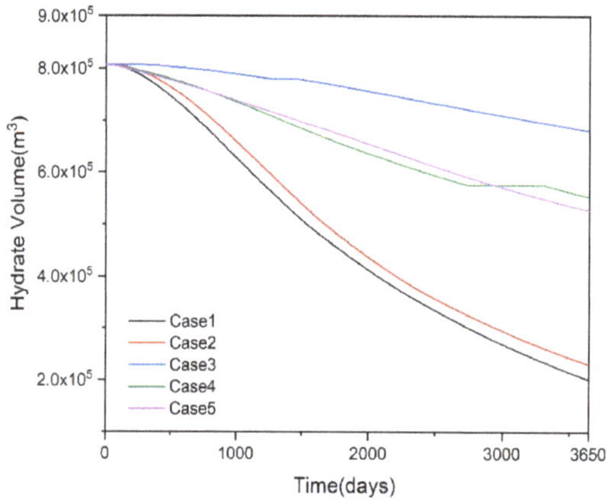

Figure 10. Hydrate volume of the reservoir for Case 1–5.

Figure 11 shows the gas volume variation patterns in reservoirs under different scenarios. From days 1 to 2000, the gas volume trends in Case 1 and Case 2 are relatively consistent. Due to continuous gas production, the gas volume in the reservoir steadily decreases. From days 2000 to 3650, the gas volume in Case 1 remains largely unchanged. Considering the hydrate decomposition pattern, this indicates that the gas production rate from hydrates equals the well's gas production rate during this period. In Case 2, the gas volume increases, suggesting that the gas production rate from hydrates exceeds the well's gas production rate, indicating a relatively low well production rate at this stage. Case 3 shows a slight decrease in gas volume, maintaining 1.77×10^8 m^3 at 3650 days. Case 4 and Case 5 exhibit similar patterns of continuous gas volume increase. Based on the aforementioned hydrate decomposition and gas production patterns, the pressure reduction strategies in Case 4 and Case 5 prove effective, enabling partial hydrate decomposition and sufficient gas supply. However, the low gas production rate of the wells impedes further gas extraction from the reservoir.

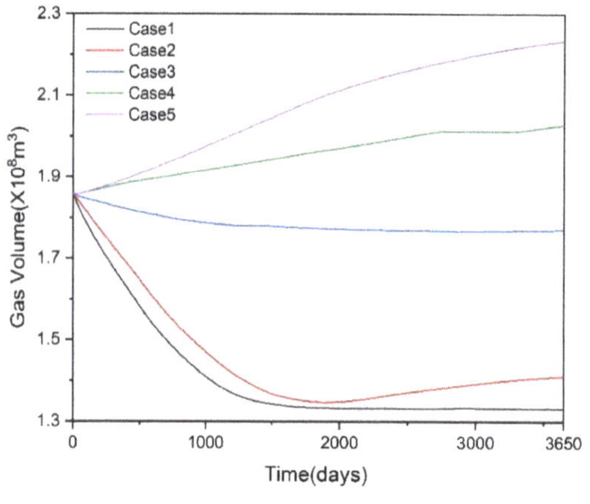

Figure 11. Gas volume of the reservoir for Case 1–5.

Based on the above analysis, a preliminary conclusion can be drawn: Case 1 and Case 2, located within the gas accumulation zone, exhibit higher gas production rates, with Case 1 demonstrating greater cumulative gas production than Case 2. It is evident that both Case 4 and Case 5 are located within the hydrate-bearing zone, where the permeability of the substrate is known to be particularly high. The hydrate-bearing reservoir is *characterised* by a relatively low level, which has a consequential effect on the reservoir's cumulative gas production. However, Case 3, with half of its perforated section situated in the gas accumulation zone, does not demonstrate a significant advantage in cumulative gas production compared to Case 4 and Case 5. This requires further analysis.

4.2. Evolution of Hydrate Distribution

As shown in Figures 12 and 13, evolution of hydrate saturation(S_h) distribution on the vertical cross-section and the horizontal cross-section. The hydrate decomposition dynamics reveal that the decomposition patterns in Case 1 and Case 2 are relatively similar. At 180 days, only partial decomposition occurred in the hydrates within the three-phase zone and near the underlying layer. By 365 days, hydrates in the three-phase zone had completely decomposed, while those near the underlying layer underwent further decomposition, with hydrate saturation decreasing by approximately 0.1. At 1825 days, significant decomposition occurred in the top and bottom parts of the hydrate reservoir, while decomposition in the middle part was relatively minor. By 3650 days, the decomposition front in the central hydrate had advanced approximately 75 m. The hydrate saturation dynamics in Case 3 reveal minimal decomposition, occurring only near the underlying formation. Concurrently, significant secondary hydrate formation is observed around the wellbore. This likely explains why half of the perforated section in Case 3 lies within the gas reservoir yet yields substantially lower production than Cases 1 and 2. The hydrate decomposition dynamics in Cases 4 and 5 are similar, with decomposition primarily occurring around the wellbore and near the underlying formation. By 3650 days, hydrates within a 20-m radius of the wellbore had largely decomposed. Concurrently, significant secondary hydrate formation was observed around the wellbore. This secondary hydrate generation likely contributed to the reduced gas production observed in Cases 4 and 5.

The variation in secondary hydrate saturation within a 2.5 m radius around the wellbore is shown in Figure 14. It was observed that the secondary hydrate saturation around the wellbore in Case 3 remained consistently high from days 1 to 3650. It has been demonstrated that from days 1 to 500, there was a significant increase in gas production in Case 3 compared with Cases 4 and 5 (as shown in Figure 8). This finding indicates the benefit of the proximity of Case 3 to the gas accumulation zone. However, the higher gas production rate induced a strong choke expansion effect, causing a rapid decrease in wellbore temperature. Simultaneously, Case 3's well is situated between the free gas and hydrate reservoirs, where temperatures are lower. Consequently, the choke expansion effect rapidly generates substantial secondary hydrates around the wellbore, reducing surrounding permeability. Between days 500 and 3650, this phenomenon decreased gas production and prevented further pressure drop propagation, thereby inhibiting hydrate decomposition. Thus, Case 3 exhibited the lowest hydrate decomposition rate and the lowest gas production. Both perforation sections in Case 1 and Case 2 are situated within the gas accumulation zone, resulting in higher reservoir temperatures. The warmer gas helps elevate temperatures around the wellbore, leading to relatively fewer secondary hydrates. Compared to Case 2, Case 1 experiences higher temperatures and fewer secondary hydrates, yielding slightly higher cumulative gas production. From days 1 to 200, gas in Case 4 primarily originates from the decomposition of hydrates around the wellbore, resulting in a lower initial gas production rate. From 200 to 1500 days, gas production in Case 4 accelerated rapidly as surrounding hydrates decomposed and permeability increased. The combination of high production rates and low surrounding temperatures led to extensive secondary hydrate formation around the well. From 1500 to 3650 days, significant secondary hydrate generation persisted, continuously suppressing Case 4's gas production rate. From 1 to 30 days, the hydrates around the Case 5 well were completely decomposed. Between 30 and 2000 days, the hydrate saturation around Case 5 remained at 0. This was due to the low gas production rate of Case 5, which resulted in negligible throttling expansion effects. Consequently, despite having the lowest wellbore temperature, no secondary hydrates formed during this phase. From 2000 to 3650 days, the gas production rate of Case 5 increased, enhancing the throttling expansion effect. Consequently, a certain amount of secondary hydrate formed around the well. This hydrate formation also caused the daily gas production to fluctuate between 6000 and 7400 m^3/d, preventing further increases in the production rate.

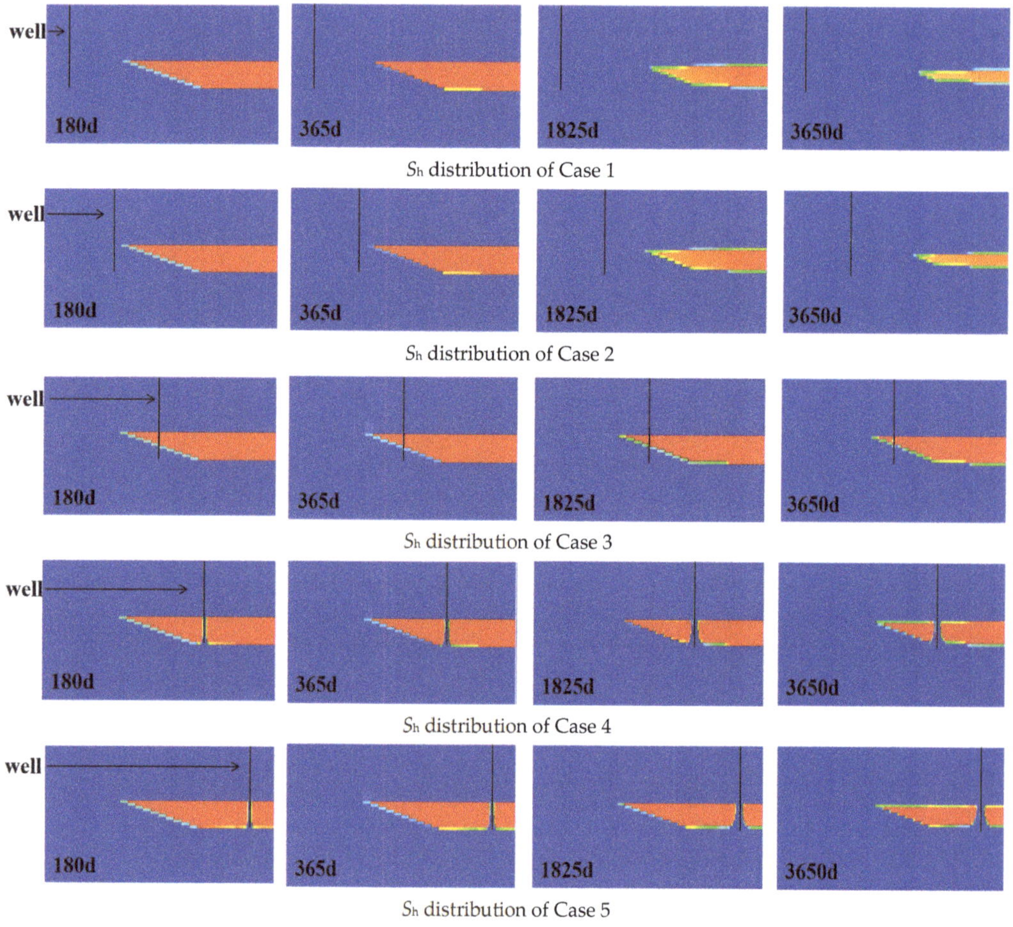

Figure 12. Evolution of S_h distribution on the vertical cross-section for Case 1–5.

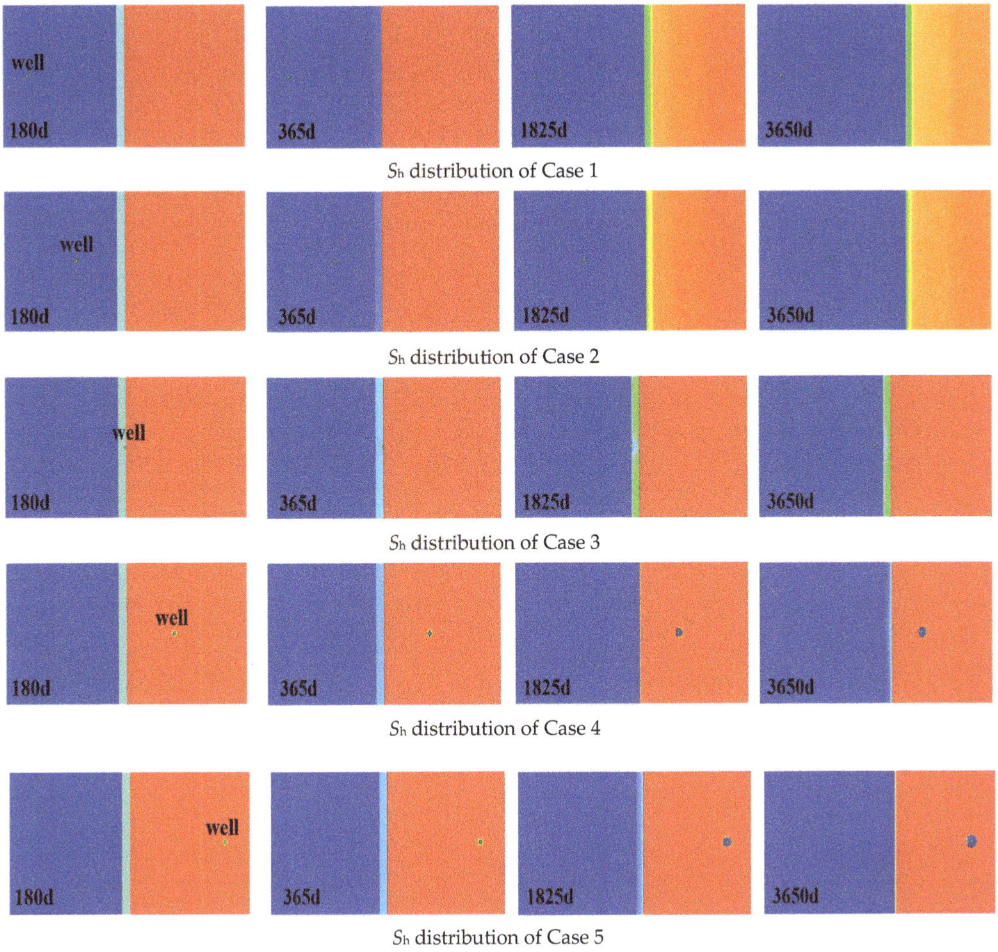

Figure 13. Evolution of S_h distribution on the horizontal cross-section.

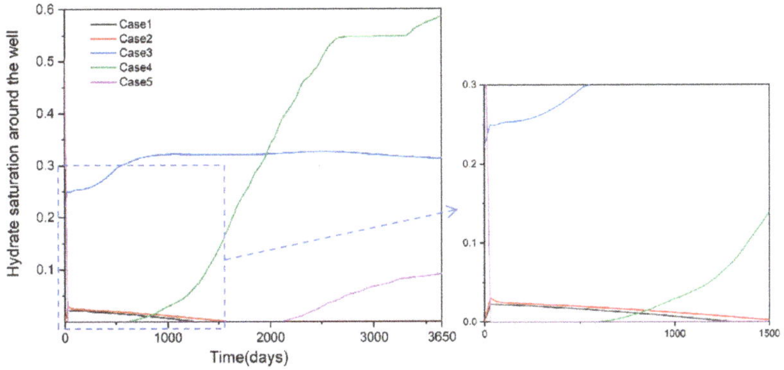

Figure 14. Hydrate saturation around the well of Case 1–5.

In summary, the saturation of hydrates around the wellbore exhibits a significant correlation with the gas production rates in Case 3, Case 4, and Case 5. The formation of secondary hydrates has been demonstrated to have a substantial impact on gas production rates in the examined scenarios. Consequently, measures such as heating are required to mitigate the formation of secondary hydrates in the surrounding area of the wellbore. In Cases 1 and 2, the perforated sections are located within the higher-temperature gas accumulation zone, where secondary hydrate formation is minimal. However, the limited well control area and reserves result in residual hydrates remaining undecomposed in the reservoir's right section. Consequently, insufficient subsequent gas supply leads to declining gas production rates. Future considerations should include increasing the number of wells to enhance recovery rates.

4.3. Effect of Wellbore Pressure on Results

The above analysis reveals that the location of the well and the formation of secondary hydrates have a significant influence on gas production behaviour. However, this conclusion was obtained at a bottomhole pressure of 10 MPa. To verify the universality of this finding, this study established four simulation scenarios with different bottom-hole pressures. These were set at 5 MPa, 7.5 MPa, 12.5 MPa and 15 MPa, respectively. All other parameters remained unchanged, consistent with the simulation scenario at a bottomhole pressure of 10 MPa. The gas production results are shown in Figure 15. Figure 16 presents the statistical analysis of hydrate saturation within a 2.5-metre radius around the wellbore.

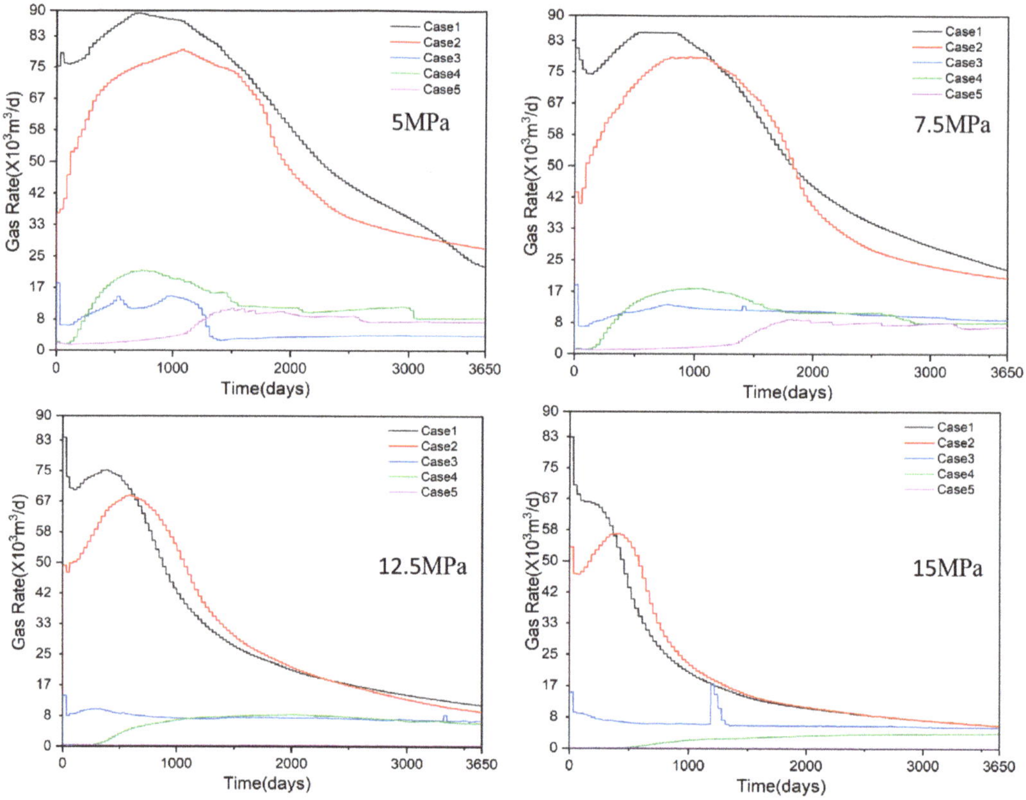

Figure 15. Gas production rates for five cases at wellhead pressures of 5, 7.5, 12.5 and 15 MPa.

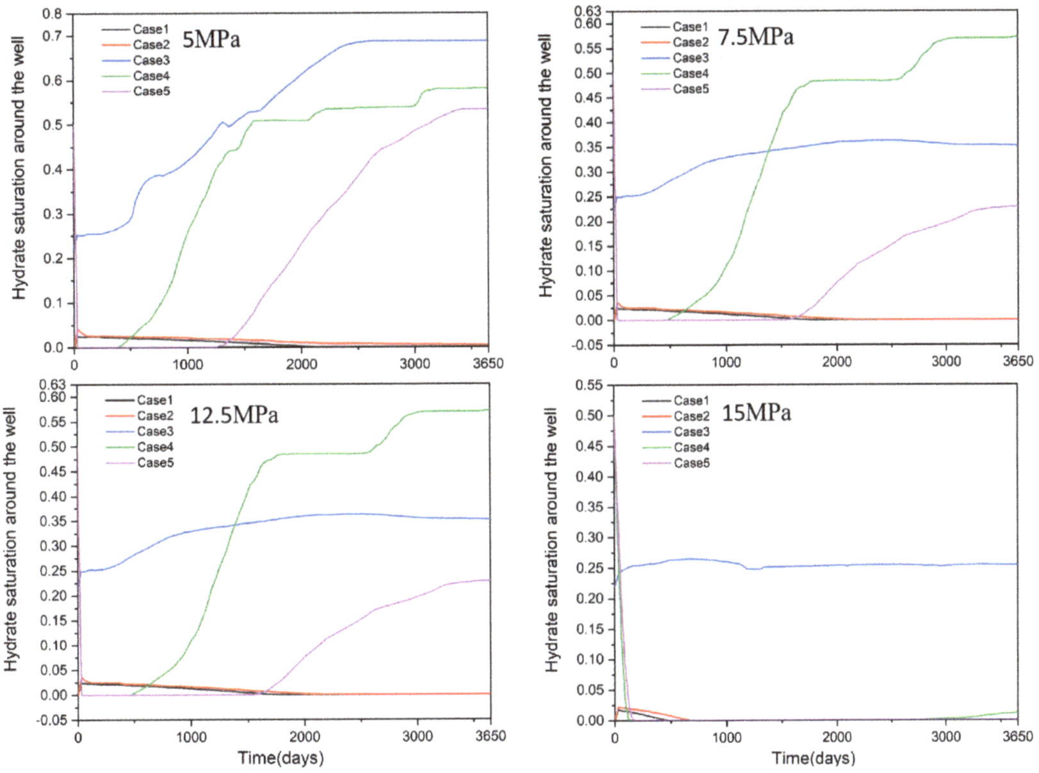

Figure 16. Hydrate saturation of Case 1–5 at wellhead pressures of 5, 7.5, 12.5 and 15 MPa.

Of the scenarios at 5 MPa, 7.5 MPa, 12.5 MPa and 15 MPa, Cases 1 and 2 exhibit significantly higher gas accumulation zone production rates than Cases 3, 4 and 5. Therefore, Case 1 within the gas accumulation zone should be prioritised as the production well. In the 5 MPa, 7.5 MPa and 12.5 MPa pressure groups, secondary hydrate formation around the wellbore was markedly higher in Cases 3, 4 and 5 than in Cases 1 and 2. This phenomenon had a significant impact on the gas production behaviour of Cases 3, 4 and 5. At 15 MPa, the smaller bottomhole pressure differential results in lower gas production in Cases 4 and 5, leading to reduced saturation of secondary hydrates around the wellbore. Therefore, under different production pressures, both well location and secondary hydrate formation significantly influence gas production behaviour.

5. Conclusions

This paper analyzes the geological background and formation mechanisms of Class 1S hydrate deposits. Based on actual data, a geological model was established and five simulation scenarios were set for different well locations, revealing the gas production characteristics of Class 1S hydrate deposits and providing development recommendations. Conclusions are as follows:

(1) The formation mechanism of Class 1S hydrate deposits involves: a robust deep gas supply system, a gas chimney-dominated transport system, and the hydrate storage system within stable domains coupled with the sealing effect of the caprock.

(2) Case 1, located within the gas accumulation zone, is recommended as the priority option to achieve the highest cumulative gas production. Additional production wells may be considered in later stages to enhance recovery rates.
(3) Secondary hydrate formation significantly impacts production in Cases 3, 4, and 5. Measures such as wellbore heating can be employed to minimize secondary hydrate formation around the wellbore.
(4) Geomechanical deformations do exert a certain influence on the results, especially in the three-phase zone. However, the article does not consider geomechanical deformation. This will be a focus of our subsequent work.

Author Contributions: Software, Z.P. and Z.Z.; Methodology, Z.P. and B.L.; Data curation, J.D.; Investigation, C.X. (Changwen Xiao); Writing—original draft, X.L.; Writing—review and editing, C.X. (Chenlu Xu); Project administration, L.Y.; Funding acquisition, H.L.; Conceptualization, L.N. and J.C. All authors have read and agreed to the published version of the manuscript.

Funding: This research was funded by the Special Project for Marine Economy Development of Guangdong (Six Marine Industries) (GDNRC [2024]48), 2025 Hainan International Science and Technology Cooperation Research and Development Project(GHYF2025017), National Natural Science Foundation of China(No.42476232), Guangzhou Science and Technology Program(No.202206050002), National Key Research and Development Program (2024YFC2814703), the Youth Research Team Project of the National Engineering Research Center of Gas Hydrate Exploration and Development (Grant No. NERC2024003), the National Natural Science Foundation of China (No. 42406232).

Data Availability Statement: The datasets generated and analyzed during the current study are available from the corresponding author by request.

Acknowledgments: Support for this work was jointly provided by the Special Project for Marine Economy Development of Guangdong (Six Marine Industries), 2025 Hainan International Science and Technology Cooperation Research and Development Project, National Natural Science Foundation of China, Guangzhou Science and Technology Program, National Key Research and Development Program, the Youth Research Team Project of the National Engineering Research Center of Gas Hydrate Exploration and Development.

Conflicts of Interest: The authors declare no conflicts of interest.

References

1. Sloan, E.D. Fundamental principles and applications of natural gas hydrates. *Nature* **2003**, *426*, 353–359. [CrossRef]
2. Kvenvolden, K.A.; Lorenson, T.D. The global occurrence of natural gas hydrate. *Geophys. Monogr.* **2001**, *124*, 87–98. [CrossRef]
3. Cherskiy, V.P.; Tsarev, V.P.; Nikitin, S.P. Investigation and prediction of conditions of accumulation of gas resource in gas-hydrate pools. *Pet. Geol.* **1984**, *21*, 84–89.
4. Liu, C.L.; Meng, Q.G.; He, X.L.; Li, C.F.; Ye, Y.G.; Lu, Z.Q.; Zhu, Y.H.; Li, Y.H.; Liang, J.Q. Comparison of the characteristics for natural gas hydrate recovered from marine and terrestrial areas in China. *J. Geochem. Explor.* **2015**, *152*, 67–74. [CrossRef]
5. Makogon, Y.F. Natural gas hydrates-a promising source of energy. *J. Nat. Gas Sci. Eng.* **2010**, *2*, 49–59. [CrossRef]
6. Makogon, Y.F.; Holditch, S.A.; Makogon, T.Y. Russian field illustrates gas-hydrate production. *Oil Gas J.* **2005**, *103*, 43–50.
7. Dallimore, S.R.; Uchida, T.; Collett, T.S. Scientific results from JAPAX/JNOC/GSC Mallik 2L-38 gas hydrate research well Mackenzie Delta, Northwest Territories, Canada. *Geol. Surv. Can. Bull.* **1999**, *544*, 295–311.
8. De Angelis, M.A.; Lilley, M.D.; Baross, J.A. Methane oxidation in deep-sea hydrothermal plumes of the endeavour segment of the Juan de Fuca Ridge. *Deep Sea Res. Part I* **1993**, *40*, 1169–1186. [CrossRef]
9. Chazallon, B.; Rodriguez, C.T.; Ruffine, Y.; Carpentier, Y.; Riboulot, Y. Characterizing the variability of natural gas hydrate composition from a selected site of the Western Black Sea, off Romania. *Mar. Pet. Geol.* **2020**, *124*, 104785. [CrossRef]
10. Klauda, J.B.; Sandler, S.I. Global distribution of methane hydrate in ocean sediment. *Energy Fuels* **2005**, *19*, 459–470. [CrossRef]
11. Takahashi, H.; Yonezawa, T.; Fercho, E. Operation Overview of the 2002 Mallik gas hydrate production research well program at the Mackenzie Delta in the Canadian Arctic. In Proceedings of the Offshore Technology Conference, Houston, TX, USA, 5 May 2003. [CrossRef]

2. Moridis, G.J.; Collett, T.S.; Dallimore, S.R.; Satoh, T.; Hancock, S.; Weatherill, B. Numerical studies of gas production from several CH4-hydrate zones at the Mallik Site, Mackenzie Delta, Canada. *J. Petrol. Sci Eng.* **2004**, *43*, 219–238. [CrossRef]
3. Alp, D.; Parlaktuna, M.; Moridis, G.J. Gas production by depressurization from hypothetical Class 1G and Class 1W hydrate reervoirs. *Energy Convers. Manag.* **2007**, *48*, 1864–1879. [CrossRef]
4. Moridis, G.J.; Kowalsky, M.B.; Pruess, K. Depressurization-induced gas production from Class 1 hydrate deposits. *SPE J. Res. Eval. Eng.* **2007**, *10*, 458–481. [CrossRef]
5. Moridis, G.J.; Sloan, E.D. Gas production potential of disperse low-saturation hydrate accumulations in oceanic sediments. *Energy Convers. Manag.* **2007**, *48*, 1834–1849. [CrossRef]
6. Moridis, G.J.; Collett, T.S.; Boswell, R.; Hancock, S.; Coh, C. *Gas Hydrates as a Potential Energy Source: State of Knowledge and Challenges*; Springer: Berlin/Heidelberg, Germany, 2013; pp. 977–1033. [CrossRef]
7. Moridis, G.J.; Collett, T.S.; Pooladi-Darvish, M.; Hancock, S.H.; Santamarina, C.; Boswell, R.; Kneafsey, T.J.; Rutqvist, J.; Kowalsky, M.B.; Reagan, M.T. Challenges, Uncertainties and Issues Facing Gas Production from Hydrate Deposits in Geologic Systems. In Proceedings of the SPE Unconventional Gas Conference, Pittsburgh, PA, USA, 23–25 February 2010. [CrossRef]
8. Moridis, G.J.; Reagan, M.T. Estimating the upper limit of gas production from Class 2 hydrate accumulations in the permafrost: 1. Concepts, system description, and the production base case. *J. Pet. Sci. Eng.* **2011**, *76*, 194–204. [CrossRef]
9. Ye, H.; Chen, J.; Yao, Y.; Chen, D.; Wu, X.; Li, D.; Zi, M. Gas recovery from low-permeability muddy silt gas hydrate reservoirs by depressurization coupled with hot water injection: Impact of hydro-lock effect. *Energy* **2025**, *316*, 134413. [CrossRef]
20. Ye, H.; Yao, Y.; Chen, D.; Duan, J.; Wu, X.; Li, D.; Zi, M. Hydrate-based CO_2 sequestration in post-exploitation natural gas hydrate reservoir: A numerical method considering mixed hydrate. *Energy* **2025**, *336*, 138406. [CrossRef]
21. Qin, F.; Sun, J.; Cao, X.; Mao, P.; Zhang, L.; Lei, G.; Jiang, G.; Ning, F. Numerical simulation on combined production of hydrate and free gas from silty clay reservoir in the South China Sea by depressurization: Formation sealing. *Appl. Energy* **2025**, *377*, 124343. [CrossRef]
22. Li, B.; Sun, Y.; Jiang, S.; Zhang, G.; Feng, J.; Shan, H.; Li, X. Production characteristics and sensitivity analysis on gas hydrate and shallow gas coexistence reservoirs: Focus on gas hydrate formation. *Appl. Energy* **2025**, *386*, 125572. [CrossRef]
23. Yu, L.; Lu, H.; Zhang, L.; Xu, C.; Kuang, Z.; Li, X.; Yu, H.; Wang, Y. Assessment of Gas Production from Complex Hydrate System in Qiongdongnan Basin of South China Sea. *Energies* **2023**, *16*, 7447. [CrossRef]
24. Zhao, Z.X.; Sun, Z.; Wang, Z.F.; Sun, Z.P.; Liu, J.B.; Zhang, C.M. The high-resolution sedimentary filling in Qiongdongnan Basin, northern south China Sea. *Mar. Geol.* **2015**, *361*, 11. [CrossRef]
25. Wei, J.G.; Liang, J.Q.; Lu, J.A.; Zhang, W.; He, Y.L. Characteristics and dynamics of gas hydrate systems in the northwestern South China Sea-Results of the fifth gas hydrate drilling expedition. *Mar. Pet. Geol.* **2019**, *110*, 287–298. [CrossRef]
26. Liang, J.Q.; Zhang, W.; Lu, J.A.; Wei, J.G.; Kuang, Z.K.; He, Y.L. Geological occurrence and accumulation mechanism of natural gas hydrates in the eastern Qiongdongnan Basin of the South China Sea: Insights from site GMGS5-W9-2018. *Mar. Geol.* **2019**, *418*, 106042. [CrossRef]
27. Meng, M.M.; Liang, J.Q.; Lu, J.A.; Zhang, W.; Kuang, Z.K.; Fang, Y.X.; He, Y.L.; Deng, W.; Huang, W. Quaternary deep-water sedimentary characteristics and their relationship with the gas hydrate accumulations in the Qiongdongnan Basin, Northwest South China Sea. *Deep Sea Res. Part I* **2021**, *177*, 103628. [CrossRef]
28. Zhang, W.; Liang, J.Q.; Lu, J.A.; Meng, M.M.; He, Y.L.; Deng, W.; Feng, J.X. Characteristics and controlling mechanism of typical leakage gas hydrate reservoir forming system in the Qiongdongnan Basin, northern South China Sea. *Nat. Gas Ind.* **2020**, *40*, 90–99, (In Chinese with English Abstract). [CrossRef]
29. He, Y.L.; Liang, J.Q.; Shi, W.Z.; Kuang, Z.K.; Deng, W.; Wang, R.; Xu, L.T.; Du, H. Influencing factors and accumulation modes of gas hydrate in the South low uplift and its surrounding area of Qiongdongnan Basin. *Earth Sci.* **2021**, *5*, 1711–1727, (In Chinese with English Abstract). [CrossRef]
30. Xie, Y.H.; Zhang, G.C.; Sun, Z.P.; Zeng, Q.; Zhao, Z.; Guo, S. Reservoir forming conditions and key exploration technologies of Lingshui 17-2 giant gas field in deepwater area of Qiongdongnan Basin. *Petrol. Res.* **2019**, *4*, 1–18. [CrossRef]
31. Chen, Z.G.; Jiang, T.; Kuang, Z.G.; Chen, C.; Xiong, P.F.; Chen, Y. Accumulation characteristics of gas hydrate-shallow gas symbiotic system in Qiongdongnan Basin. *Earth Sci.* **2022**, *47*, 1619–1634, (In Chinese with English Abstract). [CrossRef]
32. Huang, W.; Zhang, W.; Liang, J.Q.; Shang, J.J.; Meng, M.M.; Shan, C.C. Comparative study of gas hydrate accumulation system in the Qiongdongnan Basin of the South China Sea and the Ulleung Basin of Korea. *J. China Uni. Min. Tech.* **2020**, *50*, 363–380, (In Chinese with English Abstract). [CrossRef]
33. Huang, S.Z.; Li, F.; Song, P.; Hu, B.; Sun, W.Y. Study on geophysical identification technology of ultra deep water and shallow gas reservoirs in Qiongdongnan Basin. *Earth Sci.* **2022**, *339*, 1–11, (In Chinese with English Abstract). [CrossRef]
34. Xue, K.P.; Liu, Y.; Yu, T.; Yang, L.; Zhao, J.F.; Song, Y.C. Numerical simulation of gas hydrate production in shenhu area using depressurization: The effect of reservoir permeability heterogeneity. *Energy* **2023**, *271*, 126948. [CrossRef]

35. Wei, R.P.; Xia, Y.Q.; Wang, Z.F.; Li, Q.P.; Lv, X.; Leng, S.D.; Zhang, L.X.; Zhang, Y.; Xiao, B.; Yang, S.X.; et al. Long-term numerical simulation of a joint production of gas hydrate and underlying shallow gas through dual horizontal wells in the South China Sea. Appl. *Energy* **2022**, *320*, 119235. [CrossRef]
36. Zhao, E.M.; Hou, J.; Du, Q.J.; Liu, Y.G.; JI, Y.K.; Bai, Y.J. Numerical modeling of gas production from methane hydrate deposits using low-frequency electrical heating assisted depressurization method. *Fuel* **2021**, *290*, 120075. [CrossRef]
37. Gajanayake, S.; Gamage, R.P.; Wanniarachchige, P.; Zhang, D. Quantification of CO2 replacement in methane gas hydrates: A molecular dynamics perspective. *J. Nat. Gas Eng.* **2022**, *98*, 104396. [CrossRef]
38. Li, S.X.; Li, S.; Zhang, R.Y.; Li, Q.P.; Pang, W.X. Strategies for gas production from Class 2 hydrate accumulations by depressurization. *Fuel* **2021**, *286*, 119380. [CrossRef]
39. Yang, S.W.; Lang, X.M.; Wang, Y.H.; Wen, Y.G.; Fan, S.S. Numerical simulation of Class 3 hydrate reservoirs exploiting using horizontal well by depressurization and thermal co-stimulation. *Energy Convers. Manag.* **2014**, *77*, 298–305. [CrossRef]
40. Li, S.; Wang, Z.; Li, S.; Wang, X.; Hao, Y. Investigations on performance of hydrate dissociation by depressurization near the quadruple point. *J. Nat. Gas Sci. Eng.* **2021**, *90*, 103929. [CrossRef]
41. Wang, B.; Fan, Z.; Wang, P.; Liu, Y.; Zhao, J.; Song, Y. Analysis of depressurization mode on gas recovery from methane hydrate deposits and the concomitant ice generation. *Appl. Energy* **2018**, *227*, 624–633. [CrossRef]
42. Zhang, Y.; Zhang, P.; Hui, C.; Tian, S.; Zhang, B. Numerical analysis of the geomechanical responses during natural gas hydrate production by multilateral wells. *Energy* **2023**, *269*, 126810. [CrossRef]
43. Meret, S.; Al-Raoush, R.I.; Jung, J.W.; Alshibli, K.A. Comprehensive literature review on CH_4-CO_2 replacement in microscale porous media. *J. Pet. Sci. Eng.* **2018**, *178*, 48–62. [CrossRef]
44. Liu, Y.; Hou, J.; Chen, Z.; Bai, Y.; Su, H.; Zhao, E.; Li, G. Enhancing hot water flooding in hydrate bearing layers through a novel staged production method. *Energy* **2021**, *217*, 119319. [CrossRef]
45. Zhong, X.; Pan, D.; Zhu, Y.; Wang, Y.; Zhai, L.; Li, X.; Tu, G.; Chen, C. Fracture network stimulation effect on hydrate development by depressurization combined with thermal stimulation using injection-production well patterns. *Energy* **2021**, *228*, 120601. [CrossRef]
46. Uddin, M.; Coombe, D.; Law, D.; Gunter, B. Numerical studies of gas hydrate formation and decomposition in a geological reservoir. In Proceedings of the SPE Gas Technology Symposium, Calgary, AB, Canada, 15 May 2006. [CrossRef]
47. Gaddipati, M. Code Comparison of Methane Hydrate Reservoir Simulators Using CMG STARS. Ph.D. Thesis, West Virginia University, Morgantown, WV, USA, 2008.
48. Ajayi, T.; Anderson, B.J.; Seol, Y.; Boswell, R.; Myshakin, E.M. Key aspects of numerical analysis of gas hydrate reservoir performance: Alaska North Slope Prudhoe Bay Unit "L-Pad" hydrate accumulation. *J. Nat. Gas Sci. Eng.* **2018**, *51*, 37–43. [CrossRef]
49. Matsuda, H.; Yamakawa, T.; Sugai, Y.; Sasaki, K. Gas Production from Offshore Methane Hydrate Layer and Seabed Subsidence by Depressurization Method. *Engineering* **2016**, *8*, 353–364. [CrossRef]
50. Myshakin, E.M.; Gaddipati, M.; Rose, K.; Anderson, B.J. Numerical simulations of depressurization-induced gas production from gas hydrate reservoirs at the Walker Ridge 313 site, northern Gulf of Mexico. *Mar. Pet. Geol.* **2012**, *34*, 169–185. [CrossRef]
51. Vysniauskas, A.; Bishnoi, P.R. A kinetic study of methane hydrate formation. *Chem. Eng. Sci.* **1983**, *38*, 1061–1072. [CrossRef]
52. Kim, H.C.; Bishnoi, P.R.; Heidemann, R.A.; Rizvi, S.S.H. Kinetics of methane hydrate decomposition. *Chem. Eng. Sci.* **1987**, *42*, 1645–1653. [CrossRef]
53. Su, Z.; Moridis, G.J.; Zhang, K.; Wu, N. A huff-and-puff production of gas hydrate deposits in Shenhu area of South China Sea through a vertical well. *J. Pet. Sci. Eng.* **2012**, *86–87*, 54–61. [CrossRef]
54. Xiao, C.-W.; Li, X.-S.; Lu, H.-F.; Li, G.; Xu, C.-L.; Qi, R.-R.; Xing, D.-H.; Li, X.; Weng, Y.-F.; Yu, L. Experimental studies and pore-scale modeling of hydrate dissociation kinetics at different depressurization rates in a cubic hydrate simulator. *Chem. Eng. J.* **2024**, *499*, 156313. [CrossRef]

Journal of
Marine Science and Engineering

MDPI

Article

The Numerical Evaluation of Hydrate Saturation in Marine Sediment During the Injection Process of Self-Heat Generating Fluid

Kewei Zhang [1,2], Kaixiang Shen [1,2,*], Yanjiang Yu [1,2], Yingsheng Wang [1,2], Jiawei Zhou [1,2] and Jing Zeng [1,2]

1 Key Laboratory of Marine Mineral Resources, Ministry of Natural Resources, Guangzhou Marine Geological Survey, China Geological Survey, Guangzhou 511458, China; zhang3334k@163.com (K.Z.); yuyanjiang2004@163.com (Y.Y.); gmgs_wys@163.com (Y.W.); zhoujwyx@163.com (J.Z.); zeng_jing91@163.com (J.Z.)

2 National Engineering Research Center for Gas Hydrate Exploration and Development, Guangzhou 511458, China

* Correspondence: skxv@163.com

Abstract

Marine gas hydrates are recognized as a promising offshore energy resource. Self-heat fluid injection is an innovative thermal-enhanced gas recovery technique for hydrate exploitation engineering. This study numerically investigates hydrate saturation during the self-heating reagent injection process in a sub-sea hydrate reservoir, decoupled from gas production interference. This process employs two consecutive stages: reactive chemical flow stage followed by non-reactive flow stage. The simulation output parameters encompass reservoir temperature, fluid saturation, thermal conductivity, and heat flow rate. The base case demonstrates that fluid injection elevates reservoir temperature from 13.0 °C to 29.3 °C and reduces hydrate saturation from 0.40 to 0.21 through coupled heat–mass transfer mechanisms during the reactive flow stage. In the consequent non-reactive flow stage, hydrate saturation decreases to zero. Sensitivity analysis reveals that initial permeability variation governs the hydrate saturation and temperature during the non-reactive phase. The permeability range of less than 15 mD is the optimal threshold preventing hydrate reformation during fluid injection. 55–70 mD permeability triggers severe secondary hydrate generation, which decreases the fluid application feasibility. Fluid flooding demonstrates superior hydrate dissociation efficacy compared to in situ thermal stimulation. This study develops a novel simulation approach to characterize marine hydrate saturation dynamics.

Keywords: marine hydrate resource; underground fluid saturation; saturation simulation; reservoir heat transfer; reservoir permeability

1. Introduction

Natural gas hydrates (NGHs) have been acknowledged as a significant future clean energy source in marine sediment due to their enormous reserves, high energy density, and low-pollution dissociation products [1–3]. Alternating hydrate formation environment has emerged as a viable gas recovery strategy through affecting the hydrate dissociation efficiency [4]. The predominant technical challenge in the hydrate exploitation field arises from the secondary hydrate formation during the dissociation phase [5]. With the help of heat transfer in the reservoir porous media, the thermal-assisted gas production enhancement method (abbreviated as thermal stimulation method) has been widely proposed to solve the secondary hydrate generation problem through thermal compensation mechanism [6–8], which is a common issue encountered in conventional depressurization approaches, stemming from inadequate thermal energy replenishment around the wellbore region [9,10]. Following thermal fluid injection into the formation, the coupled thermo-hydrodynamic process within porous media manifests as two distinct yet interconnected phases: the heat injection followed by the gas production. From the perspective of energy exchange, heat carrier input primarily functions as supplementing sensible heat within the marine sediment [11], and the gas production predominantly involves the thermochemical endothermic process triggered by post-stimulation hydrate dissociation in reservoirs [12].

The continuum-scale multiphysics coupling that persists throughout both heat injection and gas production phases necessitates the systematic identification of key parameters governing each phase during thermal stimulation. This comprehensive parameter analysis enables coherent interpretation of system evolution from the initial heat generation phase through to the terminal gas production stage [13,14]. As the dominant fluid parameter dictating hydrate occurrence morphology and pore-scale characteristics in gas hydrate-bearing reservoirs, hydrate saturation exerts multiple control: modulating marine sedimentary mechanical properties, governing gas–water relative permeability [15,16], and serving as a thermal efficiency indicator for hydrate decomposition [17,18]. Therefore, it is crucial to assess hydrate saturation levels during both energy injection and gas production operational phases of the thermal stimulation process to ensure reservoir characterization.

From the perspective of simultaneously reducing heat loss and improving energy utilization efficiency, thermal stimulation methods have been explored with different heat carriers under various heat injection systems [19–21]. Recently, an innovative thermal-based production enhancement approach, self-heat generating fluid method (abbreviated as self-heat injection method), has been proposed to enhance the hydrate reservoir exploitation efficiency [22,23]. The key mechanism of the self-heat injection method entails thermal energy liberation from chemical heating systems to stimulate secondary hydrate dissociation within marine sediments [24]. Beyond thermal dissociation enhancement, exothermic reaction byproducts (notably saline) effectively suppress hydrate reformation [25]. While these advantages exist, the method's effectiveness fundamentally depends on initial hydrate saturation. Such dependence arises because initial hydrate saturation functions as an important petrophysical parameter that simultaneously regulates gas production capacity, thermal efficiency, and operational conditions. One example includes determining calcium-oxide-based fracturing fluid concentration in thermal stimulation [26], with higher saturation levels demonstrably extending gas production duration in marine sediment experiments [27]. However, two research gaps exist in self-heat injection methodologies: (i) the hydrate saturation evolution during the energy injection stage remains poorly characterized despite its thermo–mass–velocity coupling significance throughout the hydrate dissociation process [28–30]; and secondary hydrate reformation risks caused by localized wellbore thermal depression lack quantitative constraints during the thermal carrier injection process [31]. The absence of hydrate saturation monitoring during the thermal injection period results in uncontrolled secondary hydrate formation. This directly induces near-wellbore permeability impairment during injection, ultimately compromising the feasibility of hydrate management strategies [32].

This study bridges these gaps by establishing a hydrate saturation prediction model to: (i) quantify hydrate saturation dynamics during the self-heat fluid injection process, and (ii) develop phase equilibrium constraints against secondary hydrate nucleation in the near-wellbore area. The effects of thermal stimulation on hydrate saturation evolution, reservoir temperature distribution, and associated reservoir properties are systematically examined across baseline and sensitivity analysis cases. The $NaNO_2$-NH_4Cl exothermic system is utilized as the self-heating agent, benefiting from its thermal performance and chemical stability under hydrate-bearing sediment conditions [33]. This study emphasizes the assessment of intrinsic permeability characteristics and injection protocols on hydrate saturation patterns in marine sediments. Sensitivity analysis identifies initial permeability (1.5–100 mD) and thermal injection mode as key parameters, given permeability's dominant role in controlling thermal front advancement in hydrate reservoirs [34,35]. Additionally, in situ heating and fluid flooding constitute alternative thermal stimulation strategies for the reservoir heat loss mitigation, necessitating comparative analysis in hydrate-bearing sediments [36].

This work quantifies hydrate saturation evolution during self-heating fluid injection in gas hydrate reservoirs, employing numerical multiphysics models during the non-production period to decouple saturation estimation from produced gas flow. The injection process utilizes a vertical well within a hydrate reservoir, with porous media properties referenced from Shenhu area, South China Sea [37]. The numerical evaluation of the base case reveals a 16.3 °C average reservoir temperature increase during the "Reactive chemical flow stage" (defined in Section 2.1), induced by chemical self-heating fluid injection, while hydrate saturation declines from 40% to 21%. The self-heat generating method demonstrates superior applicability in reservoirs possessing permeability below 15 mD; however, this method exhibits limited efficacy in porous media with 55–70 mD initial permeability. The fluid flooding technique surpasses the in situ heating approach.

2. Materials and Methods

2.1. Defining Two Stages in Reservoir Energy Stimulation

This study employs numerical modeling to analyze vertical well injection systems for sodium nitrite solutions in marine hydrate reservoirs. During the initial injection phase, hydrate decomposition heat arises from two primary mechanisms: sensible heat conduction through the reservoir, and thermal convection of reactive fluids generated by the working reagent's exothermic reaction [38]. As heat and mass transfer propagate through the porous media, hydrate dissociation initiates gas–aqueous phase transitions. The dynamic evolution of three-phase saturation (gas/water/hydrate) serves as a direct indicator of thermal–fluid coupling between the heat generation system and the sedimentary reservoir. The thermochemical stimulation process involves concurrent exothermic chemical reactions and hydrate phase transformations, leading to progressive reservoir temperature increase accompanied by hydrate saturation decline. Upon completion of the exothermic period, the reservoir reaches peak temperature. The synchronous process of the chemical reaction and phase transition is described as the "reactive chemical flow stage", abbreviated as Stage 1. Equation (1) characterizes exothermic and endothermic chemical reactions for Stage 1.

$$\begin{aligned} NaNO_2 + NH_4Cl &\overset{H^+}{\Rightarrow} NaCl + N_2 + 2H_2O \\ CH_4 \cdot (H_2O)_{5.75} &\Rightarrow CH_4 + 5.75(H_2O) \end{aligned} \tag{1}$$

Upon cessation of Stage 1, the reactive chemical flow terminates while the temperature difference between the hydrate zone and the non-hydrate zone dominates the sensible heat towards hydrate decomposition within the porous media, and thus the process after Stage 1 is defined as "non-reactive flow stage" (abbreviated as Stage 2). In this stage, thermal energy influx from Stage 1 progressively attenuates due to endothermic hydrate dissociation, thereby inducing average temperature decline within the marine sediment; meanwhile, reservoir pressure dynamics exhibit complex characterization under the combined influence of injection pressure and hydrate decomposition, wherein stochastic hydrate saturation trajectories emerge: hydrates may either decompose or regenerate; consequently, their saturation within the reservoir porous media may change randomly.

2.2. Simulation Tool Introduction

COMSOL Multiphysics 5.4® is a finite element method-based software that enables the simulation of coupled physical processes, including sedimentary fluid flow, heat transfer, and hydrate phase transition within marine sediments containing gas hydrates [39]. Therefore, COMSOL Multiphysics® is employed as the numerical tool to develop the self-heating fluid injection simulation model. The assumptions and model construction are introduced as follows.

2.3. Model Construction

As shown in Figure 1, a self-heat generating fluid injection well is located in a hydrate reservoir of the South China Sea, and the well has been operated with casing perforation. The 2D simulation domain (2 m × 1 m) features initial hydrate saturation homogeneously distributed in a unilateral wellbore configuration. The sediment condition and fluid initial properties are listed in Table 1.

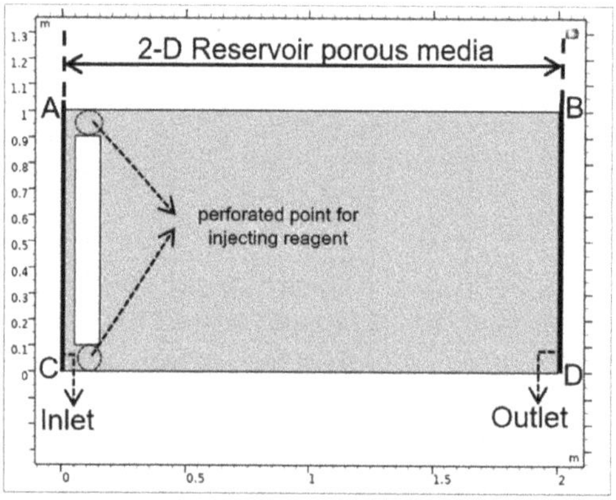

Figure 1. Hydrate reservoir porous media.

Table 1. Fundamental physical parameters of reservoir porous media and fluid.

Parameters	Physical Meaning	Values
d [m]	Reservoir length (Figure 1, AB and CD)	2.00
b [m]	Reservoir width (Figure 1, AC and BD)	1.00
k_{por} [mD]	Reservoir initial absolute permeability	1.50
ϕ	Reservoir porosity	0.33
S_{ghi}	Initial hydrate saturation	0.40
S_{wi}	Initial water saturation	0.55
S_{gi}	Initial gas saturation	0.05
M_{gh} [g/mol]	Hydrate molar mass	119.50
M_m [g/mol]	Methane molar mass	16
M_w [g/mol]	Water molar mass	18

2.4. Model of Reactive Chemical Flow Stage (Stage 1)

In Stage 1, the governing equations for flow momentum, heat transfer, and mass conservation are formulated in Equations (2)–(4). The initial parameters of Stage 1 are shown in Table 2. In this stage, the combined effects of the exothermic reaction of the self-heating system and the endothermic reaction of hydrate dissociation are evaluated.

By integrating the chemical reactions with the convection–dispersion equation, the mass of water released from hydrate dissociation and the corresponding changes in hydrate saturation have been calculated. During the process of correlating chemical reactions with fluid heat transfer, the heat exchange between the endothermic reaction of hydrate dissociation and the formation environment has been considered.

Table 2. Required physical parameters in Stage 1.

Parameters	Physical Meaning	Related Equation	Values
ρ_c [kg/m^3]	Chemical reagent fluid density	Momentum conservation equation	1.0×10^3
V_C [m/s]	Chemical reagent fluid velocity		0.19
μ_c [Pa·s]	Chemical reagent fluid dynamic viscosity		1.0×10^{-3}
g [m/s^2]	Force per unit volume of fluid		9.80
ρ_G [kg/m^3]	Hydrate density	Heat transfer equation	0.91
C_G [J/(kg·K)]	Hydrate heat capacity		1600
k_G [W/(m·K)]	Hydrate thermal conductivity		0.62
u_G [m/s]	Hydrate velocity		0
q_G [W/m^2]	Heat flux by conduction		$-k_G\nabla T$
Q_G [W/m^3]	External heat source for hydrate		0
D_i [m^2/s]	The diffusion coefficient of injected fluid molecules in porous media	Mass conservation equation	1×10^{-9}

Momentum conservation equation

The momentum conservation equation (Equation (2)) describes the flow situation of the heat generation system during the injection process of self-heat injection fluid in the reservoir porous media. The definitions of the parameters in Equation (2) are listed in Table 2.

$$\begin{aligned} \rho_c \tfrac{dV_c}{dt} &= \rho_c g - \nabla p + \mu_c \nabla^2 V_c \\ \tfrac{\partial \rho_c}{dt} &+ \nabla \cdot (\rho_c V_c) = 0 \end{aligned} \quad (2)$$

where ρ_c is the chemical reagent fluid density, kg/m3; V_C is the chemical reagent fluid velocity, m/s; μ_c is the chemical reagent fluid dynamic viscosity, Pa·s; g is force per unit volume of fluid, m/s^2.

The momentum conservation is expressed by the simultaneous formulation of the Navier–Stokes equation in the first line and the continuity equation in the second line, where the Navier–Stokes equation describes the dynamic behavior of the fluid velocity field V evolving over time. This set of equations embodies the balance of forces (inertial, body, pressure, and viscous forces) acting on the element of injected chemical reagent fluids.

$\rho_c \frac{dV_c}{dt}$ represents the inertial force per unit volume of the fluid; $\rho_c g$ is the body force per unit volume acting on the fluid, where g is the gravitational force; in this study, gravitational force is ignored. $-\nabla p$ represents the pressure gradient force per unit volume, with p being the pressure and ∇ is the gradient operator, $\mu_c \nabla^2 V_c$ is the viscous force per unit volume of fluid, describing the viscous dissipation and momentum transfer within the fluid.

Heat transfer equation at hydrate interface

Equation (3) quantifies hydrate-induced thermal variations during Stage 1 heat injection. Table 2 summarizes hydrate thermal properties and Equation (3) parameter definitions. Equation (3) is used to describe the heat transfer of hydrates considering heat convection and conduction in the reservoir porous media.

$$\begin{aligned} &\rho_G C_G u_G \cdot \nabla T + \nabla \cdot q_G = Q_G \\ &q_G = -k_G \nabla T \end{aligned} \tag{3}$$

where ρ_G is the hydrate density, kg/m^3; C_G is the hydrate heat capacity, J/(kg·K); k_G is the hydrate thermal conductivity, W/(m·K); u_G is hydrate velocity, m/s; q_G is the heat flux by conduction, W/m^2; Q_G is the external heat source for hydrate, W/m^3; D_i is the diffusion coefficient of injected fluid molecules in porous media, m^2/s.

The equations presented describe the heat transfer process in hydrate reservoirs. This set of equations is used to analyze the temporal and spatial distribution of the temperature field in hydrate reservoirs. In the first line equation (energy equation), the term $\rho_G C_G u_G \cdot \nabla T$ incorporates heat transfer (convective term) resulting from hydrate movement. The term $\nabla \cdot q_G$ accounts for the heat change due to thermal conduction, Q_G is the external heat source for hydrate. The second line equation is Fourier's law of heat conduction; this equation relates the heat flux vector q_G to the temperature gradient ∇T. By substituting Fourier's law into the energy equation, a hydrate-reservoir heat transfer governing equation that incorporates convection, conduction, and heat sources can be derived.

Mass conservation equation

The mass conservation equation governs Stage 1 aqueous-phase solute concentration distribution and porous-media reactions. The convection dispersion equation is formulated based on Fick's law Equation (4), where c represents the volume concentration (mol/m^3) of the heat generation system in the liquid phase; V_c represents the convective flow velocity (m/s); D_i represents the diffusion coefficient (m^2/s) of self-heat fluid molecules in porous media; and R_i refers to the chemical reaction rate (mol/(m^3·s)). Detailed parameters' meanings are shown in Table 2.

$$\frac{\partial c_i}{\partial t} + \nabla \cdot (-D_i \nabla c_i + c_i \cdot V_c) = R_i \tag{4}$$

where c is the volume concentration of the heat generation system in the liquid phase, mol/m^3; V_c represents the convective flow velocity, m/s; D_i refers to the diffusion coefficient of self-heat fluid molecules in porous media, m^2/s; and R_i is the chemical reaction rate, mol/(m^3·s).

The equation presented describes the mass transfer process of chemical fluid i (chemical reagent fluid) in hydrate reservoirs. It is in the form of a convection–diffusion–reaction equation. This equation comprehensively integrates diffusive, advective, and source/sink effects, serving as a governing equation for analyzing coupled mass transfer and multiphase flow in hydrate-bearing reservoirs.

The term $\frac{\partial c_i}{\partial t}$ represents the local rate of change in the concentration c_i of fluid i. The divergence term $\nabla \cdot (-D_i \nabla c_i + c_i \cdot V_c)$ characterizes the total mass flux of fluid i; $-D_i \nabla c_i$ denotes the diffusive flux driven by concentration gradients, and $c_i \cdot V_c$ represents the advective flux induced by the bulk flow of the fluid phase. The term R_i is the source/sink term, which accounts for the mass production or consumption of fluid i due to hydrate dissociation within the reservoir.

As shown in Equations (2)–(4) and Table 2, the momentum conservation constitutive parameters encompass thermally generated fluid density, viscosity, and volumetric flux during Stage 1 operational dynamics. The heat transfer equation incorporates hydrate properties including specific heat capacity and thermal conductivity. Equation (5) dynamically characterizes real-time water saturation determination through its porosity–saturation correlation [40,41].

$$S_{w\_1} = \frac{c_w \cdot 5.75 \cdot M_w}{\rho_w \phi} \tag{5}$$

where S_{w_1} is the instantaneous water saturation; c_w is the hydrate water ($5.75 \cdot H_2O$) concentration; M_w is water molar mass, g/mol; ρ_w is water density, g/cm^3; φ is porosity. By combing the instantaneous water saturation with the initial water saturation (S_{wi}), the water saturation during the hydrate dissociation process can be calculated using Equation (6).

$$\Delta S_w = (S_{w\_1} - S_{wi}) \tag{6}$$

$$Rs = \frac{\Delta S_{gh}}{\Delta S_w} = \frac{\Delta c_{gh}}{\Delta c_w} \cdot \left(\frac{M_{gh}}{\rho_{gh}\varphi}\right) \Big/ \left(\frac{5.75 \cdot M_w}{\rho_w \varphi}\right) \tag{7}$$

$$S_{gh\_1} = S_{ghi} - R_s \Delta S_w \tag{8}$$

where ΔS_w in Equation (6) is the decomposed water saturation involved in Stage 1. The instantaneous hydrate saturation can be calculated based on Equations (7) and (8). Rs is the ratio of hydrate saturation to the decomposed water saturation involved in Stage 1; subscript *gh* represents the hydrate while w refers to the water. M_{gh} is molar mass of hydrate, which equals to 119.5 g/mol. Symbol Δ represents the difference in the water or hydrate phase involved in Stage 1. S_{ghi} refers to the initial hydrate saturation. ΔS_{gh} quantifies hydrate saturation differential throughout its dissociation dynamics. S_{gh_1} is the instantaneous hydrate saturation in the porous media.

2.5. Model of Non-Reactive Flow Stage (Stage 2)

In Stage 2, the decomposition–water saturation results obtained from Stage 1 are used to describe the decomposition–water mass increment via Darcy's law, and the iterative evolution of the mass increment is triggered based on the temperature field's phase transition due to temperature differences. During the iterative calculation of decomposition–water saturation, by tracking the real-time temperature's contour map, the corresponding hydrate temperature map evolution values can be derived. This enables the quantification and analysis of hydrate saturation changes through the temperature field mapping of hydrate areas.

The thermodynamic evolution during Stage 2 is quantitatively determined through continuous monitoring of phase transition temperatures and corresponding latent heat fluxes within the sedimentary porous media. The momentum conservation framework for this stage is mathematically expressed by Equation (9), with initial boundary conditions defined in Table 3. This formulation is derived from Darcy's law, incorporating two principal parameters: (1) the intrinsic permeability of the reservoir, and (2) the average linear velocity of the aqueous phase at the end of Stage 1. The heat transfer equation maintains continuity with the Stage 1 model Equation (3). Critical parameters of Equation (3) include phase change latent heat, the prescribed temperature differential of 1 K for phase transition, and other thermophysical properties.

Momentum conservation equation

$$\begin{aligned}
&\nabla \cdot (\rho_w u_w) = Q_m \\
&u_w = -\frac{\mathrm{k_{por}}}{\mu_w}(\nabla \rho_w) \\
&Q_m = S_{w_step2}\phi(\rho_G - \rho_w)\frac{\partial S_w}{\partial t} \\
&S_{w_step2} = S_{w_res} + (1 - S_{w_res}) \cdot F \\
&F = \begin{cases} 1 & T \geq T_{pc} \\ 0 & T < T_{pc} \end{cases}
\end{aligned} \quad (9)$$

where ρ_w is the water dynamic density, kg/m^3; u_w is water velocity, m/s; Q_m is the water mass increase due to hydrate phase change, W/m^3; μ_w is the water dynamic viscosity, Pa·s.

Table 3. Required physical parameters in Stage 2.

Parameters	Physical Meaning	Related Equation	Values
C_w [J/(kg·K)]	Water heat capacity	Heat transfer equation	1600
k_w [W/(m·K)]	Water thermal conductivity		0.62
L_{pc} [J/kg]	Latent heat for phase change		43,500
Q_h [W/m^3]	External heat source		0
ρ_w [kg/m^3]	Water dynamic density	Darcy's law equation	1×10^3
u_w [m/s]	Water velocity		1.96×10^{-8}
μ_w [Pa·s]	Water dynamic viscosity		1.0×10^{-3}
Q_m [W/m^3]	Water mass increase due to hydrate phase change		Based on simulation process

2.6. Boundary Condition and Key Parameters Descriptions

2.6.1. Boundary Condition Description

Stage 1: The boundary conditions for Stage 1 are detailed as follows. At AC inlet boundary, the initial chemical reagent fluid concentration is defined as 2000 mol/m^3, while the volume flow rate is 0.19 m/s. This boundary condition is used to define the initial concentration and injection velocity of reactants to provide the primary driving condition for reactant transport in the mass and flow fields, coupled calculations with the energy field, thereby simulating the dynamic process of reactant migration from the injection point into the reservoir interior. The injection velocity of the chemical reagent is 0 at reservoir upper and lower boundaries (AB and CD shown in Figure 1). In addition, AB and CD serve as isothermal flow boundaries so as to provide stable thermodynamic constraints for the energy field, preventing solution divergence caused by fluctuating thermal boundary conditions, particularly in temperature-sensitive processes such as hydrate phase transitions. In addition, the initial chemical reagent fluid concentration in the reservoir is 0 mol/m^3 since the reservoir is initially free of reactive components. The reservoir initial temperature is set to 286.2 K (13 °C) to reflect the ambient reservoir conditions prior to fluid injection (Table 4).

Stage 2: The boundary conditions for Stage 2 are outlined below. A pressure of 20 MPa is applied at the AC boundary for chemical reagent injection while the outlet pressure is set at BD boundary with 5 MPa. The pressure gradient drives the chemical reagent flow in the reservoir, serving as the dynamic basis for coupling the energy and flow fields. The AC boundary temperature is set as the average reservoir temperature at the end of Stage 1, ensuring continuity with the initial reservoir temperature at the start of Stage 2. BD is the outlet boundary of porous heat conduction. AB and CD are isothermal flow boundaries. The initial hydrate temperature is set as 278.7 K (5.6 °C).

Table 4. Initial conditions and boundary conditions for Stages 1 and 2.

Stage	Parameters	Physical Meaning	Values
Stage 1	V_C [m/s]	AB and CD boundary velocity of chemical reagent fluid	0
	T_{1_ave} [K]	Formation ambient temperature	286.2
	$c_{\_initial}$ [mol/m^3]	Initial chemical reagent fluid concentration in the reservoir	0
	$c_{\_inlet}$ [mol/m^3]	Initial chemical reagent fluid concentration at AC inlet boundary	2000
Stage 2	$P_{\_inlet}$ [MPa]	Injected pressure for chemical reagent fluid	20
	$P_{\_outlet}$ [MPa]	Outlet pressure for chemical reagent fluid	5
	T_{2_ave} [K]	Average formation temperature at the beginning of Stage 2 (by the end of Stage 1)	303.8 (base case)
	T_{2_gh} [K]	Hydrate temperature before phase change	278.7

2.6.2. Key Parameters Description

(1) The explanation for the heat capacity

The hydrate heat capacity corresponds to the fluid-phase heat capacity in Stage 2, matching its value during the exothermic reaction period (Stage 1), while differing from the value of liquid water. The reason is as follows.

In the thermal decomposition process of hydrate reservoirs, the governing equations describe the critical state where hydrates transform into liquid water. At this critical state, hydrates remain the dominant component, with heat primarily consumed for temperature changes and latent heat absorption during phase transition. Since the equations focus on the critical transition stage, the thermal properties of hydrates dominate the system's energy balance. Hence, adopting the specific heat capacity of hydrates accurately characterizes the thermal properties under this critical condition, ensuring that the governing equations provide a relatively precise simulation of the hydrate dissociation process.

(2) The explanation for the velocity of reagent and water

The macroscopic heat transport of the working fluid in porous media is primarily considered during Stage 1, so the macroscopic average flow velocity across the porous media's cross-sectional area (0.19 m/s) has been adopted. In Stage 2, the influence of reservoir properties on flow is mainly considered. To ensure that the selection of velocity parameters in the temperature field control equation matches the physical mechanisms described by Darcy's law, the Darcy velocity (average 1.96×10^{-8} m/s) has been used to define the water flow velocity describing the physical process of fluid seepage and heat transfer during the hydrate–water phase transition.

(3) Phase transition latent heat

Phase transition latent heat (L_{pc}) refers to the energy absorbed during hydrate-to-water phase transformation at constant temperature. In this study (Table 3), L_{pc} adopts the value from reference [42], which is a validated thermodynamic parameter for methane hydrate decomposition. This reference is specifically selected because of the following: (1) It addresses similar engineering scenarios involving wellbore temperature gradients and hydrate decomposition fronts, and both studies focus on near-wellbore heat transfer during hydrate dissociation. (2) The analytical solutions of this reference's thermal field model have been verified through numerical solutions.

2.7. Mesh Configuration, Convergence Criteria and Solver Settings

Using COMSOL's automatic grid creation function to execute grid refinement for Stages 1 and 2. Resultant datasets (such as post-Stage 1 reservoir temperature) embedded as initialization parameters in the equation formulation of Stage 2. The simulation solver settings and mesh design are introduced as follows.

2.7.1. Convergence Criteria

For both Stages 1 and 2, the termination condition for iterative calculations is that the relative tolerance for calculations is maintained at 1×10^{-4}. The maximum number of iterations is limited to 200.

2.7.2. Solver Settings

Stage 1: Steady-state computational solver is employed for Stage 1 equilibrium analysis. For solver selection in parameter solving, the MUMPS (Multi-frontal Massively Parallel sparse direct Solver) solver is used to solve the pressure, while the PARDISCO (Parallel Direct Sparse Solver Interface) is employed for solving fluid flow variables, concentration, and heat transfer variables.

Stage 2: Steady-state computational solver and transient solver are employed in Stage 2. In the transient solver settings, the implicit solution method using Backward Differentiation Formulas (BDF) is employed. To ensure the solver strictly adheres to the requested output time steps and selects appropriate step sizes, minimizing numerical differences in the solved variables and improving solution stability, the transient solver uses the "strict" step size selection level instead of intermediate. For solving parameters, the MUMPS solver is used to solve pressure and temperature.

2.7.3. Mesh Configuration

Mesh configuration is shown in the validation part (Section 3.5), which represents the basic case numbered Test 1-1 in Table 5. Mesh design details will be introduced in that section where multiple grid refinement conditions are compared with each other.

Table 5. Grid information for stability validation cases (Stages 1 and 2).

Stages	Test Number	Grid Element Number	Maximum Grid Size	Minimum Grid Size	Maximum Element Growth Rate	Curvature Factor	Simulated Saturation
Stage 1	Test 1-1	3195	0.067 m	0.003 m	1.20	0.40	0.2111
	Test 1-2	3597	0.055 m	0.002 m	1.20	0.40	0.2133
	Test 1-3	5420	0.045 m	0.002 m	1.15	0.30	0.2106
Stage 2	Test 2-1	3210	0.040 m	1.5×10^{-4}	1.20	0.25	Please see Section 3.5
	Test 2-2	5774	0.030 m	1.5×10^{-4}	1.10	0.20	
	Test 2-3	7998	0.025 m	4.0×10^{-5}	1.10	0.20	

3. Results and Discussions

3.1. Base Case

3.1.1. Reactive Chemical Flow Stage (Stage 1)

In Stage 1, simulation outputs comprise reservoir temperature and three-phase saturation within the marine sediment. Hydrate saturation declines as the average reservoir temperature rises throughout Stage 1. At the end of Stage 1, the final distribution of hydrate saturation and the average value (S_{gh_1}) are shown in Figure 2, average hydrate saturation within the reservoir declines from 0.40 to 0.21 (47.5% reduction) during Stage 1. The water saturation increases from 0.55 to 0.65, while the saturation of methane climbs from residual gas saturation of 0.05 to 0.14. Average reservoir temperature rises from initial 13.0 °C to 29.3 °C (T_2 in Figure 3).

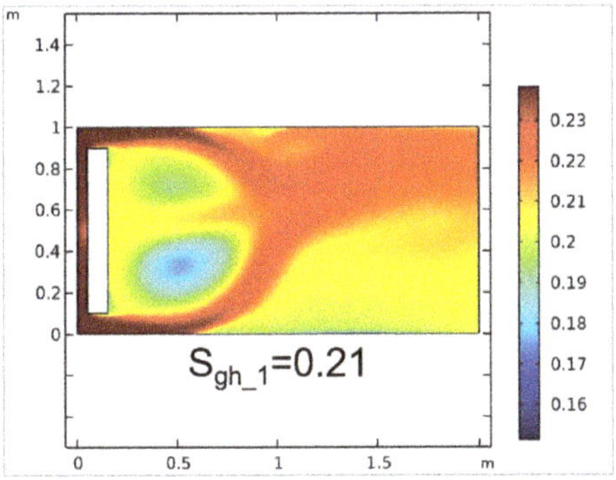

Figure 2. The reservoir hydrate saturation distribution by the end of Stage 1 (base case).

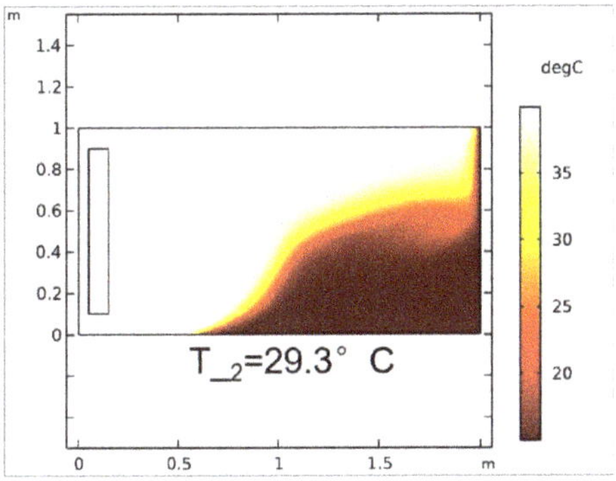

Figure 3. Average reservoir temperature by the end of Stage 1 (base case).

3.1.2. Non-Reactive Flow Stage (Stage 2)

The initial temperature and saturation conditions for Stage 2 are inherited from Stage 1 simulation outputs. Hydrate saturation exhibits continuous decay throughout Stage 2 (Figure 4), indicating full dissociation of hydrates. Correspondingly, the reservoir average temperature decreases from 29.3 °C to 27.2 °C, demonstrating strong coupling with hydrate saturation dynamics. The fluid saturation study area is delineated by the red line zone in Figure 4 (t = 0), encompassing six saturation profiles specifically designed to monitor secondary hydrate formation near the wellbore. Within this domain, white areas visualize hydrate distribution, whereas blue color maps aqueous phase saturation. The white regions inside the boundary correspond to initial hydrate saturation, whereas peripheral white areas indicate secondary hydrate accumulation.

In Stage 1, the reservoir's thermal behavior reveals that the coupled effect of fluid heat release and hydrate decomposition causes the rising average temperatures and declining hydrate saturation. During Stage 2, hydrate saturation follows an accelerated dissociation rate after 5 h compared to the 0–5 h period. This hydrate–water phase transition drives thermal absorption, resulting in slight temperature reduction.

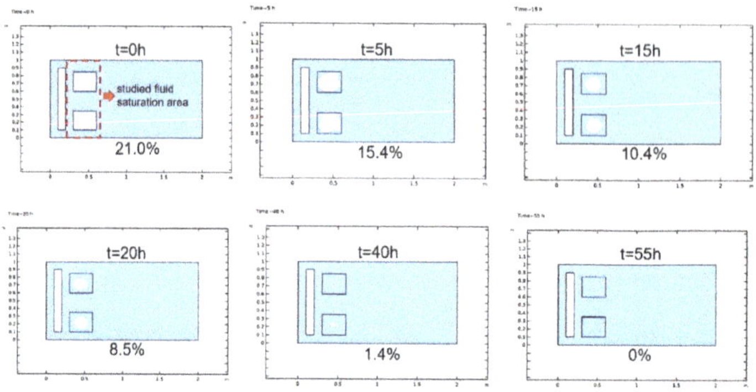

Figure 4. Hydrate saturation distribution over time for the base case (0~55 h).

3.2. *Sensitivity Analysis*

This section examines how initial absolute permeability (K) and injection modes affect hydrate saturation, reservoir temperature, and other related parameters. The sensitivity analysis indicates that, during Stage 2, both reservoir permeability and working fluid's injection mode significantly impact hydrate saturation patterns, reservoir thermal profile, hydrate-zone conductivity, and hydrate surface endothermic flux. The detailed sensitivity analysis is articulated as follows.

3.2.1. Initial Absolute Permeability

Comparative analysis of nine reservoir permeability cases (1.5–100 mD) reveals negligible impacts of initial absolute permeability on hydrate saturation and reservoir temperature in Stage 1; Stage 2, however, demonstrates pronounced permeability control over hydrate saturation distribution and temperature patterns, as detailed in Section Non-Reactive Flow Stage (Stage 2).

Permeability Study: Non-Reactive Flow Stage (Stage 2)

The hydrate-bearing media's initial permeability is classified as low-K (1.5–15 mD), medium-K (25–55 mD), and high-K (70–100 mD); hydrate saturation patterns exhibit permeability-dependent changes across these ranges. Figures 5–7, respectively, illustrate temporal hydrate saturation changes within the fluid saturation domain across each permeability range. Figure 8 compiles temporal hydrate saturation averages across all permeability cases, with Stage 2 duration divided into early, middle, and late periods. Figures 5–8 reveal heterogeneous temporal hydrate saturation evolution. Low-K domain demonstrates progressive hydrate depletion during Stage 2, whereas medium-high permeability ranges exhibit transient regeneration followed by decomposition in phase transition. Figures 6 and 7 demonstrate secondary hydrate formation via heterogeneous nucleation in porous media, aligning with the reported pattern of regenerated hydrates [43]. Secondary hydrate generation peaks are within the 55 mD permeability range. Analysis of reservoir temperature profiles (Figures 9–11) reveals that the 55 mD case exhibits significantly greater temperature decline than 5 mD and 70 mD cases during the late-stage phase (40–55 h) of Stage 2, demonstrating the most pronounced heat absorption through hydrate decomposition. Reservoir average temperature distribution is presented in Figure 12.

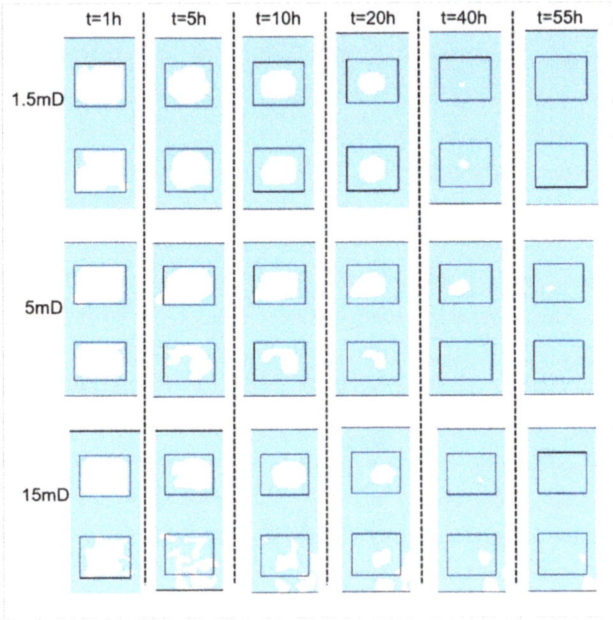

Figure 5. The evolution of hydrate saturation zone over time (1.5–15 mD).

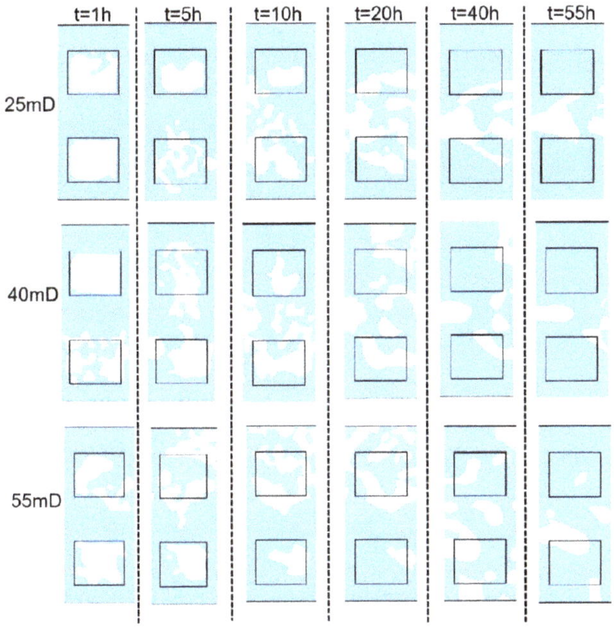

Figure 6. The evolution of hydrate saturation zone over time (25–55 mD).

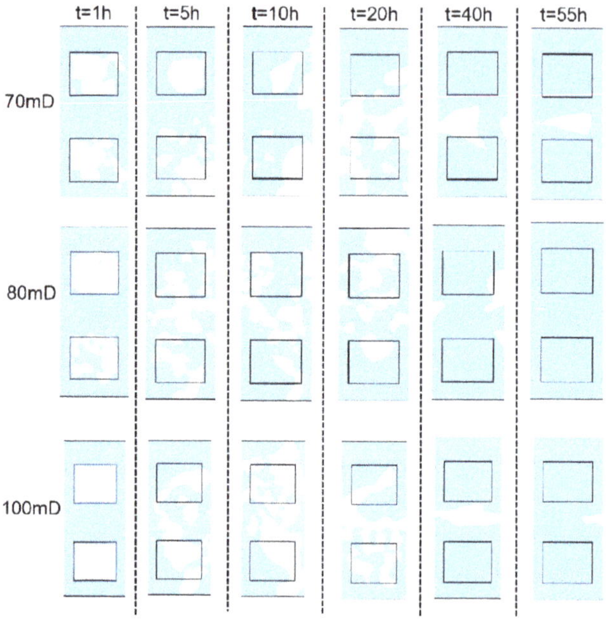

Figure 7. The evolution of hydrate saturation zone over time (70–100 mD).

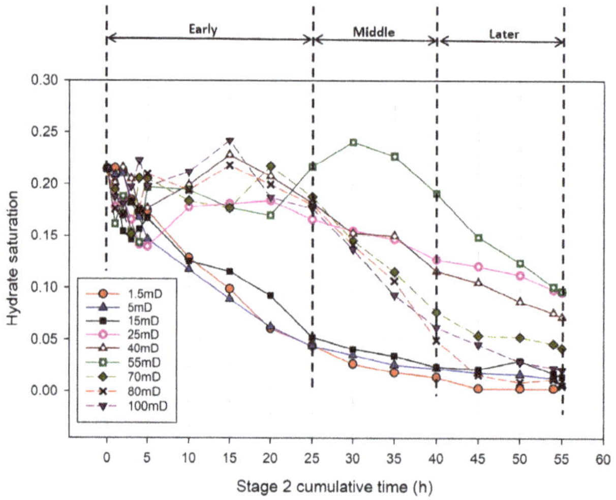

Figure 8. The estimated average hydrate saturation for various reservoir permeability conditions during Stage 2.

Thermal conductivity and endothermic heat flux govern Stage 2 temperature fluctuations in the reservoir porous media. Figure 12 reveals an anomalous temperature deviation at 25 h in the 55 mD case, exhibiting non-monotonic behavior resulting from competing thermal processes: minor thermal release via hydrate reformation and simultaneous endothermic phase transition of original hydrates [44]. This observation demonstrates marked hydrate phase transition initiation after t = 25 h. It has been confirmed that the hydrate reformation is strongly associated with the reservoir thermodynamic feedback to the attenuated heat conduction efficiency, and the thermal conductivity of porous media is positively correlated with the hydrate decomposition rate [45]. Therefore, the quantification of hydrate-bearing zone thermal indices (conductivity and endothermic flux) during Stage 2 (25–55 h) is essential, with results detailed in Figures 13 and 14. Within the low-permeability range (1.5–15 mD), hydrates exhibit progressive dissociation throughout Stage 2 without secondary formation. Both thermal conductivity (λ) and endothermic heat flow rate (Qendo) maintain depressed levels throughout 25 < t < 55 h. Within moderate-to-high permeability ranges (25–100 mD), dissociation dynamics coexist with secondary hydrate formation. The reason why hydrates regenerate and then decompose can be explained as follows.

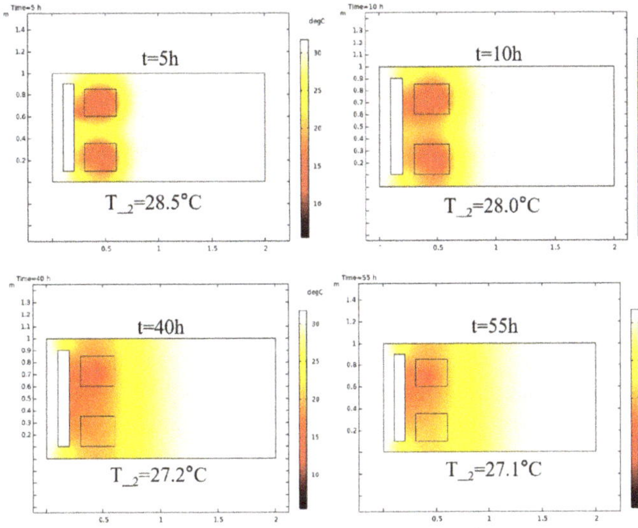

Figure 9. Reservoir temperature change profile (5 mD).

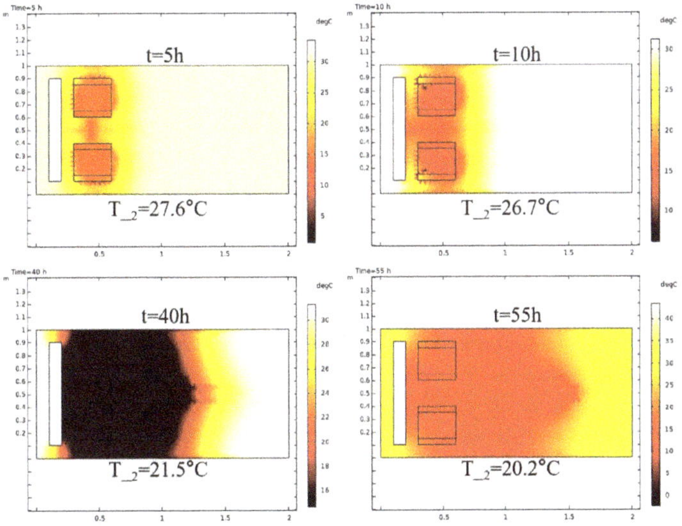

Figure 10. Reservoir temperature change profile (55 mD).

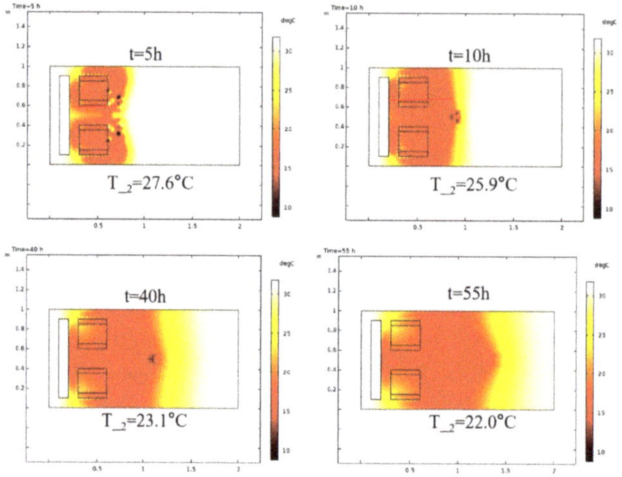

Figure 11. Reservoir temperature change profile (70 mD).

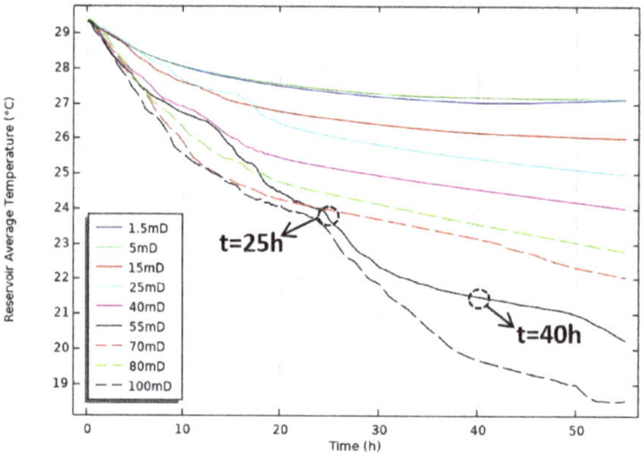

Figure 12. Average reservoir temperature for various permeability conditions.

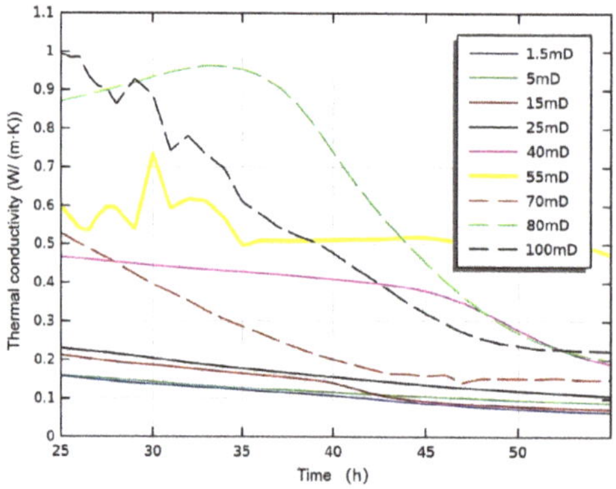

Figure 13. The thermal conductivity of the hydrate area from 25 to 55 h in various permeability conditions.

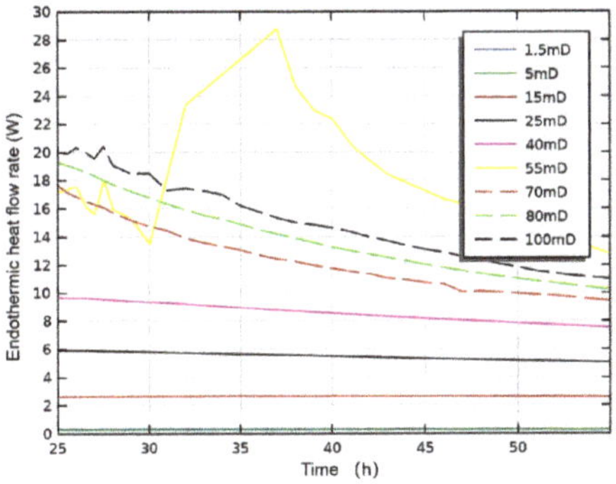

Figure 14. The endothermic heat flow rate of the hydrate area from 25 to 55 h in various permeability conditions.

In reservoirs with permeability higher than 25 mD, hydrate thermal conductivity typically exhibits twofold enhancement relative to the low-K values in 25 to 40 h duration (Figure 13). Enhanced thermal conductivity elevates heat flux, thereby augmenting the endothermic heat flow rate at hydrate interfaces. The inflow heat flow rate of the system remains fixed due to the constant properties of injected working fluid, so with the elevated endothermic heat flow rate, the net heat flow rate will be reduced. As a result, the decline of average temperature induces secondary hydrate formation in porous media. While low-K (1.5–15 mD) zones show absent secondary hydrates, medium/high-K conditions (25–100 mD) exhibit pronounced temperature difference between original and new hydrate zones with temporal progression. This finding corroborates the existing conclusion that the temperature difference drives the spatial distribution difference in secondary hydrates [31]. In addition, the observed positive correlation between thermal conductivity and reservoir permeability aligns with published studies demonstrating that enhanced gas mobility under high permeability conditions induces higher thermal conductivity in hydrate-bearing sediment [46]. During $\Delta t = 25$–55 h, the thermal conductivity (λ) in 25–100 mD cases exhibits significant decay, which is corresponding to the decline of the endothermic heat flow rate Q_{endo}. With constant Q_{in} in system thermodynamics, the net heat flow rate (Q_{net}) escalation induces attenuated reservoir cooling, triggering hydrates re-decomposition in the reservoir porous media.

In short, hydrate regeneration behavior is fundamentally controlled by initial reservoir permeability (K), with distinct regimes emerging based on permeability thresholds: when $K \leq 15$ mD, primary hydrates fully dissociate without secondary nucleation through dynamic thermal conductivity variations and endothermic heat flow rates within the hydrate zone; however, at $25 \text{ mD} \leq K < 70$ mD, secondary hydrates form during early-mid Stage 2 with peak regeneration intensity in the mid-phase; and for $K \geq 70$ mD, late-phase dissociation of secondary hydrates occurs as mass transfer dominates. These permeability-dependent patterns result from the competition between endothermic dissociation kinetics and exothermic reformation thermodynamics, where permeability governs heat distribution efficiency and fluid migration capacity, suggesting that permeability-modulation strategies should be customized to specific reservoir property classes to optimize self-heat fluid injection while minimizing secondary hydrate formation risks.

3.2.2. Injection Mode

The two injection modes, namely fluid flooding and in situ heating, are defined as follows.

Fluid flooding: Self-heat generated fluid (40 °C) flows into the 13.0 °C reservoir porous media from inlet boundary (AC of Figure 1).

In situ heating: Ambient temperature (13.0 °C) fluid flows into the reservoir with fixed 40 °C inlet boundary.

The contrast of in situ heating versus fluid flooding reveals that altered injection modes significantly influence hydrate saturation and reservoir temperature across both Stage 1 and Stage 2 processes.

Injection Mode Study: Reactive Chemical Flow Stage (Stage 1)

Under the coupled effects of chemical exothermic reactions and hydrate phase transitions within reservoir porous media, the average reservoir temperature under in situ heating rises from 13.0 °C to 24.6 °C (Figure 15), while corresponding hydrate saturation declines from 0.40 to 0.23 (Figure 16).

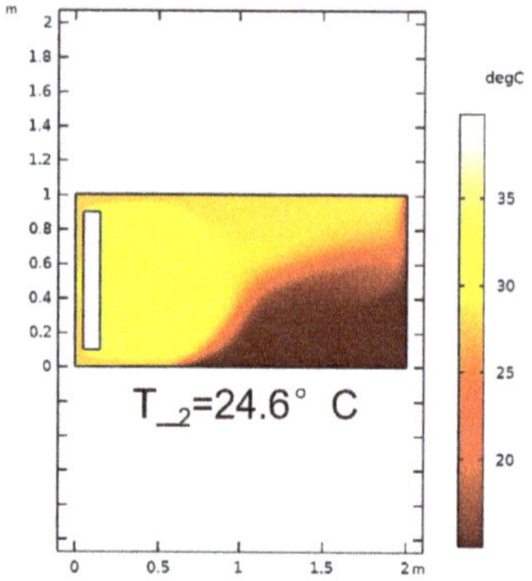

Figure 15. Reservoir temperature profile by the end of Stage 1 (in situ heating case).

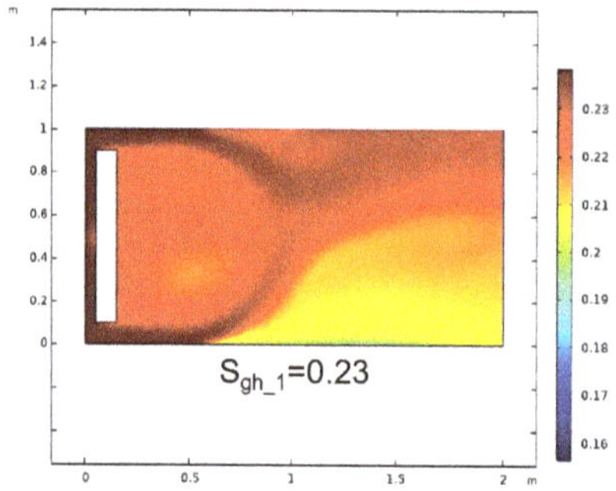

Figure 16. Hydrate saturation profile by the end of Stage 2 (in situ heating case).

Injection Mode Study: Non-Reactive Flow Stage (Stage 2)

The fluid flooding approach demonstrates superior performance to the in situ heating method in hydrate decomposition efficiency. The quantified average hydrate saturation under both injection modes is presented in Figure 17, with corresponding spatial distribution profiles illustrated in Figure 18. Figure 19 illustrates the average reservoir temperature versus cumulative duration in Stage 2 under both scenarios. The in situ heating method exhibits slower hydrate decomposition rates compared to fluid flooding, resulting in residual hydrate retention due to limited thermal propagation and fluid migration constraints.

This incomplete dissociation persists throughout the injection process, as evidenced by saturation profiles. In addition, the average reservoir temperature of the fluid flooding method is sustainably higher than that of the in situ heat generation method during Stage 2. The disparity in hydrate decomposition efficiency between both approaches stems from thermal conductivity and endothermic heat flow rate perspectives. Comparative analysis of Figures 20 and 21 with Figures 13 and 14 reveals little change in thermal conductivity in hydrate zones across the two injection modes; however, compared with the in situ heating method, the heat generation effect is better performed in the fluid flooding method (Figure 4 versus Figure 18), and thus the hydrate decomposition efficiency is elevated in the fluid flooding case throughout both Stages 1 and 2. This aligns with two recent studies: for one, solely elevating local heat source temperature insufficiently enhances hydrate decomposition efficiency due to low thermal conductivity in porous media [7]; for another, heat convection–conduction coupling effect is more contributive compared with heat conduction-dominated regimes during hydrate dissociation under thermal fluid injection [47].

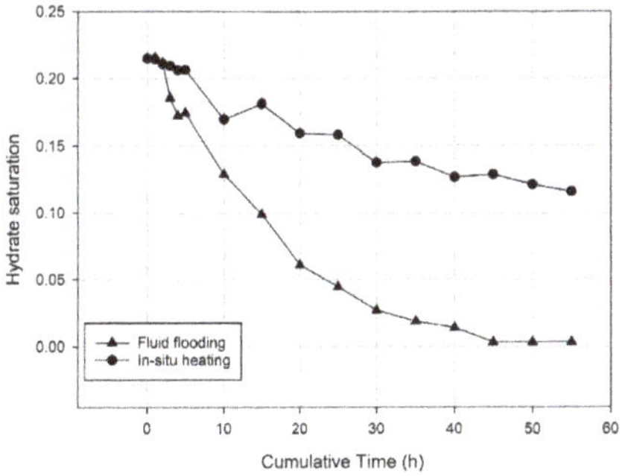

Figure 17. The estimated average hydrate saturation for two injection conditions during Stage 2.

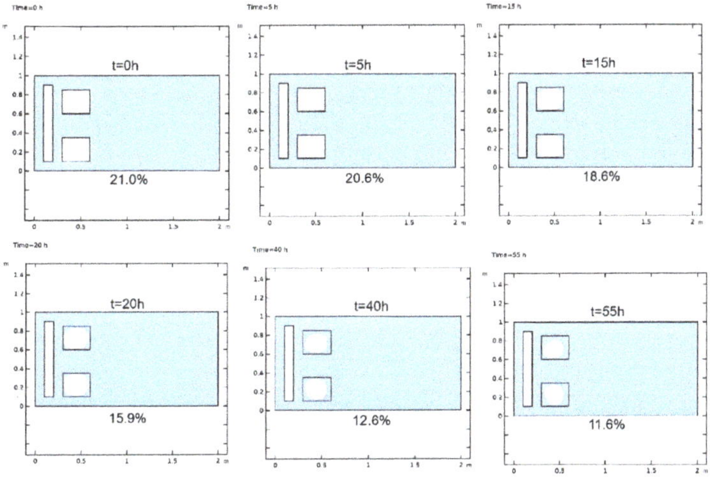

Figure 18. Hydrate saturation distribution over time for the in situ heating case (0~55 h).

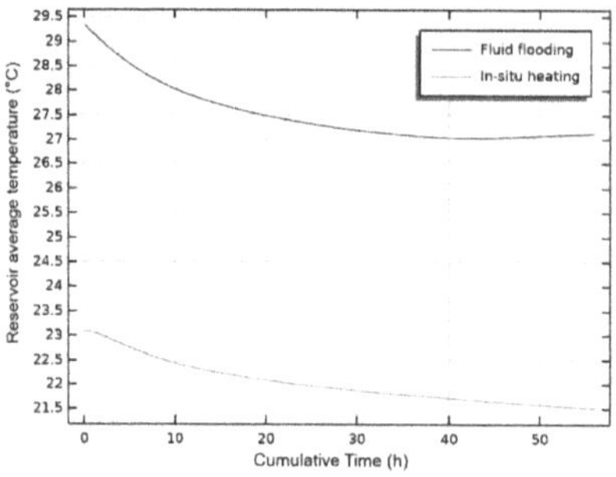

Figure 19. Average reservoir temperature of the two injection modes.

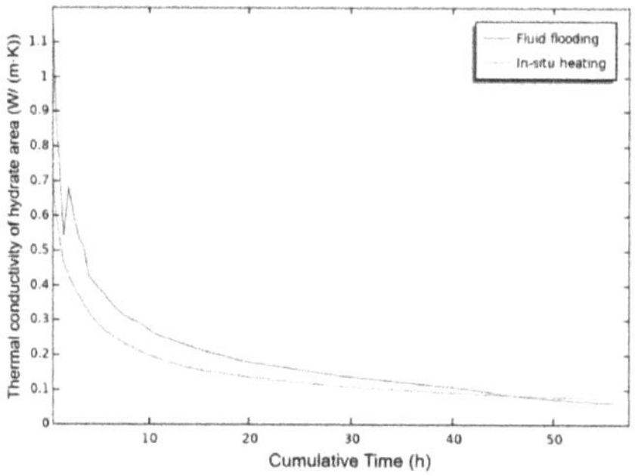

Figure 20. The thermal conductivity of the hydrate area under the two injection modes.

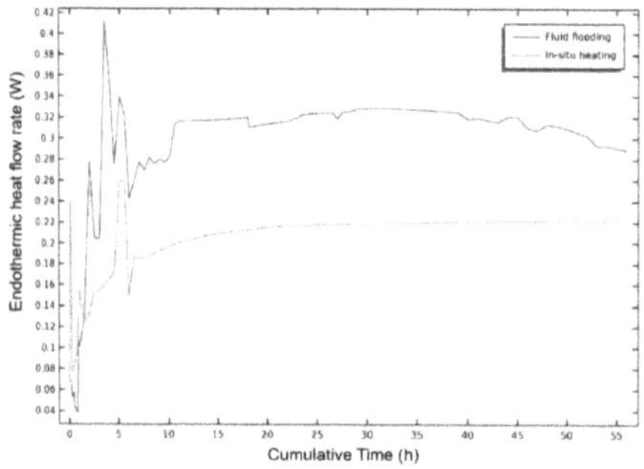

Figure 21. The endothermic heat flow rate of the hydrate area under the two injection modes.

3.3. Post-Injection Gas Production Potential

This research calculates decomposed gas volumes after the self-heating fluid injection process, examining how reservoir properties and injection modes influence post-injection gas production potential. The estimation results indicate that the gas production potential is positively correlated to the hydrate decomposition efficiency. The gas production calculation methodology and results analysis are briefly stated below.

Assuming the gas production well is the same as the fluid injection well, so the produced gas flow direction is opposite to the injected fluid pathway. By calculating the gas saturation and average reservoir temperature after Stage 2, the reservoir pressure at the beginning of the production process (immediately conducted after the injection process) is calculated based on the equation of state, with gas yield subsequently simulated through Darcy's law. Figure 22 displays gas production versus cumulative time under varying permeability conditions, while Figure 23 comparatively presents output data for two distinct fluid injection modes. The results indicate that the instantaneous gas production exhibits accelerated growth beyond the permeability condition of K > 40 mD. Comparative analysis of the two injection modes demonstrates that the fluid flooding method achieves 1.7-fold greater gas production than the in situ heating approach. Fluid flooding mode surpasses in situ thermal stimulation in gas recovery performance, primarily due to its capacity to optimize hydrate decomposition efficiency within the reservoir porous media.

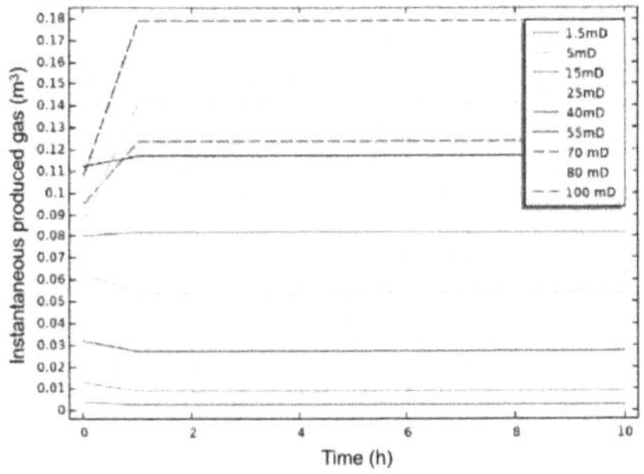

Figure 22. Instantaneous gas production of various permeability conditions in 10 h.

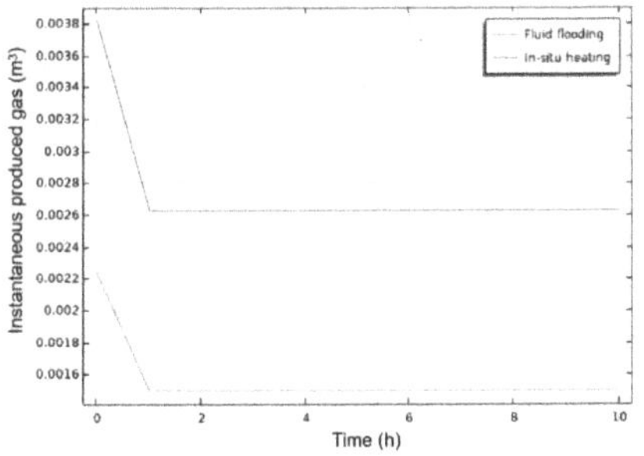

Figure 23. Instantaneous gas production of two injection modes in 10 h.

3.4. Numerical Model Validation

3.4.1. The Experimental Case Selection

It is important to validate the practicability of the simulation model through comparing simulation data with the relevant experimental test. However, as mentioned in the introduction, existing self-heat chemical injection studies predominantly emphasize gas recovery performance over injection dynamics, with hydrate saturation evolution during chemical stimulation remaining underexplored in the published experimental literature. In contrast, gas production potential in self-heat generating fluid experiments has received prior research attention [27]. Therefore, the validation is conducted based on the gas production simulation method of this study and the production data from a previous self-heat thermal stimulation experiment [48]. Validation involves applying the proposed gas prediction model to simulate gas production performance in the reference case. The key parameters dominating the validation process are described in Section 3.4.2.

3.4.2. The Validation Process

The experimental setup employs NH_4Cl + $NaNO_2$-based gas stimulation test in a cylindrical vessel to simulate formation–dissociation processes within a hydrate reservoir. In this experimental study, Case 2 (documented on reference Page 40) serves as the validation case due to production method consistency with this study's simulation scheme. The validating numerical model is constructed using the production estimation methodology outlined in Section 3.3.

The reservoir condition of the model and the production protocol are adopted from the selected experiment. The numerical simulation employs a 2D geometric model with equivalent volume (1.76 L) to the experimental reactor's effective vessel bulk volume. The model initializes three-phase saturations using experimental parameters, with injected $NH_4Cl + NaNO_2$ concentration matching the reference data (5 mol/L). The self-heating fluid injection rate is quantified via Equation (10), under the proportionality premise between flow rate and model geometric bulk. In Equation (10), Vc is the volumetric flow rate, m^3/s; BV is the model bulk volume, m^3. Since Vc_{base} = 1.33 m^3/s, BV_{base} = 2 m^3 in the base case of Section 3.3, $BV_{verified}$ is 0.88×10^{-3} m^3 due to the symmetry design of the simulation model, and the calculated $Vc_{verified}$ is 5.8×10^{-4} m^3/s. The validated model's volumetric flow rate feeds into the validation framework to compute experimental case water saturation (Equation (11), Sw).

$$\frac{Vc_{base}}{Vc_{verified}} = \frac{BV_{base}}{BV_{verified}} \tag{10}$$

The gas production simulation is conducted under experimental reservoir parameters and specified injection conditions. Simulation outputs quantify produced gas fractions within the bulk volume Equation (11). Equation (5) is used to calculate Sw. Two assumptions are made to conduct the gas fraction calculation: (1) Hydrate achieves complete dissociation in the sand-pack model upon gas production termination; (2) the ideal gas equation of state Equation (12) determines produced gas characteristics under the assumption that compressibility factor equals 1.

$$Vg_{vol} = Sg = 1 - Sw \tag{11}$$

$$n = \frac{PV}{RT} \tag{12}$$

In Equation (12), n is the amount of gaseous substance (mole), P is pressure (Pa), V is gas volume (m^3), R is molar gas constant (J/(mol·K)), and T is temperature (K). In Equation (14), subscript 1 represents condition in sand-pack model, and 2 for room condition. V_1 is calculated from Equation (13), where φ = 0.405. The calculation result of the Vg_{vol} in Equation (11) is 0.78 according to the simulation output.

$$V_1 = BV_{verified} \cdot \varphi \cdot Vg_{vol} \tag{13}$$

Equation (12) under reservoir and room conditions exhibits mutual correlation through n_1–n_2 equivalence. Equating $n_1 = n_2$ in Equation (14) yields produced gas volume V_2, which is 5.7 MPa·(0.88 L·0.405·0.78)/101 KPa·293 K/281 K = 16.36 L.

$$V_2 = \frac{P_1 \cdot V_1}{P_2} \cdot \frac{T_2}{T_1} \tag{14}$$

Experimental gas production is the value of 16.57 L, as shown in the reference, and the numerical–experimental discrepancy is 1.3%, based on the calculation result. The numerical model thus validates gas production prediction accuracy in the self-heat generation fluid stimulation. In the future, laboratory experiments will be conducted to further validate the simulation model applicability by analyzing hydrate saturation evolution during the self-heat fluid injection process.

3.5. Numerical Stability Validation

Numerical stability validation is conducted using different refining grids. Based on the computational discrepancies of hydrate saturation values under different grid densities, the numerical model's calculation errors are analyzed and the relevant results are presented in Table 5. For Stage 1, hydrate saturation profiles for Test 1-1 are shown in Figure 2, while those for Tests 1-2 and 1-3 are shown in Figure 24. For Stage 2, hydrate saturation comparing results are shown in Figures 25 and 26. As shown in the grid density distribution map (Figure 27), the grid density distribution provides a visualization of the mesh configuration.

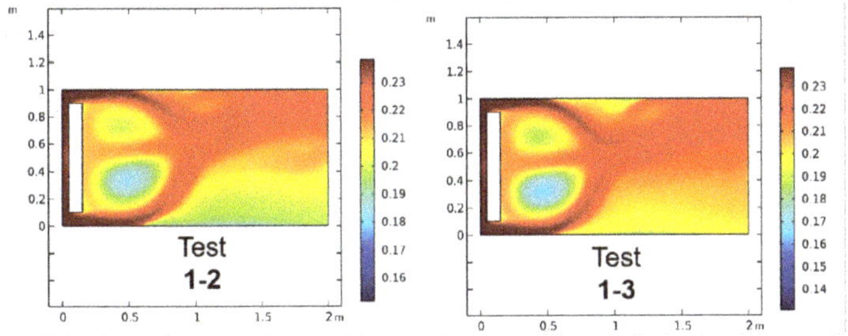

Figure 24. Hydrate saturation profile under various refining grids.

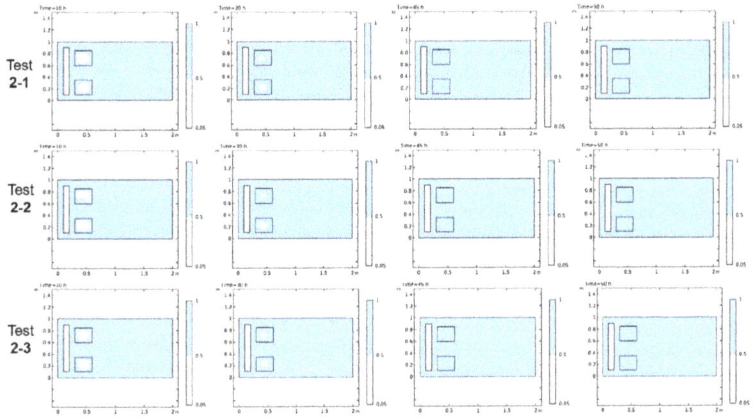

Figure 25. Hydrate saturation profile under various refining grids (Stage 2).

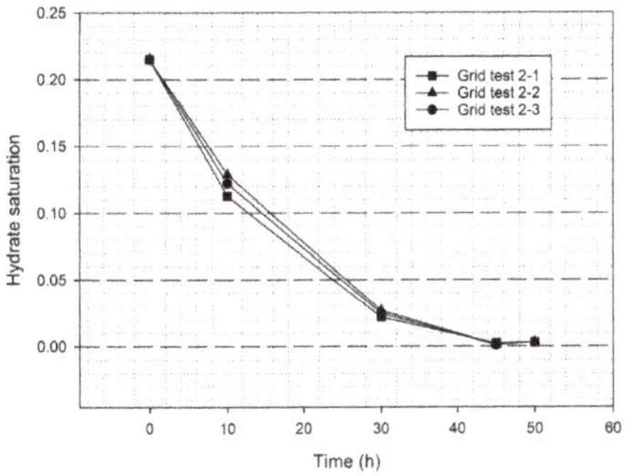

Figure 26. Hydrate saturation record under various refining grids (Stage 2).

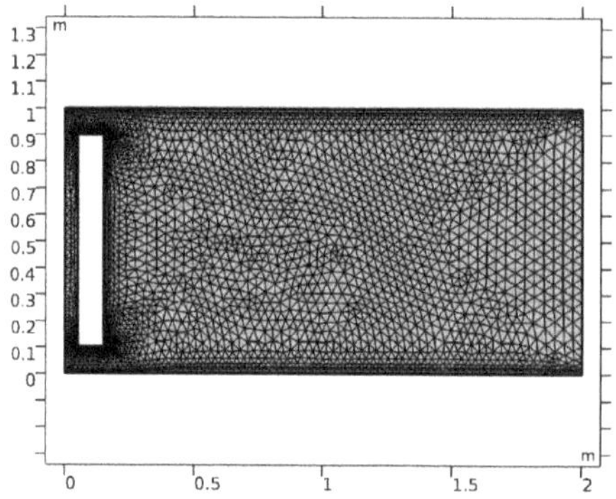

Figure 27. Grid density distribution map.

For Stage 1, the hydrate saturation calculation results represent saturation calculation accuracy. As shown in Table 5, three grid refining cases show relatively negligible changes. The mean absolute deviation (MAD) of each test number of data relative to the baseline (Test 1-1) is 0.0022 for Test 1-2 and 0.0005 Test 1-3; the MAD between the data of two compared tests and the baseline is extremely low (<1.1%).

For Stage 2, all data maintain changes within the same order of magnitude. The MAD of each test number of data relative to the baseline (Test 2-1) is calculated at different times, and the results show the MADs of Test 2-2 and 2-3 are 0.0018 and 0.0012, respectively, which are both lower than 0.002.

In short, all deviations of Stage 1 and 2 are within the commonly accepted 5% MAD engineering tolerance threshold. The validation demonstrates that the computed hydrate saturation values exhibit minimal deviations under different grid refinement parameter conditions. Therefore, the insensitivity of hydrate saturation variation to grid refinement parameters (including element size, maximum growth rate, and curvature factor) validates the model's stability, as it eliminates potential artifacts caused by mesh-related parameters.

3.6. Comparative Analysis with Prior Studies

3.6.1. Enhancing the Permeability-Secondary Hydrates Response Mechanism Study

Under the circumstance of employing different numerical simulation methodologies, varying research scales, and distinct hydrate decomposition mechanisms, this study has reached similar conclusions to previous research regarding the permeability threshold effect and provided complementary explanations. Specifically, previous studies utilized a large-scale reservoir model based on depressurization-induced decomposition of thermal–hydraulic–chemical (THC), indicating that high permeability (≥75 mD) accelerates fluid flow, leading to rapid heat dissipation and promoting secondary hydrate formation [49]. This study adopts a heat–mass transfer coupled near-wellbore model driven by self-heating to facilitate decomposition, proposing that secondary hydrates in the 55–70 mD range tend to form extensively due to the evolution of heat flow and thermal conductivity. Both studies demonstrate that high permeability intervals promote secondary hydrate formation. Additionally, this study reveals that under relatively low permeability conditions (<15 mD), localized heat accumulation occurs, which helps maintain high temperatures and inhibit secondary hydrate formation. This conclusion expands the understanding of permeability's impact on secondary hydrate formation during hydrate decomposition.

In short, this study and previous research jointly reveal the controlling role of permeability on secondary hydrate dynamics from different perspectives (thermodynamics vs.

fluid dynamics), enhancing mechanistic understanding and supporting the argument of this study that heating near the wellbore is recommended for mitigating secondary hydrates.

3.6.2. Complementing Existing Research on Heat Transfer Mechanisms

Although this study suggests that the fluid flooding mode of self-heat fluid can achieve higher hydrate decomposition efficiency compared to in situ injection, it only discusses one heating mode of continuous local heat release in in situ wellbore; however, previous research indicates that the in situ heat generation mode under huff-puff injection can yield higher hydrate decomposition gas production than heat flooding [50]. The inspiration from this study for our research is that in terms of the heat generation mechanism of self-heating system injection into hydrate reservoirs, a combination of flooding and huff-puff in situ heating can be designed to potentially maximize the advantages of both injection modes. The combination of the two can possibly create a composite thermal transfer mechanism of "macro-convection + micro-reaction". Additionally, the huff-puff in situ heating method can be attempted to reduce the risk of residual secondary hydrate formation near the wellbore under 55–70 mD conditions.

3.6.3. Synergistic Findings in Formation Heating Modes with Other Studies

The large-scale numerical simulation of seawater flooding and the self-heating fluid flooding around the wellbore in this study jointly demonstrate that the combination of thermal convection and thermal conduction is more efficient than static thermal conduction in terms of heat transfer efficiency [51]. Regarding the inhibition of secondary hydrate formation, due to the low thermal conductivity of sediments [52], both the in situ heating method in this study and wellbore electrical heating exhibit localized and delayed thermal energy diffusion [47]. Therefore, this study, along with current electrical heating studies, reveals the role of external heat source types in controlling hydrate dissociation and secondary formation.

3.7. Evaluation of Self-Heat System's Impact on Methane Hydrate Decomposition: Quantitative Heat Release and Chemical Product Effects

3.7.1. Energy Efficiency Analysis: Self-Heat System Driven CH_4 Hydrate Decomposition

For reaction equation (Equation (1)), the reaction enthalpy of 1 mol of the chemical heating system (1 mol $NaNO_2$ + 1 mol NH_4Cl) is −332.58 kJ/mol. The decomposition of 1 mol of methane hydrate requires 55 kJ of energy [53]. Under the assumption of neglecting heat transfer in the porous medium and fluids of the formation, the heat released by the self-heat system reaction can be totally utilized for the decomposition of methane hydrates. Therefore, the amount of methane hydrate decomposed by 1 mol of exothermic reactants is calculated by dividing the total energy released by the energy required to decompose 1 mol of methane hydrate, which is 332.58 kJ/55 kJ/mol = 6.05 mol.

3.7.2. $NaNO_2/NH_4Cl$ Heating Effect and Hydrate Dissociation Performance

In order to assess the heating performance of the $NaNO_2$ + NH_4Cl exothermic system under different reactant mass concentration conditions, a series of temperature rise experiments has been conducted with varying reactant mass concentrations. The main experimental instrument is an insulated reactor with temperature-acquisition capability (Chengdu, China) (Figure 28). The key step is to inject reactants (chemical products from Chengdu, China) into the sealed reactor via the liquid injection while collecting temperature data. By altering the mass concentration ratio of the reactants, the temperature changes have been obtained in the reactor under different concentration ratios, while keeping the dosage of the heating control agent (hydrochloric acid concentration) constant during the experiment. The results show that the peak temperature increase in the system rises with higher reactant dosage. The post-reaction system temperatures under different reactant mass concentration ratios are shown in Figure 29. Considering both the heating effect and cost, the optimal reactant concentration ratio combination is 6% $NaNO_2$+ 7% NH_4Cl. More experimental details will be published in future works.

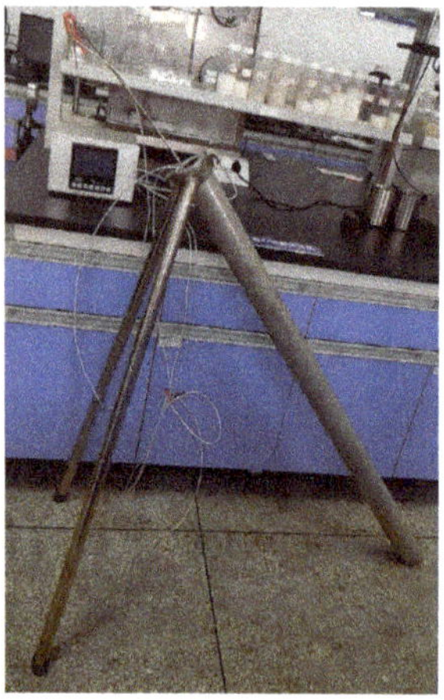

Figure 28. Temperature-acquisition insulated reactor.

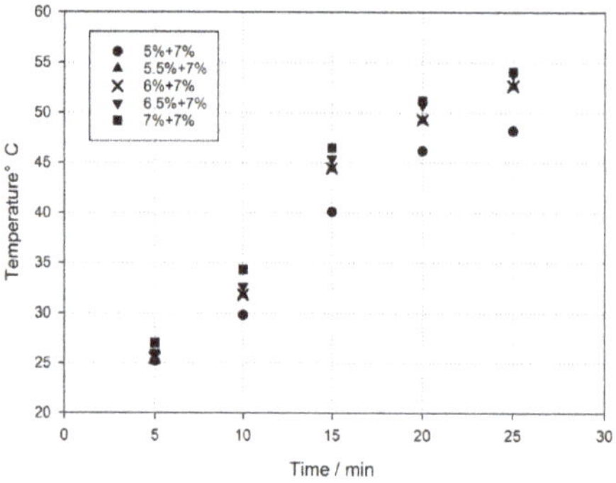

Figure 29. Post-reaction system temperatures under various reactant mass concentration ratios.

The calculation method for heat release and decomposed gas molar number can be outlined as follows: assuming the total mass of the solution is 1 kg, different reactant concentration ratios correspond to varying heat release amounts of the heating system. Based on the molar ratio of the two reactants, the reactant that determines the heat release can be identified. By combining the molar mass, mass concentration, and total mass of the relevant reactants, the molar number of the reactant reaction can be calculated. As the reaction enthalpy value of the chemical heating system per 1 mol is confirmed (332.58 kJ), the heat release of the reaction system under this molar number can be determined. By combining the energy required for the decomposition of 1 mol of hydrate (55 kJ/mol), the molar number of methane hydrate that can be decomposed under this heat release is calculated as shown in Table 6.

Table 6. Heat release capacity, hydrate decomposition capacity, and reaction product yield under different reactant mass concentration ratios (X wt%$NaNO_2$ + Y wt%NH_4Cl).

Physical Parameters	Unit	5.0% + 7.0%	5.5% + 7.0%	6.0% + 7.0%	6.5% + 7.0%	7.0% + 7.0%
Heat release	KJ	241.0	265.1	289.2	313.3	337.40
Decomposable CH_4 hydrate molar number	mol	4.38	4.82	5.26	5.70	6.13
Generated N_2	mol	0.73	0.80	0.87	0.94	1.01
Generated NaCl	mol/L	0.74	0.82	0.89	0.97	1.04

3.7.3. Quantity Comparison Between Reaction-Generated N_2 and Hydrate-Dissociated CH_4

In order to compare the amount of nitrogen (N_2) produced in the exothermic reaction with the methane released from hydrate decomposition, the molar quantities of N_2 generated and methane decomposed from hydrates under the aforementioned different reactant mass concentration ratios are calculated. The calculation method for the moles of N_2 is as follows: First, calculating the mass of the reactants based on the mass concentration of the reactants and then determining the moles of the reactants using the molar mass. Next, considering the reactant with the smaller number of moles as the basis for determining the moles of N_2, and then calculating the moles of N_2 based on the stoichiometric ratio.

The influence level of nitrogen gas (N_2) on methane hydrate decomposition is relatively insignificant. According to the N_2 mass calculation under various reactant mass ratios (Table 6), the N_2 mass is obviously lower than that of the hydrate-dissociated gaseous CH_4, indicating that the partial pressure of N_2 may slightly reduce the phase equilibrium pressure of methane hydrate, yet the effect is restricted.

The quantified correlation between injected N_2 condition and methane hydrate extent has been studied recently in the hydrate research field. N_2 injection has been experimentally proved to promote methane hydrate replacement kinetics, as evidenced by NMR (Nuclear Magnetic Resonance) data, revealing that N_2 is selectively trapped in small cages (5^{12}) and enhances guest molecule exchange; however, the research of the pure N_2 gas replacement effect on methane hydrate decomposition is conducted within the temperature range of 268.8–278.2 K [54], which is beyond the temperature range of this work (>286.2 K for simulation and 298.2 K for lab test), so the corresponding conclusion that the impact of N_2 on methane hydrate dissociation increases with higher temperature and lower injection pressure can hardly explain the influence of N_2 existence on this study's hydrate decomposition, which means the clear effect of produced N_2 on methane needs to be validated under more practical reservoir conditions in the future.

3.7.4. The NaCl Effect and Crystallization Tendency

As a thermodynamic inhibitor, the presence of NaCl critically influences methane hydrate stability by shifting the phase equilibrium condition, which necessitates the evaluation of the NaCl effect on methane hydrate dissociation [55,56]. A previous experimental study indicates that the impact of NaCl aqueous solution on methane hydrate crystallization demonstrates concentration-dependent behavior, and a concentration of 1.2 mol/L NaCl represents the critical threshold where its influence on methane hydrate formation shifts from promotion to inhibition [57]. In this study, the concentration of 1.2 mol/L NaCl serves as a critical threshold for determining whether in situ generated NaCl facilitates hydrate dissociation. As shown in Table 6, the calculated NaCl concentrations for each concentration ratio fall within the range that does not induce hydrate deformation. However, further calculation implies that 8.5 wt%$NaNO_2$ + 7.0 wt%NH_4Cl represents the critical reactant mass ratio threshold for hydrate inhibition, since its NaCl molar density reaches 1.28 mol/L.

The solubility of sodium chloride increases slightly with rising temperature. Therefore, the solubility under the lowest temperature (room temperature) condition during the experiment is used as the critical indicator for investigating sodium chloride crystallization. At 25 °C, the solubility of sodium chloride is 36.0 g/100 g water, which converts to a molar concentration of 5.4 mol/L. Additionally, the solubility of sodium chloride under actual hydrate reservoir conditions (15 °C) is 35.8 g/100 g water, equivalent to a molar concentration of 5.1 mol/L. Under the current reaction conditions studied, the molar concentration of the reaction product NaCl is less than 1.0 mol/L. Therefore, neither under experimental conditions nor at the actual hydrate reservoir temperature will sodium chloride crystallization or precipitation occur. Although the concentration of sodium chloride theoretically remains within the crystallization-free range in this investigation, the potential for NaCl crystallization remains a noteworthy concern in self-heat-fluid-based hydrate reservoir extraction methods, warranting continued attention and research in the future.

In short, NaCl exhibits concentration-dependent effects on methane hydrate stability, with 1.2 mol/L as the critical threshold where its role changes from enhancing formation to accelerating deformation, while reaction-generated NaCl concentrations (<1.28 mol/L) remain below thresholds, inducing hydrate growth promotion under reactant mass ratio conditions listed in Table 6. With 7.0 wt%NH_4Cl, the percentage of $NaNO_2$ needs to be higher than 8.5 wt% to trigger the NaCl hydrate inhibition effect. Theoretical calculations confirm that NaCl concentrations (basically below 1.0 mol/L) remain far lower than solubility thresholds (5.1–5.4 mol/L) under all tested conditions, preventing NaCl crystallization.

3.8. *Recommendations for Field Application*

From the perspectives of engineering design (Section 3.8.1) and field operation optimization (Sections 3.8.2 and 3.8.3), the following discussion explores the practical implications of numerical simulation results for in situ applications.

3.8.1. Permeability-Graded Design for Secondary Hydrate Control

(1) Prioritize low-permeability reservoirs (K < 15 mD)

Simulations indicate that in this range, self-heating fluid injection achieves complete hydrate dissociation without secondary generation (as discussed in Section 3.2). It is recommended to prioritize such reservoirs during target screening, based on in situ logging data.

(2) For reservoirs with permeability of 15–55 mD

It is recommended to inject hydrate inhibitors before self-heating working fluid to prevent secondary hydrate formation during the later injection phase.

(3) Avoid high-risk permeability zones (55–70 mD)

Severe secondary hydrate formation might occur in this range. Such reservoirs should be excluded in engineering design, or alternative composite techniques should be employed to reduce risks.

3.8.2. Field Operation Optimization: Thermal Front Dynamic Control

Thermal front propagation speed: It is suggested to dynamically adjust the thermal front's advance speed based on reservoir permeability. Low-permeability reservoirs (K < 15 mD): a constant high flow rate is recommended to rapidly exceed the dissociation threshold. Medium-permeability reservoirs (15–55 mD): adjusting injection volume flexibly to increase flow rate early to establish thermal fields and reduce it during mid-phase to prevent secondary generation.

3.8.3. Field Operation Optimization: Real-Time Monitoring and Feedback Control

In self-heat fluid-assisted hydrate extraction, it is advised to deploy DTS (Distributed Temperature Sensing) to track thermal fronts monitoring near-wellbore temperature gradients in real time; additionally, using numerical simulations to refine injection parameters, which prevents premature heat dissipation. In addition, adding pressure monitoring points to dynamically adjust injection parameters and to warn of secondary hydrate blockage risks based on pressure changes.

3.9. Limitations and Future Work Directions

This study numerically demonstrates that the self-heating fluid injection method represents an effective strategy for inducing controlled gas hydrate dissociation through thermally stimulated processes; in addition, this investigation employs numerical simulation methods to quantitatively analyze the evolution of hydrate saturation in marine sediments characterized by multi-phase fluid flow dynamics. The computational results provide insights for designing laboratory experiments by identifying optimal reservoir parameters and thermal stimulation scenarios through self-heating fluid injection techniques; however, several limitations and relevant future research directions should be acknowledged, as detailed below.

3.9.1. The Assumption of Isotropic and Homogeneous Porous Media

The central research problem of this study investigates the variation patterns of hydrate saturation during self-heating fluid injection in the near-wellbore region. As the first foundational model exploring this injection process, addressing the primary research question necessitates engineering simplifications regarding other factors. The influence of heterogeneity and anisotropy within the 2 m near-wellbore zone is provisionally treated as a secondary consideration, with the associated reason, potential impacts, and current offsets detailed below.

(1) The reason for the simplification

Based on hydrate dissociation sensitivity study focusing on sample uniformity, the hydrate dissociation rate constant is highly sensitive to the medium permeability K of hydrate-bearing sediments [58]. Therefore, the homogeneous reservoir assumption in this study can decouple the cross-influences between reservoir permeability heterogeneity and chemical reaction kinetics on gas–aqueous phase transformation, thereby meeting the requirement for engineering simplification. Additionally, a strong coupling relationship exists between hydrate saturation and permeability anisotropy [59]. To reduce computational costs for the primary mechanistic indicator (hydrate saturation) and clarify dominant variables, this study avoids the interference of reservoir anisotropy on the main control mechanisms of hydrate saturation.

(2) Potential implications for the results

The advance velocity of the thermal front is overestimated: the lower vertical permeability in actual reservoirs may lead to redistribution of thermal fluids, causing the thermal compensation advantage in fluid displacement mode to remain unverified in heterogeneous formations.

The degree of hydrate saturation reduction is overestimated: vertical permeability attenuation caused by sedimentary bedding may inhibit the vertical diffusion of thermal fluids, leading to an overestimation of hydrate saturation reduction in strongly anisotropic ormations.

Increased uncertainty in the threshold for secondary hydrate formation: under the same macroscopic permeability, isolated pores and connected pores have different inhibitory effects on secondary hydrate nucleation. Therefore, conclusions about the threshold for secondary hydrate formation may deviate due to variations in pore structure.

(3) Current offset strategies

The current offset strategy is the sensitivity analysis regarding intrinsic permeability in this study. The sensitivity analysis sets a permeability variation range of 1.5–100 mD, which in fact covers the fluctuation amplitude of permeability after equivalent treatment of heterogeneous and anisotropic reservoirs. This approach compensates for the limitations of heterogeneity and anisotropy simplification to some extent.

(4) Future plans

Future research will focus on developing multi-scale models that incorporate heterogeneous and anisotropic properties of marine hydrate-bearing sediments. Advanced layer design techniques will be considered to characterize reservoir heterogeneity for the improved hydrate saturation prediction.

3.9.2. Hydrate Saturation-Geomechanical Assumption

(1) The rationality of geomechanical effects simplification

To facilitate the initial analysis of hydrate saturation variation characteristics during the self-heat fluid injection process, while simplifying computational complexity and emphasizing the influence of formation properties on the study objectives, this research adopts a decoupling approach between mechanical properties and hydrate saturation. The specific analysis proceeds as follows.

The current model focuses on "coupled heat-mass transfer mechanisms" and serves to validate its feasibility during the injection process specifically, which requires prioritizing the solution of the fundamental scientific issue of thermal–fluid coupling. For gas hydrate reservoirs in the Shenhu area of the South China Sea, which is primarily composed of clayey silt, thermal conduction dominates the "coupled heat-mass transfer" process during hydrate dissociation, while mechanical responses can be considered secondary factors. Additionally, the current model aims to rapidly validate the decomposition efficiency of self-generating thermal fluids. The incorporation of mechanical coupling would significantly increase model complexity and potentially obscure the dominant thermal activation-driven decomposition mechanism, as the mechanical strength of hydrate reservoirs remains highly correlated with multiple fundamental formation properties. For example: (1) The linear positive correlation between hydrate saturation and sedimentary shear strength [60]. (2) The nonlinear negative correlation between permeability and effective confining pressure [61]; the dominant effect of confining/overburden stress on the sedimentary porosity and permeability evolution [62]. (3) The dependence of pore structures and sedimentary particle heterogeneity on effective stress [63]. (4) The impact of stress change on sedimentary creep effect through pore destruction [64]. The aforementioned studies indicate that geomechanical effect comprehensively controls various reservoir properties simultaneously, which means porosity, permeability, and pore structure are supposed to be considered as dependent variables once the mechanical characteristics are considered in the simulation model. This leads to model construction challenges in two aspects: 1. Nonlinear Equation Coupling: Temperature field (heat conduction equation) and saturation field (fluid transport equation) are affected by pore deformation feedback, forming strong nonlinear terms. 2. The alteration of reservoir geomechanical properties induced by hydrate decomposition necessitates continuous updates to permeability and porosity values. Consequently, it becomes particularly challenging to isolate a single independent variable during reservoir permeability sensitivity analysis. Those are the computational complexity reasons for the simplification.

(2) Field Application Risks and Mitigation Strategies Analysis

The potential field-scale risks arising from decoupling hydrate saturation from reservoir geomechanical effects are outlined below, with each item incorporating corresponding preventive and compensatory engineering designs:

(1) Local overpressure during self-heat thermal fluid injection may induce microfractures. Operational pressure should be appropriately reduced to prevent excessive rock deformation.

(2) For field applications, to mitigate sediment strength reduction and potential geological collapse caused by hydrate saturation decline after thermal decomposition, it is recommended to inject rock consolidants during the later stage of self-heat thermal fluid operation to enhance sedimentary mechanical strength [65].

(3) For medium-high permeability reservoirs (15–55 mD), secondary hydrates during mid-stage injection initially increase then decrease, potentially triggering mechanical strength fluctuations and localized dynamic evolution of formation porosity/permeability. A real-time pore pressure feedback system must be deployed to monitor formation strain, so as to prevent wellbore shear failure, and dynamically adjust injection rates based on pore pressure changes.

3.9.3. Operational Parameter Limitations in Sensitivity Analysis

While the study evaluates the impacts of reservoir permeability and injection mode as dominant control parameters, it serves as a simplified condition because it does not explicitly address other critical operational variables. The relevant parameters and their potential influences are: (1) Reagent concentration: Influences reaction kinetics and heat generation potential. (2) Injection rate: Affects fluid distribution and thermal front propagation. (3) Initial temperature gradient: Modulates baseline heat transfer efficiency.

The simplification is adopted to prioritize the primary permeability-driven dissociation mechanisms in the first numerical exploration of self-heating fluid injection. Future work should expand sensitivity analyses to quantify the interplay of these parameters, particularly their combined effects on secondary hydrate reformation risks.

3.9.4. Validation Scope Limitation and Future Plans

(1) Validation scale limitation

The validation is restricted to gas production data from a single experimental case ($NH_4Cl + NaNO_2$ stimulation) due to the scarcity of published experimental studies directly monitoring hydrate saturation evolution during self-heating fluid injection, and thus the current validation scope does not encompass direct measurement of spatial–temporal hydrate saturation dynamics, which represents a research gap for characterizing hydrate dissociation/reformation mechanisms.

(2) Lab and field saturation validation plans

To address this limitation, we propose the following experimental and field-related research plans in the future:

(1) Lab-scale saturation monitoring application: designing reactor-based self-heat injection experiments with embedded NMR sensors to quantify real-time hydrate saturation distributions under controlled injection conditions (e.g., injection rate and reagent concentration) and to validate the conclusions of current numerical studies.

(2) Field data application: Collaborating numerical findings with offshore hydrate drilling projects data to calibrate the numerical model using in situ saturation logs from pilot thermal stimulation tests, leveraging well-log interpretations that quantify porosity–permeability evolution as functions of hydrate saturation.

3.9.5. Limitation of Phase-Specific Hydrate Regeneration Modeling

In the self-heat fluid injection model, the secondary hydrate regeneration issue is considered only in Stage 2. This simplification neglects potential hydrate reformation throughout the entire injection process, which may lead to underestimation of flow impedance and overestimation of injection efficiency. This limitation suggests the need for more comprehensive modeling approaches that account for continuous hydrate reformation dynamics during the whole injection process.

4. Conclusions

In this study, the hydrate saturation change during self-heating chemical fluid injection was investigated, while coupled heat-mass transfer was examined with a focus on fluid-phase behavior in the porous media. The sensitivity analysis was conducted to assess how initial reservoir permeability and fluid injection mode impact hydrate saturation and average reservoir temperature. This work contributes to the specific application details of self-heat-fluid-assisted injection technique, providing valuable parameter references for gas hydrate exploitation experimental/numerical investigations. The main conclusions are summarized below.

(1) The stimulation process involving self-heating fluid injection into hydrate-bearing marine sediment is conceptualized as the reactive chemical flow stage (Stage 1), followed by the non-reactive flow stage (Stage 2). A near-wellbore 2D three-phase reservoir model is constructed to perform numerical simulations for both stages. In the base case, reservoir temperature rises from 13.0 °C to 29.3 °C with a 47.5% hydrate saturation reduction in Stage 1, followed by full dissociation in Stage 2. The evolution of hydrate saturation and reservoir temperature demonstrates that the self-heat fluid method provides a viable approach for effectively enhancing hydrate decomposition in the thermal stimulation process.

(2) During the self-heating fluid injection process, variations in reservoir permeability critically affect thermal conductivity and heat flow rate in hydrate-bearing zones, resulting in non-uniform temperature distributions and hydrate saturation patterns within marine porous media. In low-permeability reservoirs ($K < 15$ mD), complete hydrate dissociation occurs during Stage 2, without secondary hydrate formation, whereas in the 15–55 mD permeability range, a dual-phase behavior is observed—initial hydrate regeneration in early-to-mid Stage 2 followed by secondary decomposition until the end of Stage 2. To minimize secondary hydrate formation, the self-heating fluid method should be applied only in reservoirs with permeability below 15 mD, while its use is strongly discouraged in the 55–70 mD range due to significant secondary hydrate generation.

(3) Fluid flooding injection exhibits robust heat generation within the hydrate reservoir, evidenced by Stage 1 reservoir temperature analysis and Stage 2 hydrate saturation monitoring. This injection mode outperforms in situ heating mode by enhancing

hydrate dissociation efficiency during the injection process of self-heat generating fluid, yielding higher post-injection gas recovery potential.

(4) For the research domain of secondary hydrate formation suppression, by comparing with other research, this investigation supplements and extends the preceding studies on hydrate decomposition mechanisms: it concurrently demonstrates the regulatory influence of reservoir intrinsic permeability in secondary hydrate dynamics through thermodynamic and fluid dynamic analyses, corroborated by prior large-scale reservoir research. The study further suggests that integrating self-heat fluid flooding with huff-puff in situ heating could potentially establish a macro-convection + micro-reaction composite thermal transfer modality, thereby optimizing decomposition efficacy while alleviating secondary hydrate risks in 55–70 mD reservoirs.

Author Contributions: Conceptualization, K.Z. and K.S.; Methodology, K.Z.; Data Processing, K.Z.; Formal Analysis, K.Z. and Y.W.; Writing—Original draft, K.Z.; Writing—Review and editing, K.Z. and Y.W.; Funding Acquisition, K.Z. and K.S.; Investigation, K.S.; Project Administration, Y.Y.; Resources, Y.Y.; Validation, J.Z. (Jiawei Zhou); Visualization, J.Z. (Jing Zeng). All authors have read and agreed to the published version of the manuscript.

Funding: This work was funded by the National Key Research and Development Program of China (No. 2023YFC2811002) and (No. 2023YFC2811005).

Data Availability Statement: The original contributions presented in this study are included in the article. Further inquiries can be directed to the corresponding author.

Acknowledgments: The authors would like to acknowledge the experimental resources support from Liangjun Du from Sichuan Shenhe New Materials Technology Co., Ltd.

Conflicts of Interest: The authors declare that they have no competing interests or personal relationships that could have appeared to influence the work reported in this paper. The authors declare no conflicts of interest.

References

1. Wei, R.P.; Xia, Y.Q.; Wang, Z.F.; Li, Q.P.; Lv, X.; Leng, S.D.; Zhang, L.X.; Zhang, Y.; Xiao, B.; Yang, S.X.; et al. Long-Term Numerical Simulation of a Joint Production of Gas Hydrate and Underlying Shallow Gas Through Dual Horizontal Wells in the South China Sea. *Appl. Energy* **2022**, *320*, 119235. [CrossRef]
2. Liang, H.; Duan, Y.G.; Pei, J.; Wei, N. Study on Hydrate Phase Equilibrium Diagram of Methane Containing System Based on Thermodynamic Model. *Front. Energy Res.* **2021**, *9*, 743296. [CrossRef]
3. Ye, J.L.; Qin, X.W.; Xie, W.W.; Lu, H.L.; Ma, B.J.; Qiu, H.J.; Liang, J.Q.; Lu, J.A.; Kuang, Z.G.; Lu, C.; et al. The Second Natural Gas Hydrate Production Test in the South China Sea. *China Geol.* **2020**, *3*, 197–209. [CrossRef]
4. Wang, L.Y.; Dou, M.Y.; Wang, Y.; Xu, Y.L.; Li, Y.; Chen, Y.; Li, L.S. A Review of the Effect of Porous Media on Gas Hydrate Formation. *ACS Omega* **2022**, *7*, 33666–33679. [CrossRef]
5. Xiao, C.W.; Li, X.S.; Li, G.; Yu, Y.S.; Lv, Q.N.; Yu, Y.; Weng, Y.F.; Liu, J.W.; Yu, J.X. Kinetic Studies of the Secondary Hydrate Formation in Porous Media Based on Experiments in a Cubic Hydrate Simulator and a New Kinetic Model. *Fuel* **2024**, *358*, 130168. [CrossRef]
6. Ma, X.L.; Sun, Y.H.; Liu, B.C.; Guo, W.; Jia, R.; Li, B.; Li, S.L. Numerical Study of Depressurization and Hot Water Injection for Gas Hydrate Production in China's First Offshore Test Site. *J. Nat. Gas Sci. Eng.* **2020**, *83*, 103530. [CrossRef]
7. Wu, N.Y.; Li, Y.L.; Wan, Y.Z.; Sun, J.Y.; Huang, L.; Mao, P.X. Prospect of Marine Natural Gas Hydrate Stimulation Theory and Technology System. *Nat. Gas Ind. B* **2021**, *8*, 173–187. [CrossRef]
8. Wu, H.Z.; Zhang, X.; Yang, Z.H.; Liu, D. Numerical Simulation on Depressurization Extraction of Natural Gas Hydrate Based on a Coupled Thermal-Hydraulic-Chemical Model. *Energy Rep.* **2022**, *8*, 269–278. [CrossRef]
9. Qin, X.W.; Lu, C.; Wang, P.K.; Liang, Q.Y. Hydrate Phase Transition and Seepage Mechanism during Natural Gas Hydrates Production Tests in the South China Sea: A Review and Prospect. *China Geol.* **2022**, *5*, 201–217. [CrossRef]
10. Pan, D.B.; Ma, X.L.; Chen, C.; Zhong, X.P. Numerical Analysis of Hydrate Exploitation Performances of Depressurization Combined with Thermal Fluid Injection in Marine Hydrate Reservoir with Impermeable Boundary. *Gas Sci. Eng.* **2023**, *110*, 204894. [CrossRef]

11. Mao, M.; Yan, K.; Li, X.; Chen, Z.; Wang, Y.; Feng, J.; Chen, C. Review of Heat Transfer Characteristics of Natural Gas Hydrate *Energies* **2024**, *17*, 717. [CrossRef]
12. Yu, Q.N.; Tang, H.M.; Li, C.L.; Chen, Z.J.; Zhang, K.; Yu, Y.; Liang, S. Numerical Simulation of Natural-Gas-Hydrate Decomposition in Process of Heat-Injection Production. *Processes* **2023**, *11*, 2349. [CrossRef]
13. Han, S.; Deng, X.J.; Li, W.C.; Yu, L.; Guan, L.F. A Study on the Coupling Mechanism of Hydrate Dissociation-Seepage-Heat Transfer Induced by Suction Anchor Installation in Marine Sediments. *Ocean Eng.* **2023**, *288*, 116083. [CrossRef]
14. Chen, Y.; Gao, Y.H.; Xin, G.; Yao, W.; Gao, D.Z.; Chen, L.T.; Sun, B.J. Theoretical Analysis and Numerical Simulation of Multi-Fields Coupled Variation During Deepwater Hydrate-Bearing Reservoir Exploitation. *Int. Conf. Comput. Exp. Eng. Sci.* **2019**, *21*, 88–89 [CrossRef]
15. Shen, S.; Li, Y.H.; Sun, X.; Wang, L.; Song, Y.C. Stress Behavior of Hydrate-bearing Sands with Changing Temperature and Hydrate Saturation. *J. Nat. Gas Sci. Eng.* **2022**, *98*, 104389. [CrossRef]
16. Chen, B.B.; Sun, H.R.; Zheng, J.J.; Yang, M.J. New Insights on Water-Gas Flow and Hydrate Decomposition Behaviors in Natural Gas Hydrates Deposits with Various Saturations. *Appl. Energy* **2020**, *259*, 114185. [CrossRef]
17. Wang, S.Y.; Zhang, X.H.; Lu, X.B.; Nie, X.Y. Laboratory Study on Physical and Mechanical Properties of Hydrate Sediment Samples. In Proceedings of the Twenty-Fourth International Ocean and Polar Engineering Conference, Busan, Republic of Korea, 15–20 June 2014; p. ISOPE-I-14-098.
18. Li, Q.C.; Cheng, Y.F.; Zhang, H.W.; Yan, C.L.; Liu, Y.W. Simulating the Effect of Hydrate Dissociation on Wellhead Stability during Oil and Gas Development in Deepwater. *J. Ocean Univ. China* **2018**, *17*, 35–45. [CrossRef]
19. Wu, S.T.; Zhou, X.B.; Lu, J.S.; Liang, D.Q.; Li, D.L. Experimental Study on CH4 Hydrate Dissociation by the Injection of Hot Water, Brine, and Ionic Liquids. *J. Mar. Sci. Eng.* **2023**, *11*, 713. [CrossRef]
20. Zhu, Y.; Li, X.H.; Zhao, C.; Zhang, J.; Bondarenko, V.; Dreus, A.; Liang, J.Q.; Li, X.Y.; Zhu, J.L.; Wang, P.F.; et al. Decomposition Characteristics of Methane Hydrate in Porous Media under Continuous and Intermittent Microwave Heating. *Fuel* **2023**, *332*, 126230. [CrossRef]
21. Hou, J.X.; Zhao, E.M.; Ji, Y.K. Numerical Simulation of Gas Production from Class III Hydrate Reservoirs Using Low-Frequency Electric Heating-Assisted Depressurization with Horizontal Wells. *Fuel* **2024**, *357*, 129906. [CrossRef]
22. Wei, N.; Pei, J.; Li, H.T.; Sun, W.T.; Xue, J. Application of In-Situ Heat Generation Plugging Removal Agents in Removing Gas Hydrate: A Numerical Study. *Fuel* **2022**, *323*, 124397. [CrossRef]
23. Xu, T.; Zhang, Z.B.; Li, S.D.; Li, X.; Lu, C. Numerical Evaluation of Gas Hydrate Production Performance of the Depressurization and Backfilling with an In Situ Supplemental Heat Method. *ACS Omega* **2021**, *6*, 12274–12286. [CrossRef]
24. Ping, X.L.; Han, G.Q.; Zhang, J.Q.; Chang, J.H.; Cen, X.Q.; Tang, H. Study on the Flow Behavior of Wellbore Fluids of a Natural Gas Hydrate Well with the Combined Depressurization and Heat Injection Method. *Processes* **2023**, *11*, 1625. [CrossRef]
25. Ahmadpanah, S.J.; Manteghian, M.; Ganji, H. Effect of Cyclopentane and Sodium Chloride on the Formation Kinetics of Natural Gas Hydrate in Different Operating Conditions. *J. Nat. Gas Sci. Eng.* **2022**, *99*, 104417. [CrossRef]
26. Wang, J.T.; Shao, Z.H.; Bai, Y.J.; He, G.L.; Wang, X.D.; Liu, L.; Liao, B.; Sun, X.H.; Lv, K.H.; Sun, J.S. Fabrication and Property Evaluation of Calcium-Oxide-Loaded Microcapsules during Supplemental Heat-Based Exploitation of Natural Gas Hydrates *RSC Adv.* **2023**, *13*, 7212–7221. [CrossRef] [PubMed]
27. Liu, S.; Zhang, Y.Y.; Luo, Y.J.; Liang, Y.P.; Li, B. Analysis of Hydrate Exploitation by a New In-Situ Heat Generation Method with Chemical Reagents Based on Heat Utilization. *J. Clean. Prod.* **2020**, *249*, 119399. [CrossRef]
28. Chibura, P.E.; Zhang, W.; Luo, A.J.; Wang, J.J. A Review on Gas Hydrate Production Feasibility for Permafrost and Marine Hydrates. *J. Nat. Gas Sci. Eng.* **2022**, *100*, 104441. [CrossRef]
29. Wu, T.W.; Wan, K.; Li, X.S.; Wang, Y.; Chen, Z.Y. Heat Utilization Efficiency Analysis of Gas Production from Hydrate Reservoir by Depressurization in Conjunction with Heat Stimulation. *Energy* **2023**, *263*, 125625. [CrossRef]
30. Ren, L.L.; Jiang, M.; Wang, L.B.; Zhu, Y.J.; Li, Z.; Sun, C.Y.; Chen, G.J. Gas Hydrate Exploitation and Carbon Dioxide Sequestration under Maintaining the Stiffness of Hydrate-Bearing Sediments. *Energy* **2020**, *194*, 116869. [CrossRef]
31. Xiao, C.W.; Li, X.S.; Li, G.; Yu, Y.; Weng, Y.F.; Lv, Q.N.; Yu, J.X. Key Factors Controlling the Kinetics of Secondary Hydrate Formation in the Porous Media. *Gas Sci. Eng.* **2023**, *110*, 204911. [CrossRef]
32. Liao, T.; Yuan, L.; Li, W.; Kan, J.Y.; Luo, W.; Xiong, X.Q.; Li, N. One-Dimensional Numerical Simulation on Removal of CO_2 Hydrate Blockage around Wellbore by N2 Injection. *Processess* **2024**, *12*, 204. [CrossRef]
33. Wang, Y.N.; Sun, Z.X.; Li, Q.P.; Lv, X.; Ge, Y. A Thermal Chemical Reaction System for Natural Gas Hydrates Exploitation. *Front. Energy Res.* **2022**, *9*, 804498. [CrossRef]
34. Li, Q.P.; Lv, X.; Pang, W.X.; Yao, H.Y.; Ge, Y.; Wang, J.A. Numerical Simulation of the Effect of Porous Media Permeability on the Decomposition Characteristics of Natural Gas Hydrate at the Core Scale. *Energy Fuels* **2021**, *35*, 5843–5852. [CrossRef]
35. Vikram, V.; David, L.; Samardeep, S.; Sharma, A.; Ranjith, P.G. Sensitivity Analysis of Methane Hydrate Bearing Class 3 Reservoirs during Thermal Injection. *J. Pet. Sci. Eng.* **2020**, *195*, 107575. [CrossRef]

36. Liang, Y.P.; Liu, S.; Zhao, W.T.; Li, B.; Wan, Q.C.; Li, G. Effects of Vertical Center Well and Side Well on Hydrate Exploitation by Depressurization and Combination Method with Wellbore Heating. *J. Nat. Gas Sci. Eng.* **2018**, *55*, 154–164. [CrossRef]
37. Li, J.F.; Ye, J.L.; Qin, X.W.; Qiu, H.J.; Wu, N.Y.; Lu, H.L.; Xie, W.W.; Lu, J.A.; Peng, F.; Xu, Z.Q.; et al. The First Offshore Natural Gas Hydrate Production Test in South China Sea. *China Geol.* **2018**, *1*, 5–16. [CrossRef]
38. Falser, S.; Uchida, S.; Palmer, A.C.; Soga, K.; Tan, T.S. Increased Gas Production from Hydrates by Combining Depressurization with Heating of the Wellbore. *Energy Fuels* **2012**, *26*, 6259–6267. [CrossRef]
39. Sun, X.; Luo, H.; Soga, K. A Coupled Thermal-Hydraulic-Mechanical-Chemical (THMC) Model for Methane Hydrate Bearing Sediments Using Comsol Multiphysics. *J. Zhejiang Univ.-Sci. A* **2018**, *8*, 600–623. [CrossRef]
40. Comsol Search. All Research Results About Products, COMSOL Blog, and Application Gallery, etc. Available online: https://www.comsol.com/search/ (accessed on 12 May 2022).
41. Datta, A.K. Porous Media Approaches to Studying Simultaneous Heat and Mass Transfer in Food Processes. I: Problem Formulations. *J. Food Eng.* **2007**, *80*, 80–95. [CrossRef]
42. Zhao, Z.W. Thermodynamic Analysis for the Gas Production from Gas Hydrate by Heating. *China Min. Mag.* **2013**, *8*, 125–128. (In Chinese)
43. Kou, X.; Feng, J.C.; Li, X.S.; Wang, Y.; Chen, Z.Y. Visualization of Interactions between Depressurization-Induced Hydrate Decomposition and Heat/Mass Transfer. *Energy* **2022**, *239*, 122230. [CrossRef]
44. Kou, X.; Wang, Y.; Li, X.S.; Zhang, Y.; Chen, Z.Y. Influence of Heat Conduction and Heat Convection on Hydrate Dissociation by Depressurization in a Pilot-Scale Hydrate Simulator. *Appl. Energy* **2019**, *251*, 113405. [CrossRef]
45. Li, G.; Moridis, G.J.; Zhang, K.N.; Li, X.S. The Use of Huff and Puff Method in a Single Horizontal Well in Gas Production from Marine Gas Hydrate Deposits in the Shenhu Area of South China Sea. *J. Pet. Sci. Eng.* **2011**, *77*, 49–68. [CrossRef]
46. Malagar, B.R.; Lijith, K.P.; Singh, D.N. Formation & Dissociation of Methane Gas Hydrates in Sediments: A Critical Review. *J. Nat. Gas Sci. Eng.* **2019**, *65*, 168–184. [CrossRef]
47. Wei, W.N.; Li, B.; Li, G.; Zhang, T.T.; Wan, Q.C.; Zhou, J.H. Investigating Optimal Thermal-Assisted Depressurization Regimes for Enhanced Methane Hydrate Dissolution in a Multiple Well System. *Energy Fuels* **2023**, *37*, 15795–15807. [CrossRef]
48. Zhang, Y.Y. Experimental Study on Gas Production from Hydrate Reservoir Using Self-Heating System Injection Method. Master's Thesis, Chongqing University, Chongqing, China, 2019. (In Chinese).
49. Wang, Z.Q.; Liu, S.Y.; Li, H.Y.; Li, S.X.; Xu, J.C.; Wang, X.P. A Numerical Simulation Study of Methane Hydrate Reformation During the Dissociation Process Induced by Depressurization. *Fuel* **2022**, *313*, 122983. [CrossRef]
50. Zhang, Y.Y.; Wang, Z.Y.; Chen, L.Q.; Li, H.; Zhang, J.B.; Yang, H.M. Effects of Hydrate Saturation and Sand-Filling Content on Hydrate Exploitation Using In-Situ Heat Supply with Chemical Reagents. *Lect. Notes Civ. Eng.* **2024**, *472*, 568–589. [CrossRef]
51. Yu, T.; Chen, B.B.; Jiang, L.L.; Zhang, L.X.; Yang, L.; Yang, M.J.; Song, Y.C.; Abudula, A. Feasibility Evaluation of a New Approach of Seawater Flooding for Offshore Natural Gas Hydrate Exploitation. *Energy Fuels* **2023**, *37*, 4349–4364. [CrossRef]
52. Zhang, Q.; Zhang, Q.; Zhang, Q.; Wang, Y.F.; Wang, Y.F.; Wang, Y.F. Numerical Simulations of Combined Brine Flooding With Electrical Heating–Assisted Depressurization for Exploitation of Natural Gas Hydrate in the Shenhu Area of the South China Sea. *Front. Earth Sci.* **2022**, *10*, 843521. [CrossRef]
53. Sloan, E.D.; Koh, C.A. *Clathrate Hydrates of Natural Gases*, 3rd ed.; CRC Press: Boca Raton, FL, USA, 2007.
54. Mok, J.H.; Choi, W.J.; Seo, Y.W. The Dual-Functional Roles of N_2 Gas for the Exploitation of Natural Gas Hydrates: An Inhibitor for Dissociation and an External Guest for Replacement. *Energy* **2021**, *232*, 121054. [CrossRef]
55. Wei, N.; Zheng, H.R.; Guo, B.Y.; Hu, H.Y.; Li, C. Acceleration Mechanism of Methane Hydrate Dissociation in Inorganic Salt Solutions: Experimental and Molecular Dynamics Study. *Geoenergy Sci. Eng.* **2025**, *252*, 213913. [CrossRef]
56. Xing, Q.Z.; Wan, L.; Zhang, J.Y.; Zhang, S.F.; Cui, X.Y. Experimental Study of Methane Hydrate Decomposition Kinetics in NaCl, KCl, and PVP Solutions. *ACS Omega* **2025**, *10*, 22647–22656. [CrossRef]
57. Zeng, H.P.; Zhang, Y.; Zhang, L.; Chen, Z.Y.; Li, X.S. Effects of the NaCl Concentration and Montmorillonite Content on Formation Kinetics of Methane Hydrate. *J. Mar. Sci. Eng.* **2022**, *10*, 548. [CrossRef]
58. Yin, Z.Y.; Moridis, G.; Linga, P. On the Importance of Phase Saturation Heterogeneity in the Analysis of Laboratory Studies of Hydrate Dissociation. *Appl. Energy* **2019**, *255*, 113861. [CrossRef]
59. Li, R.R.; Han, Z.H.; Zhang, L.Q.; Zhou, J.; Wang, S. Effect of Hydrate Saturation on Permeability Anisotropy for Hydrate-Bearing Turbidite Sediments Based on Pore-Scale Seepage Simulation. *J. Mar. Sci. Eng.* **2024**, *12*, 1079. [CrossRef]
60. Chen, H.; Du, H.; Shi, B.; Shan, W.C.; Hou, J.Q. Mechanical Properties and Strength Criterion of Clayey Sand Reservoirs During Natural Gas Hydrate Extraction. *Energy* **2021**, *242*, 122526. [CrossRef]
61. Wang, Y.H.; Wu, P.; Huang, L.; Song, Y.C.; Li, Y.H. Stress-Induced Permeability Evolution Mechanism of Hydrate-Bearing Sediments from Pore-Scale Perspective. *Energy* **2025**, *319*, 134977. [CrossRef]
62. Zhang, N.; Wang, H.N.; Jiang, M.J. A Mechanistic Model for Porosity and Permeability in Deformable Hydrate-Bearing Sediments with Various Hydrate Growth Patterns. *Acta Geotech.* **2023**, *19*, 1–26. [CrossRef]

63. Shang, Y.H.; Lei, G.; Tang, J.D.; Qu, J.T.; Ning, F.L. Permeability of Marine Gas Hydrate-Bearing Sediments Considering Effective Stress and Particle Size Distribution: A Theoretical Model. *SSRN J.* **2024**. *preprint*. [CrossRef]
64. Lei, G.; Qu, J.T.; Zhao, Y.J.; Wu, Q.; Liu, T.L. Permeability Behavior of Hydrate-Bearing Clayey-Silty Sediments During Creeping. *J. Rock Mech. Geotech. Eng.* **2024**, *17*, 2492–2510. [CrossRef]
65. Xu, Z.Q.; Shen, K.X.; Wang, Y.S.; Wu, J.; Liu, P.L.; Du, J.; Huang, Q.S.; Chen, C. Consolidation-Acidizing Experiments on Methane Natural Gas Sediment Skeleton. *Front. Energy Res.* **2024**, *12*, 1332495. [CrossRef]

Journal of
Marine Science and Engineering

MDPI

rticle

High-Precision Small-Scale 3D Seismic Technology for Natural Gas Hydrate Exploration in the Northern South China Sea

Dasen Zhou [1], Siqing Liu [1,*], Xianjun Zeng [1,*], Limin Gou [2], Jing Li [1], Jingjing Zhang [1], Xiaozhu Hao [1], Qingxian Zhao [1], Qingwang Yao [3], Jiafa Zhang [1], Jiaqi Shen [1], Zelin Mu [1] and Zelin He [1]

1 Guangzhou Marine Geological Survey, Guangzhou 511458, China; zhoudasen@mail.cgs.gov.cn (D.Z.); ljingcugb@163.com (J.L.); zhangjingjing@mail.cgs.gov.cn (J.Z.); dinbuyi@126.com (X.H.); zqingxian@mail.cgs.gov.cn (Q.Z.); jfzhang0532@163.com (J.Z.); shenjiaqi@mail.cgs.gov.cn (J.S.); mouzelin@mail.cgs.gov.cn (Z.M.); hekuuuu@sina.com (Z.H.)
2 College of Marine Science, China University of Geosciences (Beijing), Beijing 100083, China; goulm@163.com
3 State Key Laboratory of Precision Measurement Technology and Instruments, Tianjin University, Tianjin 300072, China; yaoqingwang@tju.edu.cn
* Correspondence: liusiqing@mail.cgs.gov.cn (S.L.); 13660343708@139.com (X.Z.)

Abstract

To address the demand for high-precision exploration of natural gas hydrates in the northern South China Sea, this paper presents a novel high-precision small-scale 3D seismic exploration technology. The research team independently developed a seismic acquisition system, incorporating innovative designs such as a narrow trace spacing of 3.125 m and a short streamer length of 150 m. By integrating advanced processing techniques, including pre-stack noise suppression, spectral broadening, and refined velocity analysis, the system significantly enhances the precision and spatial resolution of shallow seismic data. During field trials in the Qiongdongnan basin, the system successfully acquired 3D seismic data over an area of 50 km^2, enabling fine-scale imaging of sub-seabed strata within the upper 300 m. This represents a notable improvement in resolution compared to conventional 3D seismic technologies. When benchmarked against international counterparts such as P-cable, our system demonstrates distinct advantages in terms of exploration depth (reaching 1800 m) and dominant frequency range (spanning 10~390 Hz). The research findings provide a reliable technical approach for the detailed characterization of natural gas hydrates and the inversion of reservoir parameters, thereby holding significant practical value for advancing the industrial development of natural gas hydrates in China's offshore areas.

Keywords: high-precision small-scale 3D seismic exploration; data processing; natural gas hydrate; Qiongdongnan basin

1. Introduction

Marine streamer seismic exploration is essential for resource exploration [1,2]. Although cost-effective, 2D seismic struggles to capture complex subsurface heterogeneities [3–6]. In contrast, 3D seismic, with its areal acquisition geometry, is better at identifying small-scale structures and reservoirs [7–11]. As a result, 3D seismic has become the primary method for characterizing natural-gas hydrate systems, shifting from initial anomaly detection with 2D to delineating internal architectures and boundaries. However, traditional 3D systems are limited by their wide channel spacing (≥12.5 m) and long streamers (>1 km), which lead to insufficient resolution for imaging shallow strata (<300 m) and capturing hydrate reservoir details. This highlights the urgent need to develop a high-precision, small-scale 3D seismic system, where the streamer trace spacing is smaller and the streamer separation is shorter compared to conventional streamers.

Internationally, progress is evident. Norway, along with the UK and Germany, developed the P-cable system [12], which was successfully deployed in the Gulf of Mexico (2008) [13,14] and the Arctic (2013) [14]. Available from Geometrics, it tows 6–24 short streamers (25–200 m long) with narrow streamer spacing (1.5625–12.5 m) and fine trace spacing (3.125–12.5 m), providing a more straightforward, more efficient high-resolution solution that can be operated by a small crew.

In China, hydrate exploration mainly relies on high-resolution 2D or quasi-3D methods (single/dual-source with a single streamer) [15,16]. There is a heavy dependence on imported equipment for conventional 3D surveys, which limits the flexibility in streamer and trace spacing configuration. While techniques like pre-stack noise attenuation and spectral broadening (e.g., deconvolution) are well established in conventional seismic processing, their use in addressing the unique challenges of small-scale, high-density 3D acquisition systems remains underexplored. The main innovation of our processing approach is not in inventing new algorithms but in developing an integrated workflow specifically designed to overcome the inherent limitations of compact systems, such as reduced noise immunity and limited spatial coverage. Therefore, creating autonomous equipment and adapting key technologies are crucial for China's advanced hydrate exploration.

To address this, the Guangzhou Marine Geological Survey and domestic institutions developed a high-precision, small-scale 3D seismic system. Its main innovations include a very narrow channel spacing (3.125 m), short streamers (150 m), and a dual-source, four-streamer alternating shooting mode. This setup greatly improves lateral and vertical resolution for shallow targets, closing the capability gap of conventional 3D systems. The system provides more flexibility and accuracy in data acquisition but requires more advanced data processing. It is especially suitable for detailed localized exploration. Using multi-element piezoelectric streamers, the system was tested in the Qiongdongnan basin, collecting data over 50 km^2. After optimized processing tailored to the data, it produced high-quality imaging results.

2. Geological Background

The study area is located in the Qiongdongnan basin of the northern South China Sea, specifically on the Songnan low uplift within the central depression of the basin. It is surrounded by the Ledong–Lingshui sag, the Songnan–Baodao sag, and the Changchang sag, forming a typical "multi-sag encirclement" tectonic framework. This area serves as a primary target for hydrocarbon migration and accumulation [17–20]. The tectonic activity of the Songnan low uplift is controlled by the overall tectonic movements of the Qiongdongnan basin, having undergone successive stages, including a rift phase from the Paleocene to the early Oligocene, a rift-to-sag transition phase during the late Oligocene, and a sag phase from the Miocene to the Quaternary. From bottom to top, the strata developed in sequence include the Yacheng formation, Lingshui formation, Sanya formation, Meishan formation, Huangliu formation, Yinggehai formation, and Pleistocene strata [17,21–25].

Gas hydrates in the Qiongdongnan basin mainly form in deep-water turbidite sand bodies within the Miocene Huangliu and Meishan formations. The Yinggehai formation itself generally is not the primary reservoir for natural gas hydrates in the basin; however, it acts as an important source rock and seal layer. The hydrate deposits in this region show significant dynamic and active features and are closely linked to deep petroleum systems. Seismic profiles across the area often show extensive vertical migration pathways, like gas chimneys, mud diapirs, and fault systems. These features usually appear as chaotic reflections, pull-downs, or acoustically blurred zones, indicating a continuous upward flow of fluids and gases. Direct hydrocarbon indicators such as acoustic blanking and bright spots are commonly observed in the sections, pointing to the presence of substantial free gas along with gas hydrates in shallow formations. Due to active gas migration, the bottom simulating reflector (BSR) usually appears as discontinuous with weak amplitude, showing much less continuity than in the Shenhu area. The distribution of gas hydrate reservoirs strongly correlates with deep structures like diapirs and gas chimneys, supporting a "lower-generation, upper-accumulation" reservoir formation model.

In 2018 and 2020, the Guangzhou Marine Geological Survey deployed a quasi-3D seismic survey grid in the Qiongdongnan basin to conduct detailed investigations of favorable areas for gas hydrates (Figure 1). Additionally, three gas hydrate drilling and coring operations were carried out in the region in 2018, 2019, and 2021, yielding physical samples of gas hydrates [26–32]. This paper focuses on the processing and analysis of line01, within the high-precision small-scale 3D seismic exploration area, passing through the W09 drilling site (Figure 1b).

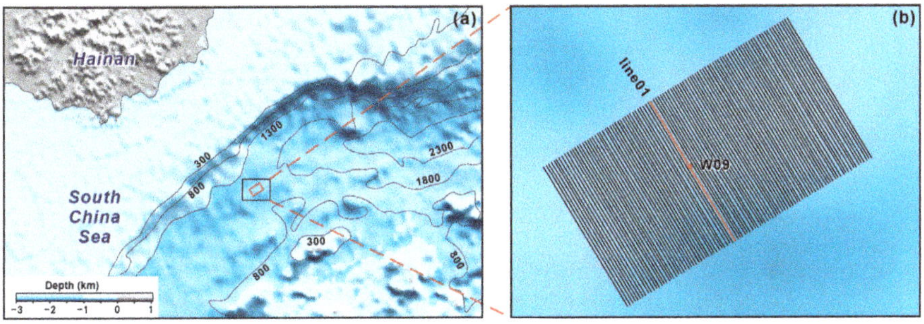

Figure 1. (**a**) Location map of the studyarea. The red rectangle represents the location of the 3D seismic study area. (**b**) Seismic survey lines. W09 is the drilling and coring site for gas hydrate exploration conducted in 2019.

The primary target layer for high-precision small-scale 3D seismic exploration in this study area is the Quaternary strata. The seafloor reflection appears as a continuous, strong-amplitude peak reflection across the entire survey area on seismic profiles. The bottom boundary of the Quaternary, which is the top interface of the Yinggehai formation and the bottom interface of the Ledong formation, shows highly continuous, medium-amplitude peak reflections in the shallow-marine to semi-deep-marine uplift regions. In deep-sea areas, this reflection remains visible but shows significant variations in continuity and amplitude. Between these two interfaces, multiple stable-block, flow-interface reflection events are consistently observed throughout the entire study area, indicating favorable seismic–geological conditions for the implementation and application of seismic methods.

3. High-Precision Small-Scale 3D Seismic System

3.1. Acquisition System

The high-precision small-scale 3D seismic exploration system is designed to achieve high-resolution imaging of shallow subsurface structures, particularly for natural-gas hydrate exploration. In contrast to conventional 3D seismic systems, the key innovations of this system lie in its acquisition geometry and instrumental configuration—notably, the use of short streamers (150 m in length) and an ultra-dense channel spacing of 3.125 m. These features significantly enhance horizontal and vertical resolution for targets within 300 m below the seafloor.

The system employs a dual-source and four-streamer acquisition configuration (Figure 2). Each streamer contains 48 hydrophones grouped into operational sections, with a total active section of 150 m. A single leading section (400 m long) is used to connect the streamers to the vessel, which simplifies deployment, reduces tangling risks, and offers greater operational flexibility compared to multi-lead systems. This design also provides sufficient clearance for seismic source operation.

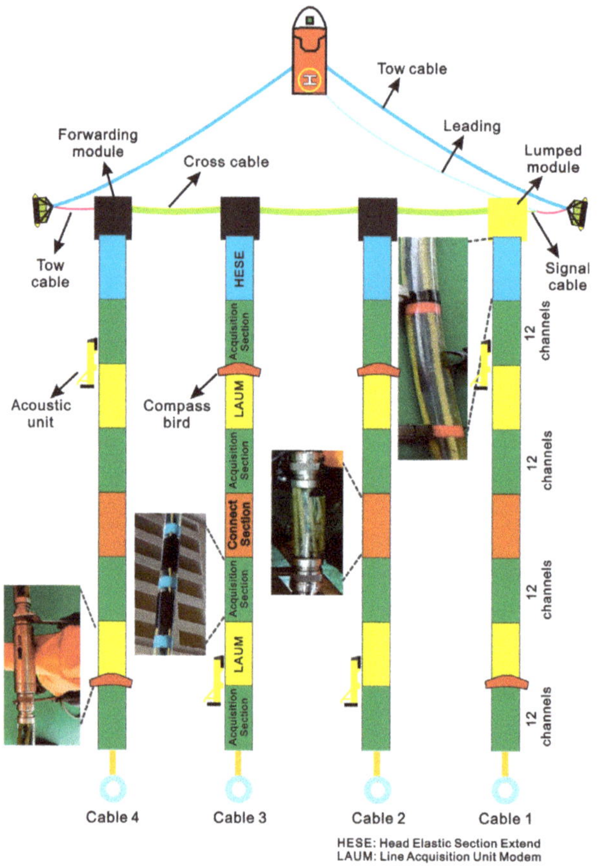

Figure 2. Schematic diagram of underwater equipment connections for seismic acquisition system.

Signal acquisition and transmission are optimized for high-channel-count data. Each acquisition node services multiple hydrophone channels, reducing the number of underwater electronic units. Data are digitized near the sensors and transmitted via a hierarchical network of acquisition section, LAUM, and the leading section back to the onboard recording system. The system uses advanced positioning and depth control devices (acoustic units and compass birds) (Figure 2) to maintain accurate streamer geometry. This streamlined and high-density acquisition approach differentiates the system from conventional 3D towed-streamer configurations and enables high-resolution imaging suited to the detailed characterization of gas hydrate reservoirs.

3.2. Observation System

The high-precision small-scale 3D seismic exploration system utilizes various positioning and depth control devices. These include RGPS, acoustic units, and compass birds. The acoustic units function as simultaneous multichannel transceiver nodes within the acoustic network. The compass birds supply depth adjustment and measurement for the streamers. The system also features spacing control devices such as spacing ropes and bridles. Together, these components enable real-time monitoring of positions and precise control of the distance between the streamer and the seismic source. Compared to conventional 3D seismic surveys, it has shorter receiver streamer lengths, narrower trace intervals, and smaller streamer separations. The system's overall parameters include transverse streamer spacing, source spacing, number of sources, firing interval, bin coverage, and others.

To maximize coverage and ensure data quality, the high-precision small-scale 3D seismic survey uses a dual-source alternating shooting acquisition method, achieving sixfold bin coverage. Under the dual-source and four-streamer acquisition mode, to prevent "blank zones" in the bins during single-pass acquisition, the transverse streamer spacing must be at least twice the source spacing, and the bin size is directly proportional to both streamer and source spacing. From a field implementation difficulty perspective, a larger transverse streamer spacing reduces the risk of streamer entanglement, enhancing operational feasibility and improving field efficiency. However, the lateral resolution of field acquisition is inversely proportional to bin size, necessitating smaller bins for precise imaging in 3D seismic surveys, which creates a trade-off between these two factors.

A practical combination of streamer and source deployment depths can reduce the impact of virtual reflections (ghost effects) from the seawater surface. Cable-source depth simulations (Figure 3) showed that when the source depth is 2 m and the streamer depth ranges from 2 to 3 m, there are no notches in the frequency band (Figure 3a). When the source depth is 3 m and the streamer depth ranges from 2 to 3 m, only a single notch appears at high frequencies (Figure 3b). As the streamer depth increases, multiple notches emerge within the frequency band (Figure 3a,b). With increasing source depth, notches shift to lower frequencies and become deeper (Figure 3c,d). In summary, shallower source and streamer depths generally reduce the impact of notches on seismic signals. However, shallower deployments are more susceptible to external disturbances such as waves, leading to poorer signal-to-noise ratios (SNRs) in the seismic data. Therefore, considering sea conditions and noise at the survey site, a source depth of 3 m and a streamer depth of 5 m were selected [33].

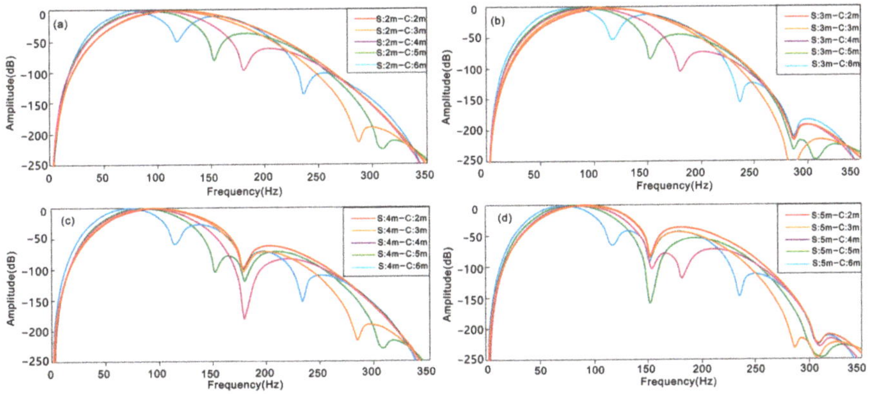

Figure 3. Comparison of the superimposed responses for cable and source depth combination (**a**–**d**): (**a**) source depth: 2 m; cable depths: 2~6 m; (**b**) source depth: 3 m; cable depths: 2~6 m; (**c**) source depth: 4 m; cable depths: 2~6 m; (**d**) source depth: 5 m; cable depths: 2~6 m.

Considering the water depth of 1600 to 1800 m in the study area and the need for precise imaging of subsurface layers shallower than 300 m, along with operational challenges, efficiency, and the safety of underwater equipment, we proposed acquisition parameters for a dual-source and four-streamer high-precision small-scale 3D seismic survey. A comparison with conventional 3D seismic parameters (taking R/V Hai Yang Di Zhi Ba Hao as an example) is presented in Table 1.

Table 1. 3D seismic acquisition parameters.

System Type	High-Precision Small-Scale 3D	Conventional 3D
Record Length	5 s	7–10 s
Sampling Rate	1 ms	2 ms
Channels per Streamer	48 channels (3.125 m × 48 = 150 m)	12 channels (12.5 m × 12 = 150 m)
Trace Spacing	3.125 m	12.5 m
Number of Streamers	4 × 1	4 × 20
Streamer Separation	50 m	100 m
Streamer Depth	5 m	8 m
Shot Interval	12.5 m	25 m
Source Depth	3 m	7 m
Source Volume	545 in^3	4365 in^3
Number of Sources	2	2
Source Firing Mode	Alternating Left/Right Sources	Alternating Left/Right Sources
Bin Size	3.125 m × 12.5 m	12.5 m × 25 m
Fold	6	90

4. Methods

Compared to conventional 3D seismic acquisition methods, the high-precision small-scale 3D seismic exploration technology uses high-frequency sources and small trace intervals, resulting in higher horizontal and vertical resolution in the acquired data. During the data processing for this operation, various measures were taken to use high-resolution data to obtain detailed subsurface structures beneath the seafloor without losing resolution. These include applying pre-stack noise attenuation to suppress swell noise; using a constant velocity-scanning method based on regional velocities for short-spread 3D data; researching spectral-broadening techniques and optimizing seismic data migration and processing parameters with commercial surface-consistent, reflection-wave, and residual-static-correction method software (paradigm2017, Houston, TX, USA) tailored to the characteristics of 3D seismic data; and applying to pick-up, decompose, and apply minor residual-time-shift, static-correction methods, eliminating static time differences and improving the SNR and resolution.

4.1. Pre-Stack Noise-Attenuation Technology

In the study area, ghost waves are not well-developed, and the primary noise types are swell noise and linear noise with stable apparent velocity, showing significant differences in signal and noise characteristics. Since swell noise has low frequency (<10 Hz) and high energy, this study applies high-pass filtering to suppress it. During processing, line01 was processed with high-pass filtering using parameters set to HP (8, 10), which achieved satisfactory results, as shown in Figure 4. As the linear interference noise exhibits a stable apparent velocity (1500 m/s) that significantly differs from that of the signal, spherical divergence compensation was first applied to address overall amplitude attenuation caused by geometric spreading and restore seismic wave energy. Subsequently, F-K dip filtering was used to suppress the energy of this linear noise. Comparison of shot gathers and stacked sections before and after processing demonstrates the effective removal of linear interference (Figure 5).

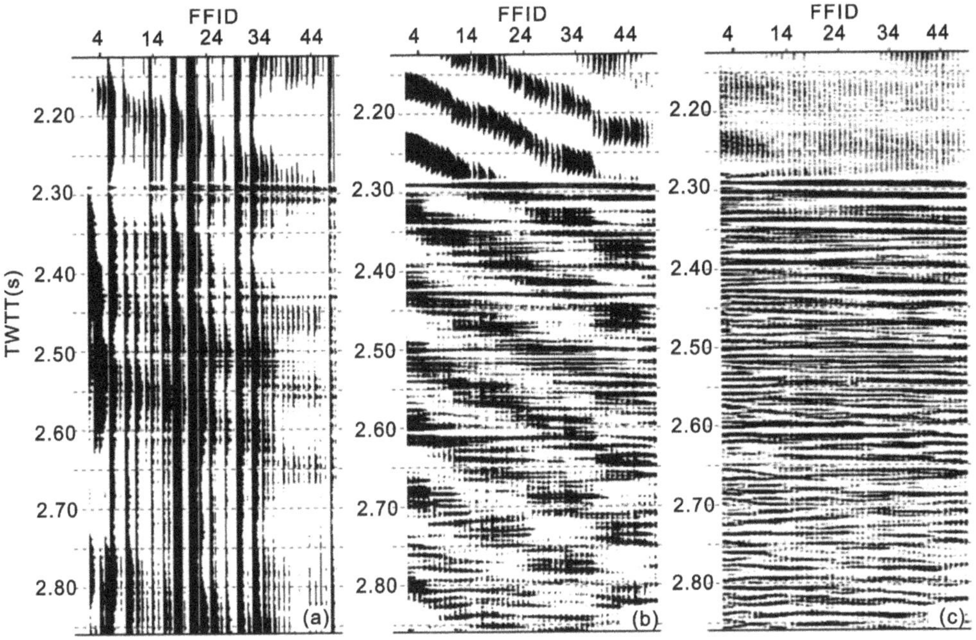

Figure 4. Shot gather records at different noise attenuation stages. The raw data (**a**) contain intense swell noise and linear interference, which are effectively suppressed in (**b**) and (**c**), respectively.

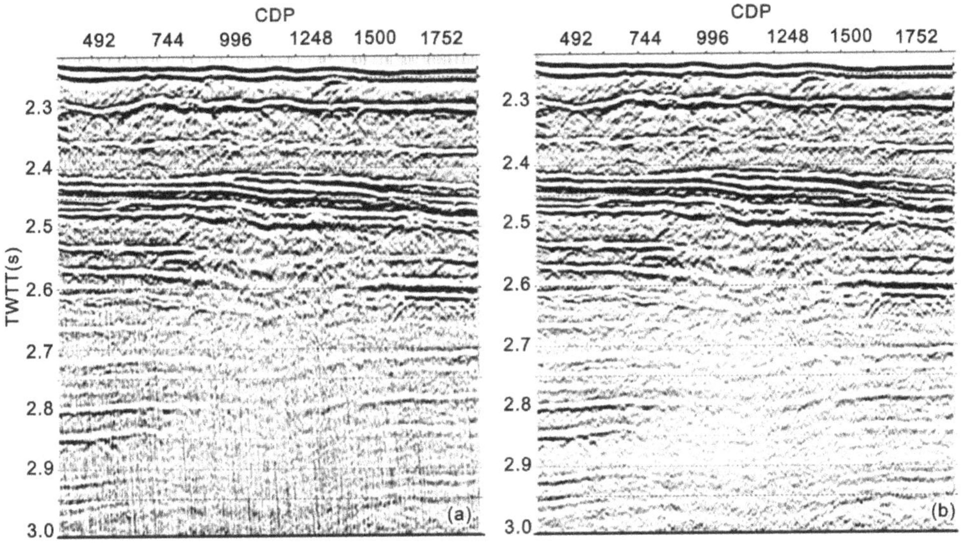

Figure 5. Stacked sections before (**a**) and after (**b**) F-K dip filtering.

4.2. Spectral-Broadening Technique

The raw data from the high-precision small-scale 3D seismic exploration system (Dynageo Systems Co., Ltd., Beijing, China) show a relatively high SNR, providing a solid foundation for enhancing data resolution through spectral broadening. This paper employs a one-step statistical wavelet-deconvolution method to process the line01 survey line. The core of this process is that the wavelet is statistically estimated from the seismic data within a designated time window. Specifically, statistical wavelet deconvolution is applied solely in the shot domain to broaden the spectrum, adjust the wavelet shape, and improve wave group characteristics, thereby fully leveraging and exploring the resolution advantages and potential of 3D seismic data.

Comparisons of shot gathers, shot gather spectra, stacked sections, and wavelet autocorrelation sections before and after processing show that wavelet consistency improves significantly after processing at a 150 Hz dominant frequency (Figure 6b). This effectively addresses the issue of wavelet shape variability, resulting in largely consistent waveforms and frequencies in the seismic records, a notable increase in dominant frequency, and better wave group characteristics. For the processing results at a 220 Hz dominant frequency, while the wavelet is further compressed and noise energy increases, leading to a slight decrease in the SNR (as indicated by the yellow box in Figure 6), the wavelet consistency remains good (Figure 6c). This provides richer interlayer information and highlights small-scale variations in strong reflection events more prominently (as shown by the red arrows in Figure 6).

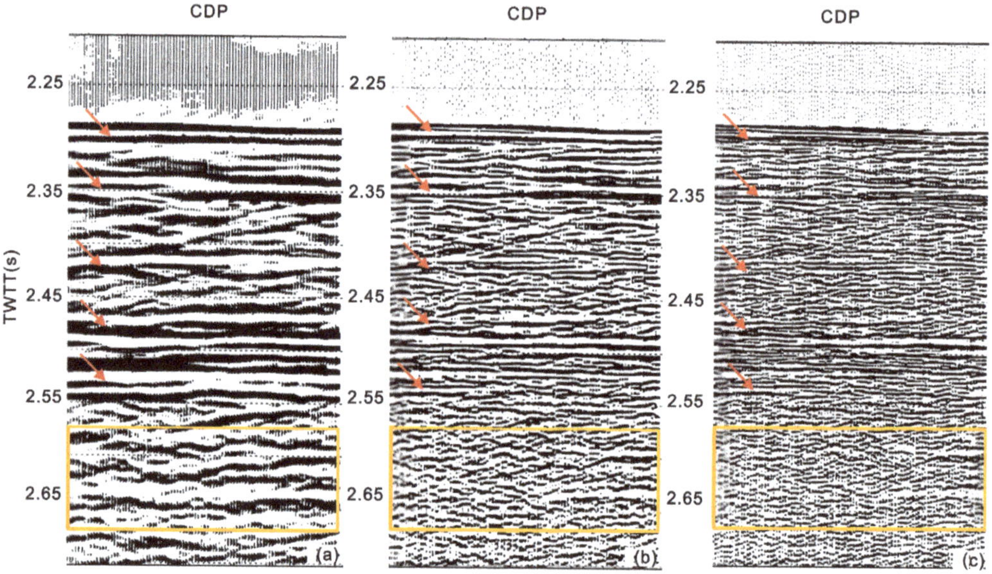

Figure 6. One-step statistical wavelet deconvolution results (stacked sections): (**a**) pre-deconvolution; (**b**) post-deconvolution with a dominant frequency of 150 Hz; (**c**) post-deconvolution with a dominant frequency of 220 Hz.

4.3. Post-Stack Phase-Shift Time Migration

By referencing the migration methods used in both domestic and international P-cable 3D and short-streamer 3D seismic data processing, the phase-shift migration method, which is widely adopted in production practices, was selected for post-stack time migration. Following migration, diffracted waves and inclined events were largely accurately repositioned, with distinct breakpoints. Meanwhile, characteristics in the section, such as disordered phases of events and irregular variations in amplitude and frequency caused by the presence of natural gas, were also made highly visible (Figure 7).

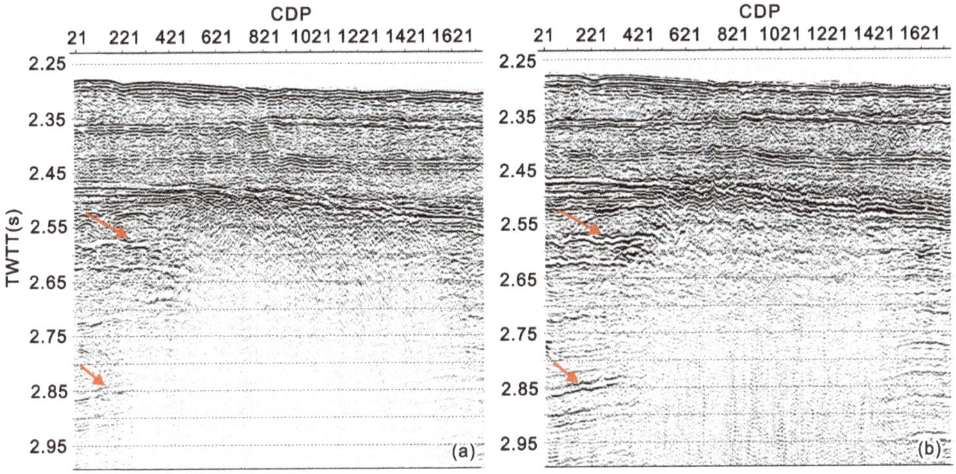

Figure 7. Comparison of seismic sections before (**a**) and after (**b**) migration.

4.4. Velocity Analysis Technique

The high-precision small-scale 3D seismic exploration system uses short streamers, resulting in less distinct hyperbolic reflection wave characteristics in CDP (common depth point) gathers. Consequently, conventional stacking velocity spectra show poor focusing, making it challenging to obtain subsurface velocity information effectively. This limitation necessitates a tailored approach to velocity analysis. While the use of regional velocities from historical 3D data as a reference could be potentially risky if the geological conditions were not consistent, this risk was mitigated in our study. The historical data were acquired from the immediate vicinity of our current survey area, ensuring geological representativeness. The regional stacking velocities, which ranged from approximately 1500 m/s at the seabed to around 1900–2100 m/s at the target depth, served as a robust initial model rather than a direct substitute. This model was geographically calibrated to our survey coordinates and formed the basis for our subsequent constant-velocity scanning, ensuring that a geologically reasonable starting point constrained the derived velocities. For this processing task, regional velocities from historical 3D data were used as reference velocities, and a constant-velocity scanning method based on these reference velocities was adopted for single-point velocity analysis to provide a velocity field for normal moveout (NMO) correction. Single-point velocity analysis is suitable for research areas with relatively simple structural conditions. When the stratigraphic dip is significant or the structure is complex, multi-point velocity analysis is used, with multi-point velocity curves interpolated along the spatial direction to establish a stacking velocity field. Given the relatively simple structural conditions of the target horizons in the operational area of this project, single-point velocity analysis was selected. By repeatedly comparing CMP (common middle point) gathers and stacked profiles, the optimal velocity was determined based on the criteria of flattened event axes in the gathers and maximum signal energy in the stacked profiles. After NMO correction, the event axes in the CMP gathers of line01 were essentially flattened, and the signal energy in the stacked profiles was enhanced, indicating high-velocity accuracy (Figure 8). Ultimately, a stacking velocity field with stable lateral velocities and vertically varying velocities was developed.

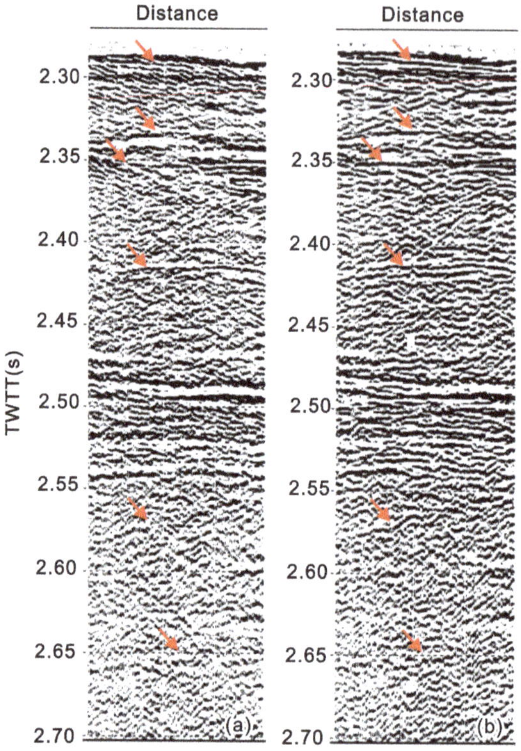

Figure 8. Comparison of stacked sections before (**a**) and after (**b**) NMO correction.

4.5. Residual-Static Correction Technique

Residual-static correction processing for high-precision small-scale 3D seismic exploration is a vital step performed after NMO correction. Its purpose is to eliminate time differences among traces within CDP gathers, align the reflection wave event phases across all traces in the gather, enhance stacking effects, improve the SNR and resolution, and highlight signal characteristics. For high-precision small-scale 3D seismic data, the datum surface is approximately a plane located 2 m below the sea surface. Time differences caused by factors such as streamer bending, depth variations from sea breezes and swells, variations in source depth, and errors in stacking velocities are relatively small. They can be treated as residual-static-correction time differences.

To address small-amplitude time differences, the high-precision small-scale 3D seismic exploration technology employs a surface-consistent, reflection-wave, and residual-static correction method. Satisfactory processing results are achieved when the correlation analysis time window is set to 2100–2800 ms, and the time shift is set to twice the sampling interval (−2–2 ms). The surface-consistent, reflection-wave, and residual-static correction method is a commonly used approach for eliminating short-wavelength static-correction time differences. Its specific implementation involves the following three stages: time difference picking, decomposition, and correction. This method utilizes CDP gather records after NMO correction, taking the stacked record trace as the model trace. Based on the cross-correlation method, it statistically calculates the time differences between each trace within the CDP gather and the model trace. According to surface-consistent conditions, these time differences are decomposed into shot point and receiver point components, thereby correcting the static-correction time differences. After processing, the SNR, resolution, and wave group characteristics of the stacked profile are improved (Figure 9).

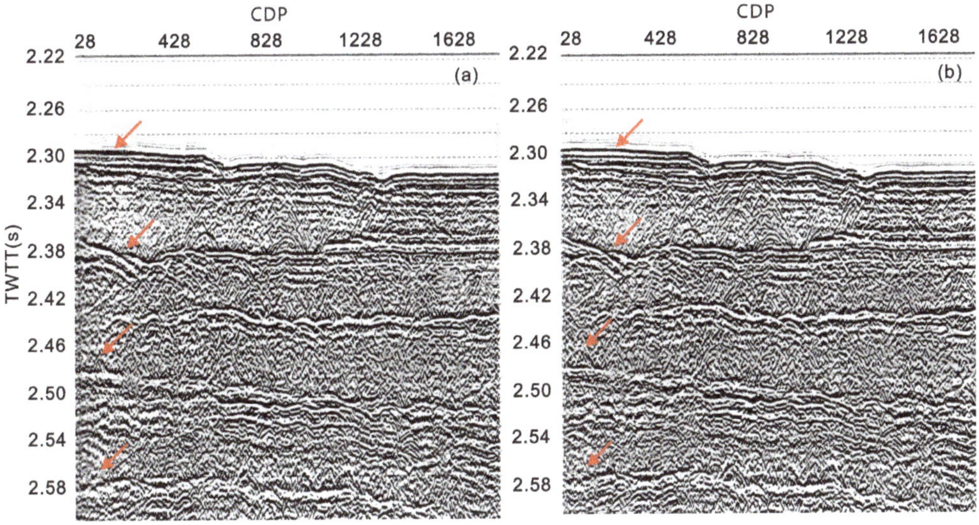

Figure 9. Comparison of stacked sections before (**a**) and after (**b**) residual-static correction.

5. Results

Figure 10 presents the seismic profile of line01 passing through station W09. The high-precision small-scale 3D seismic data exhibit excellent continuity, strong energy, and clearly highlight micro-topographic structures in shallow seafloor reflections. Features, such as mass transport deposits, gas cloud zones, bright spots, faults, and locally enhanced reflections, are readily identifiable. Previous hydrate drilling and comprehensive studies have confirmed that hydrate enrichment is closely related to these mass transport deposits and shallow deformation structures. The high-resolution imaging of shallow sedimentary geological features provides critical support for understanding deep-water sedimentary dynamics and assessing hydrocarbon resource potential.

Compared to conventional 3D seismic data (Figure 10b) acquired at the same location, the high-precision small-scale 3D profile (Figure 10a) shows richer stratigraphic information with more precise boundaries and more detailed structures. Assuming a velocity of 2000 m/s, the high-precision 3D data (with a dominant frequency of 150 Hz, for example) provides a theoretical resolution of about 3.3 m, whereas conventional data (60 Hz) resolves at roughly 8.3 m. In practice, the high-precision system can distinguish layers as thin as 3–4 m, much better than the 8–9 m possible with conventional methods. Small-scale variations at reflection interfaces are highlighted, gas-related blur zones are more distinct, and fault features are more clearly defined. Although both datasets have similar SNRs, the high-precision data reveal more detailed fluid-dynamic, superior-resolution information. Conversely, the conventional processing results lack clarity and do not preserve fine structural and stratigraphic details.

The high-precision small-scale 3D seismic data significantly improve the dynamic and kinematic characteristics of reflected waves, such as amplitude, frequency, and phase. This provides a strong basis for high-resolution geophysical inversion and resource assessment. By establishing a reasonably low-frequency model, extracting high-quality seismic wavelets, and selecting suitable inversion parameters, systematic seismic attribute analysis and acoustic impedance inversion were performed. Using the relationship between hydrate saturation from well logs and velocity from cross-plot analysis, key parameters, including gas hydrate saturation and porosity in the target reservoir, were calculated (Supplementary Figure S1). Integrating geological interpretation and geophysical inversion results, key evaluation parameters such as BSR distribution area and adequate thickness were obtained. Quantitative assessments of both gas hydrate resources and free-gas resources within the survey area were then conducted using the volumetric method.

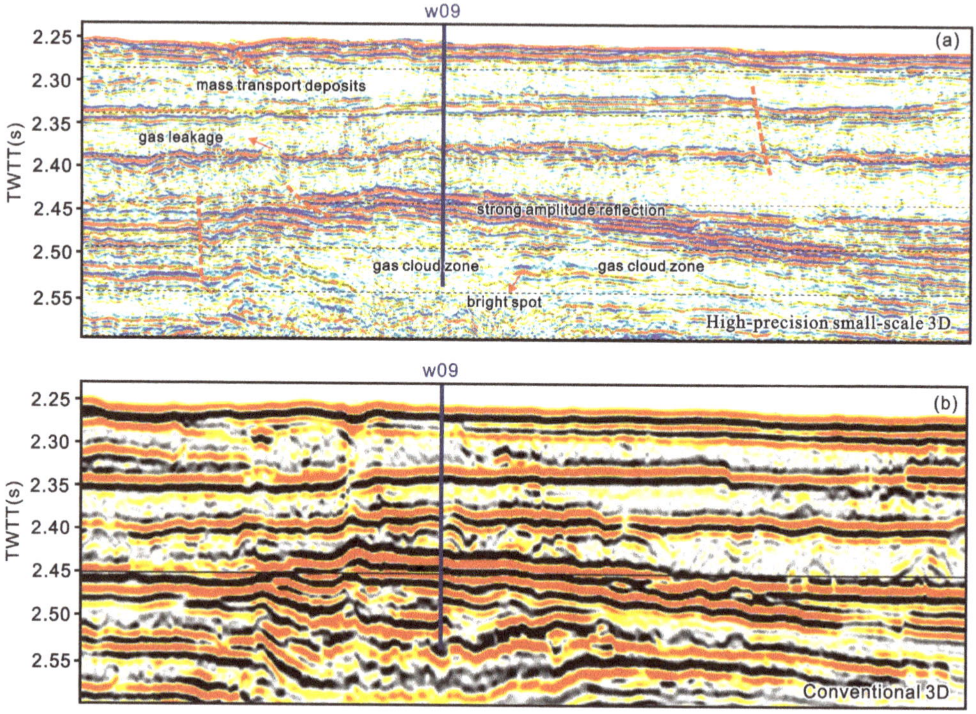

Figure 10. Comparison of high-precision small-scale 3D processing results for line01 (**a**) and conventional processing results (**b**). W09 is a drilling and coring well used for gas hydrate exploration.

6. Discussion

6.1. Comparison Between High-Precision Small-Scale 3D Seismic Data and P-Cable Data

According to Petersen (2010) [12], P-cable technology can also be used for high-resolution 3D seismic exploration of natural gas hydrates. The P-cable system utilizes shorter streamer arrays, typically consisting of 12 to 24 compact receiver cables measuring from 25 to 50 m. These cables feature hydrophone intervals of about 3.125 m and are deployed with cross-cable spacing of 6 to 12 m. The operational water depth for P-cable surveys is generally less than 1300 m, with target frequency ranges of 10 to 300 Hz (expandable to 10 to 400 Hz) and dominant frequencies concentrated between 120 and 170 Hz. In contrast, the high-precision small-scale 3D seismic survey in this study was conducted in water depths of 1500 to 1800 m, achieving a broader target frequency range of 10 to 390 Hz with dominant frequencies spanning from 120 to 220 Hz. This wider frequency bandwidth demonstrates superior vertical resolution, enabling detailed characterization of small-scale geological features (Figure 11).

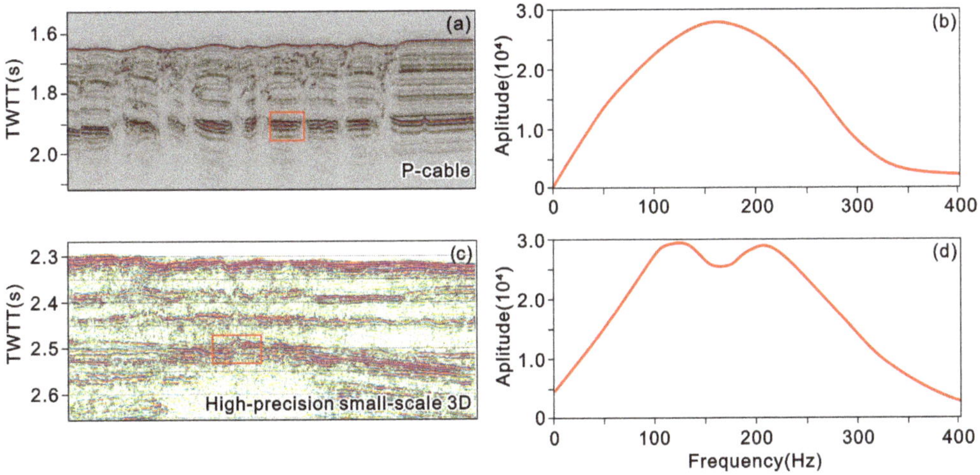

Figure 11. Comparison of processing results between the high-precision small-scale 3D seismic system and the foreign P-cable system: (**a**) P-cable seismic profile [34]; (**b**) spectrum of the P-cable seismic profile [34]; (**c**) high-precision small-scale 3D seismic profile; (**d**) spectrum of a high-precision small-scale 3D seismic profile.

This configuration, however, entails inherent trade-offs. The lower fold and narrower cross-line aperture, compared to larger P-cable arrays, may limit the signal-to-noise ratio and spatial sampling for larger-scale subsurface features in the cross-line direction. Consequently, system selection requires balancing the pursuit of ultra-high vertical resolution against the benefits of broader azimuth coverage and higher fold afforded by systems with more streamers.

6.2. Considerations for Data-Processing Parameter Selection

Spectral broadening technology holds significant scientific and practical value in seismic exploration. Its core mechanism involves expanding the frequency bandwidth of seismic signals—particularly by recovering or enhancing high-frequency components—to improve data resolution and geological interpretation accuracy substantially. This technology reduces the seismic wavelet size, allowing for the identification of thinner strata and surpassing the traditional resolution limits to achieve meter-scale or even finer precision. Simultaneously, it clearly reveals subtle structural features such as minor faults and micro-fractures, significantly improving reservoir characterization and providing solid evidence for accurately mapping internal reservoir architectures. Based on the principles mentioned above, this study employs line01 as a case example, utilizing one-step statistical wavelet deconvolution for spectral broadening. During processing, two sets of output wavelet-dominant frequencies were configured—150 Hz and 220 Hz—yielding two distinct processing results—a high-SNR dataset (150 Hz) and a high-resolution dataset (220 Hz).

The final high-SNR result profile is shown in Figure 12a. This profile displays strong-amplitude, well-continuous reflection events with high-SNR qualities. The target formation exhibits a frequency range from 6 to 400 Hz, a dominant bandwidth from 10 to 360 Hz, and a dominant frequency of 140 Hz (Figure 12c). The lateral resolution along the main survey line is 3.125 m, while the vertical resolution ranges from 2 to 3 m. The final high-resolution result profile is presented in Figure 12b. For the target formation (shallower than 300 m below the seafloor), the dominant bandwidth spans from 10 to 390 Hz with a dominant frequency of 200 Hz (Figure 12d). The lateral resolution remains 3.125 m, while the vertical resolution improves to 1–2 m. This processing outcome features higher dominant frequencies, broader bandwidths, and greater wavelet compression, resulting in decomposed composite waves and attenuated individual reflection amplitudes. To preserve gas-induced reflection anomalies and small-scale structural features, post-stack noise attenuation was intentionally minimized. Nevertheless, the primary reflection events maintain strong amplitudes and good continuity, indicating an overall moderate SNR for this dataset.

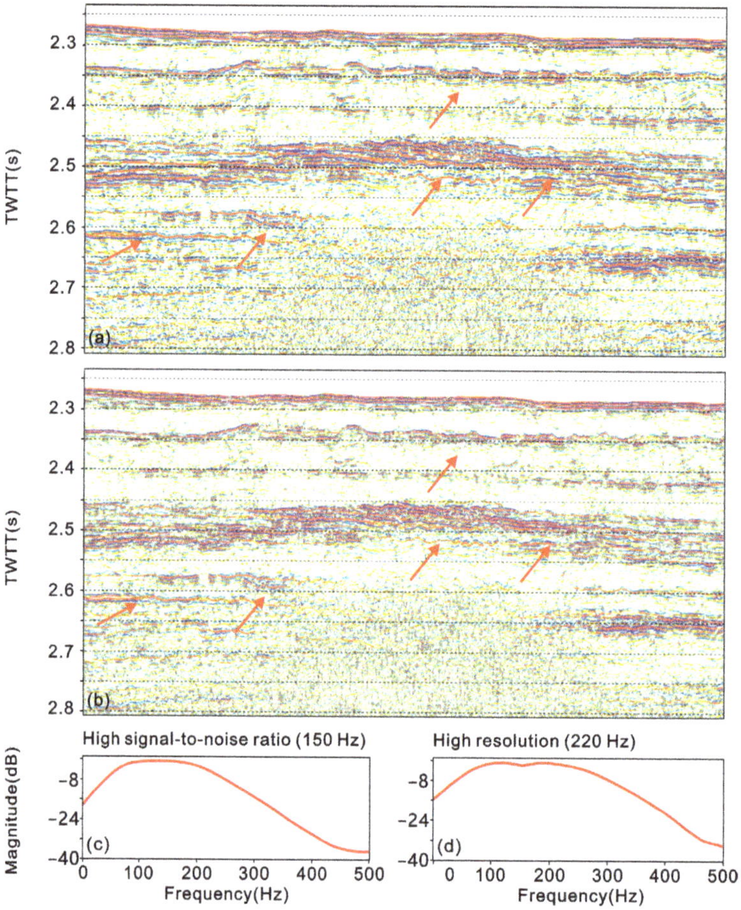

Figure 12. Comparative analysis of the high-SNR processed seismic profile (**a**) and the high-resolution processed seismic profile (**b**). Panels (**c**,**d**) show the spectral diagrams corresponding to the high-SNR processed profile and the high-resolution processed profile, respectively.

Regarding the target strata, a comparison of the two processed seismic profile results reveals the following:

1. In terms of SNR, the high-SNR processing result is superior. The continuity of strong-amplitude events is better, making them easier to trace across the entire profile. Amplitude and phase anomalies caused by fluids are more pronounced. Due to the improved resolution, composite events in the high-resolution profile are separated, with weaker amplitudes and less continuity, which makes accurate identification, interpretation, and tracing more challenging.

2. In terms of resolution, the high-resolution processing result is superior. The processed profile provides a more detailed depiction of the seafloor's characteristics and primary reflective interfaces within the target layer, providing richer information on faults and fracture systems. The correlation between fluids and fracture systems is more prominent.

7. Conclusions

This paper presents an independent research and development project for a high-precision small-scale 3D seismic exploration system. By optimizing acquisition parameters (such as small trace spacing and short streamers) and data-processing methods (including pre-stack noise attenuation and spectral broadening), the system significantly enhances the vertical and horizontal resolution, as well as the imaging quality, of seismic data. Compared to conventional seismic techniques, this system demonstrates higher precision and flexibility in the exploration of shallow gas hydrates, providing reliable technical support for natural-gas hydrate exploration.

The research team completed a 50 km^2 3D seismic acquisition operation in the Qiongdongnan basin, northern South China Sea, successfully achieving precise imaging of sub-seafloor strata within 300 m depth. Through comparison with conventional 3D seismic data, the high-precision small-scale 3D seismic data exhibit superior advantages in terms of resolution and detailed depiction, offering crucial evidence for the precise identification of gas hydrates and the acquisition of reservoir parameters.

Although the current system is equipped with only four streamers, its low cost, high precision, and flexible operation indicate promising application prospects. In the future, by increasing the number of streamers, the system can further enhance exploration efficiency, providing technical assurance for the industrial development of natural gas hydrates. Additionally, the successful application of this technology offers new solutions for deep-water resource exploration and marine geological research.

Supplementary Materials: The following supporting information can be downloaded at: https://www.mdpi.com/article/10.3390/jmse13091703/s1, Figure S1. (a) P-wave impedance profile. (b) Gas saturation in hydrate. (c) Gas hydrate saturation. (d) Hydrate saturation.

Author Contributions: Conceptualization, D.Z., S.L., X.Z., Q.Z., Q.Y., and Z.M.; methodology, L.G., S.L., and D.Z.; software, L.G., J.L., J.Z. (Jingjing Zhang), and J.Z. (Jiafa Zhang); validation, S.L.; formal analysis, S.L. and Q.Z.; investigation, D.Z., X.H., Q.Y., J.S., and Z.H.; resources, X.Z.; data curation, D.Z.; writing—original draft preparation, D.Z., S.L., L.G., J.L., J.Z. (Jingjing Zhang), J.Z. (Jiafa Zhang), J.S., Z.M., and Z.H.; writing—review and editing, S.L., D.Z., and X.Z.; visualization, S.L. and L.G.; supervision, S.L., X.Z., and X.H.; project administration, X.H.; funding acquisition, X.H. and Q.Z. All authors have read and agreed to the published version of the manuscript.

Funding: This work is supported by the National Key Research and Development Program of China (grant no. 2024YFC2813303).

Data Availability Statement: The original contributions presented in this study are included in the article. Further inquiries can be directed to the corresponding author.

Acknowledgments: We are grateful to the crew of the R/V Hai Yang Di Zhi Ba Hao, as well as the scientists and technicians, for their valuable contributions during the HYDZ8-202205 and HYDZ8-202301 cruises. We are deeply grateful to Pengfei Wen for his expert guidance and insightful suggestions for this paper.

Conflicts of Interest: The authors declare no conflicts of interest.

References

1. Du, L.Z.; Qiu, J.H.; Zhang, Q.; Zhang, X.P. Development and application of a high-fidelity and high-resolution telemetry seismic data acquisition system. *Chin. J. Geophys.* **2019**, *62*, 3964–3973, (In Chinese with English abstract).
2. Zhou, D.S.; Zeng, X.J.; Yang, C.; Zhai, J.F. Technical Analysis of the P-cable High Resolution 3D Marine Acquisition System. *J. Ocean Technol.* **2022**, *41*, 51–60.
3. Markl, R.; Bryan, G.; Ewing, J. Structure of the Blake-Bahama Outer Ridge. *J. Geophys. Res. Atmos.* **1970**, *75*, 4539–4555. [CrossRef]
4. Shipley, T.; Houston, M.; Buffler, R.; Shaub, F.; McMillen, K.; Ladd, J.; Worzel, J. Seismic Evidence for Widespread Possible Gas Hydrate Horizons on Continental Slopes and Rises. *Aapg Bull.* **1979**, *63*, 2204–2213.
5. Holbrook, W.; Hartley, H.; Wood, W.; Stephen, R.; Daniel, L. Methane Hydrate and Free Gas on the Blake Ridge from Vertical Seismic Profiling. *Science* **1996**, *273*, 1840–1843.
6. Hornbach, M.; Demian, M.; Holbrook, W.; Avendonk, H.; Gorman, A. Three-dimensional seismic imaging of the Blake Ridge methane hydrate province: Evidence for large, concentrated zones of gas hydrate and morphologically driven advection. *J. Geophys. Res.* **2008**, *113*, B07101. [CrossRef]
7. Hyndman, R.; Spence, G.A. seismic study of methane hydrate marine bottom simulating reflectors. *J. Geophys. Res.* **1992**, *97*, 6683–6698. [CrossRef]
8. Hobro, J.; Minshull, T.; Singh, S. Tomographic seismic studies of the methane hydrate stability zone in the Cascadia Margin. *Geol. Soc. Lond. Spec. Publ.* **1998**, *137*, 133–140. [CrossRef]
9. Riedel, M. Three-Dimensional Seismic Investigations of Northern Cascadia Marine Gas Hydrates. Doctoral Dissertation, University of Victoria, Victoria, BC, Canada, 2001.
10. Rensbergen, P.; Batist, M.; Klerkx, J.; Hus, R.; Poort, J.; Vanneste, M.; Granin, N.; Khlystov, O.; Krinitsky, P. Sub-lacustrine mud volcanoes and methane seeps caused by dissociation of gas hydrates in Lake Baikal. *Geology* **2002**, *30*, 631–634. [CrossRef]
11. Popescu, I.; Gilles, L.; Nicolae, P.; Batist, M.; Gillet, H. Seismic expression of gas and gas hydrates across the western Black Sea. *Geo-Mar. Lett.* **2007**, *27*, 173–183. [CrossRef]
12. Petersen, C.J.; Bünz, S.; Hustoft, S.; Mienert, J.; Klaeschen, D. High-resolution P-Cable 3D seismic imaging of gas chimney structures in gas hydrated sediments of an Arctic sediment drift. *Mar. Pet. Geol.* **2010**, *27*, 1981–1994. [CrossRef]
13. Bangs, N.L.; Hornbach, M.J.; Berndt, C. The mechanics of intermittent methane venting at South Hydrate Ridge inferred from 4D seismic surveying. *Earth Planet. Sci. Lett.* **2011**, *310*, 105–112. [CrossRef]
14. Bünz, S.; Polyanov, S.; Vadakkepuliyambatta, S. Active gas venting through hydrate-bearing sediments on the Vestnesa Ridge, offshore W-Svalbard. *Mar. Geol.* **2012**, *332–334*, 189–197. [CrossRef]
15. Xu, H.N.; Shu, H.; Li, L.Q.; Zhang, M.; Zhang, G.X. 3-D seismic data processing techniques for gas hydrate by the single-source and single-cable acquisition method. *Prog. Geophys.* **2009**, *24*, 1801–1806.
16. Zhang, X.D.; Wen, P.F.; Xu, Y.X.; Zhang, B.J. Footprint analysis and supperss processing of dual-source and single cable 3D seismic data for gas hydrate. *Mar. Geol. Front.* **2015**, *31*, 55–61.
17. Zhu, Y.M.; Sun, L.T.; Hao, F.; Tuo, L. Geochemical composition and origin of tertiary oils in the Yinggehai and Qiongdongnan basins, offshore South China Sea. *Mar. Pet. Geol.* **2018**, *96*, 139–153. [CrossRef]
18. Huang, B.J.; Tian, H.; Li, X.S.; Wang, Z.F.; Xiao, X.M. Geochemistry, origin and accumulation of natural gases in the deep water area of the Qiongdongnan Basin, South China Sea. *Mar. Pet. Geol.* **2016**, *72*, 254–267. [CrossRef]
19. Li, Y.Y.; Pang, L.; Wang, Z.D.; Meng, Q.X.; Guan, P.; Xu, X.M.; Fang, Y.X.; Lu, H.L.; Ye, J.L.; Xie, W.W. Geochemical characteristics and significance of organic matter in hydrate-bearing sediments from Shenhu area, South China Sea. *Molecules* **2022**, *27*, 2533. [CrossRef]
20. Cui, H.P.; Su, X.; Chen, F.; Melanie, H.; Yang, S.X.; Liang, J.Q.; Su, P.B.; Dong, H.L.; Hou, W.G. Microbial diversity of two cold seep systems in gas hydrate-bearing sediments in the South China Sea. *Mar. Environ. Res.* **2019**, *144*, 230–239. [CrossRef]
21. Liang, J.Q.; Zhang, W.; Lu, J.A.; Wei, J.G.; Kuang, Z.G.; He, Y.L. Geological occurrence and accumulation mechanism of natural gas hydrates in the eastern Qiongdongnan Basin of the South China Sea: Insights from site GMGS5-W9-2018. *Mar. Geol.* **2019**, *418*, 106042. [CrossRef]
22. Zhu, W.L.; Huang, B.J.; Mi, L.J. Geochemistry, origin, and deep-water exploration potential of natural gases in the Pearl River Mouth and Qiongdongnan basins, South China Sea. *AAPG Bull.* **2009**, *93*, 741–761. [CrossRef]
23. Liang, Q.Y.; Hu, Y.; Feng, D.; Peckmann, J.; Chen, L.Y.; Yang, S.X.; Liang, J.Q.; Tao, J.; Chen, D.F. Authigenic carbonates from newly discovered active cold seeps on the northwestern slope of the South China Sea: Constraints on fluid sources, formation environments, and seepage dynamics. *Deep Sea Res. Part I Oceanogr. Res. Pap.* **2017**, *124*, 31–41. [CrossRef]
24. Ma, M.; Li, C.; Lv, C.F.; Chen, G.J.; Yang, F.; Yan, Y.K.; Yin, N.; Zhang, G.C. Geochemistry and provenance of a multiple-stage fan in the Upper Miocene to the Pliocene in the Yinggehai and Qiongdongnan basins, offshore South China Sea. *Mar. Pet. Geol.* **2017**, *79*, 64–80. [CrossRef]

5. Zhang, W.; Liang, J.Q.; Wei, J.G.; Lu, J.A.; Su, P.B.; Lin, L.; Huang, W.; Guo, Y.Q.; Deng, W.; Yang, X.L.; et al. Geological and geophysical features of and controls on occurrence and accumulation of gas hydrates in the first offshore gas-hydrate production test region in the Shenhu area, northern South China Sea. *Mar. Pet. Geol.* **2020**, *114*, 104191. [CrossRef]
6. Lei, C.; Ren, J.L.; Pei, J.X.; Lin, H.T.; Yin, X.Y.; Tong, D.J. Tectonic framework and multiple episode tectonic evolution in deepwater area of Qiongdongnan Basin, northern continental margin of South China Sea. *Earth Sci.* **2011**, *36*, 151–162. [CrossRef]
7. Wei, J.G.; Liang, J.Q.; Lu, J.A.; Zhang, W.; He, Y.L. Characteristics and dynamics of gas hydrate systems in the northwestern South China Sea: Results of the fifth gas hydrate drilling expedition. *Mar. Pet. Geol.* **2019**, *110*, 287–298. [CrossRef]
8. Lai, H.F.; Fang, Y.X.; Kuang, Z.G.; Ren, J.F.; Liang, J.Q.; Lu, J.A.; Wang, G.L.; Xing, C.Z. Geochemistry, origin and accumulation of natural gas hydrates in the Qiongdongnan Basin, South China Sea: Implications from site GMGS5-W08. *Mar. Pet. Geol.* **2021**, *123*, 104774. [CrossRef]
9. Lai, H.F.; Qiu, H.J.; Liang, J.Q.; Kuang, Z.G.; Fang, Y.X.; Ren, J.F.; Lu, J.A. Geochemical characteristics and gas-to-gas correlation of two leakage-type gas hydrate accumulations in the western Qiongdongnan Basin, South China Sea. *Acta Geol. Sin.* **2022**, *96*, 680–690. [CrossRef]
0. Fang, Y.X.; Wei, J.G.; Lu, H.L.; Liang, J.Q.; Lu, J.A.; Fu, J.; Cao, J. Chemical and structural characteristics of gas hydrates from the Haima cold seeps in the Qiongdongnan Basin of the South China Sea. *J. Asian Earth Sci.* **2019**, *182*, 103924. [CrossRef]
1. He, Y.L.; Liang, J.Q.; Kuang, Z.G.; Deng, W.; Ren, J.F.; Lai, H.F.; Meng, M.M.; Zhang, W. Migration and accumulation characteristics of natural gas hydrates in the uplifts and their slope zones in the Qiongdongnan Basin, China. *China Geol.* **2022**, *5*, 234–250. [CrossRef]
2. Klauda, J.; Sandler, S. Global distribution of methane hydrate in ocean sediment. *Energy Fuels* **2005**, *19*, 459–470. [CrossRef]
3. Zhou, D.S.; Yang, C.; Zeng, X.J.; Gou, L.M.; Yao, Q.W.; Hu, T.Y. Application of high-precision small 3D seismic acquisition technology in deep water gas hydrate exploration. *Acta Geol. Sin.* **2024**, *98*, 2678–2685.
4. Singhroha, S.; Bünz, S.; Plaza-Faverola, A. Gas hydrate and free gas detection using seismic quality factor estimates from high-resolution P-cable 3D seismic data. *Interpretation* **2016**, *4*, SA39–SA54. [CrossRef]

Journal of
Marine Science and Engineering

MDPI

Article

Gas Hydrate Exploration Using Deep-Towed Controlled-Source Electromagnetics in the Shenhu Area, South China Sea

Jianping Li [1,2], Zhongliang Wu [1,2,*], Xi Chen [1,2], Jian'en Jing [3], Ping Yu [1,2], Xianhu Luo [1,2], Mingming Wen [1,2], Pibo Su [1,2], Kai Chen [3], Meng Wang [3], Yan Gao [1,2] and Yao Zhang [1,2]

1 Guangzhou Marine Geological Survey, China Geological Survey, Guangzhou 511458, China; lijianping14@126.com (J.L.)
2 Key Laboratory of Marine Mineral Resources, Ministry of Natural Resources, Guangzhou 511458, China
3 School of Geophysics and Information Technology, China University of Geosciences, Beijing 100083, China
* Correspondence: wzllzw1997@163.com

Abstract

This study presents the first application of a deep-towed transmitter–receiver marine controlled-source electromagnetic (TTR-MCSEM) system for gas hydrate exploration in the Shenhu area of the South China Sea. High-resolution electromagnetic data were acquired along a 13 km transect using dynamic source–receiver offsets and a 500 A transmitter. The results reveal the following: (1) unprecedented near-seafloor resolution (20~100 m) for the precise delineation of hydrate-bearing caprock, surpassing conventional ocean-bottom electromagnetic systems; (2) laterally continuous high-resistivity anomalies (~10 Ω·m) extending from the base of the gas hydrate stability zone to the seafloor, which correlate with seismic bottom-simulating reflector (BSR) distributions and suggest heterogeneous hydrate saturation; and (3) fault-controlled fluid migration pathways that supply hydrate reservoirs and lead to seabed methane seepage at structural highs. Through 2D inversion, we show that the inverted resistivity values (~10 Ω·m) are slightly higher than those obtained from resistivity logs (~5 Ω·m). Saturation values derived from inverted resistivity exhibit remarkable consistency with well-log-based measurements. The high efficiency of the system confirms its potential for the transformative quantitative assessment of hydrate systems, seafloor massive sulfides, and marine geohazards.

Keywords: marine controlled-source electromagnetics (CSEM); gas hydrate; Shenhu; electrical resistivity structure; deep-towed system; South China Sea

1. Introduction

Natural gas hydrates are ice-like compounds composed of methane and water, formed under low-temperature and high-pressure conditions. They are widely distributed in continental margin sediments and permafrost regions worldwide [1]. As a significant potential energy resource and critical environmental factor, gas hydrates have attracted global interest. The Shenhu area, located on the northern slope of the South China Sea and structurally adjacent to the Zhu II Depression of the Pearl River Mouth Basin, is a key target for marine gas hydrate exploration in China (Figure 1A). Due to rapid neogene sedimentation, high hydrocarbon generation potential, and active deep fluid migration along faults, the Shenhu area hosts a unique gas hydrate accumulation system [2,3]. Since 2007, the Guangzhou Marine Geological Survey (GMGS, Guangzhou, China) has conducted three drilling expeditions (GMGS1, GMGS3, and GMGS4), recovering high-saturation hydrate samples. In 2018, China's first offshore gas hydrate production test was conducted here in silt-clay sequences [4].

Geophysical techniques—including seismic, controlled-source electromagnetic (CSEM) and gravity methods—provide critical constraints for mapping sub-seafloor sediment architectures and identifying zones with anomalous physical properties [5]. Among these, the seismic-derived bottom-simulating reflector (BSR) is a key indicator of gas hydrate presence, typically marking the base of the gas hydrate stability zone (GHSZ) within sedimentary sequences [6]. However, extensive well-log data from global hydrate exploration campaigns reveal that BSR occurrence alone is insufficient to confirm hydrate presence, as it exhibits false-positive rates exceeding 40% on passive continental margins [7].

Seismic methods have dominated gas hydrate exploration in the South China Sea, providing high-resolution structural images and identifying key indicators such as BSRs, amplitude blanking zones, and velocity anomalies [8–11]. However, under complex geological conditions, seismic data often struggle to resolve shallow hydrate reservoirs and fluid migration pathways simultaneously. In contrast, controlled-source electromagnetics (CSEM) exhibits high sensitivity to resistive anomalies (e.g., hydrates) and offers superior resolution compared to potential-field methods, making it a vital non-seismic tool for marine exploration [12–16]. CSEM is particularly effective in delineating the upper boundary of the hydrate stability zone [17] and vertical hydrate distribution [18], with demonstrated success in hydrate surveys [19–24]. Significant progress has been made in China regarding the application of marine CSEM for gas hydrate exploration and structural imaging [25–29].

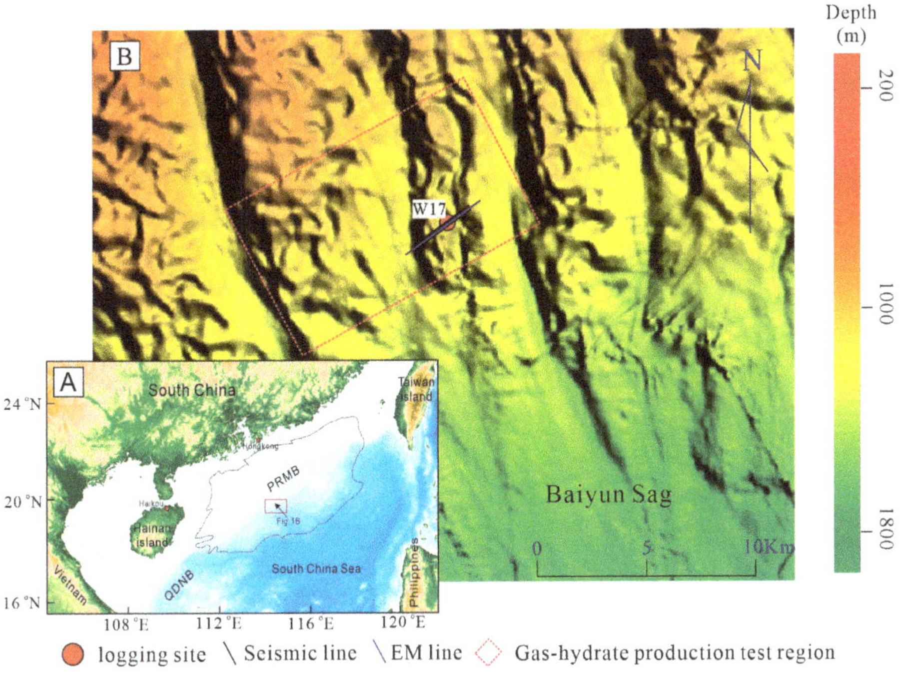

Figure 1. (**A**) Geographical location of the Shenhu area in the Pearl River Mouth Basin (PRMB) of the northern South China Sea (SCS); (**B**) submarine geomorphology and locations of boreholes drilled in the trial production gas hydrate reservoir in the Shenhu area. The dashed box indicates the gas hydrate production test site, the black line represents the seismic line, and the blue line indicates the marine CSEM line (modified from Zhang, 2020 [9]).

There are two primary marine CSEM approaches: (1) traditional ocean-bottom electromagnetics (OBEM), in which a horizontal electric dipole source is towed 20–100 m above the seafloor and receivers are deployed on the seabed; and (2) fully towed marine CSEM (TTR-MCSEM), in which both transmitters and receivers are towed near the seafloor. TTR-MCSEM offers two key advantages: (i) enhanced navigation accuracy that reduces positional uncertainty—a major limitation in long-offset CSEM inversion [30]—and improves survey efficiency [31–34]; and (ii) higher near-seafloor resolution, despite shallower penetration compared to OBEM, as demonstrated in hydrate explorations in the Black Sea and North Sea [19,35,36]. Beyond deepwater applications, towed CSEM systems also show significant advantages in shallow-water exploration [37]. Their high-resolution capabilities have enabled the successful detection of submarine freshwater reservoirs, particularly through resistivity contrasts exceeding 20 Ω·m between saline pore fluids and freshwater lenses [38,39].

2. Geologic Setting

The study area is situated within the Baiyun Sag in the southern part of the Pearl River Mouth Basin (PRMB) (Figure 1B), a major hydrocarbon-bearing depocenter on the northern slope of the South China Sea. Tectonically, the sag has evolved through two main phases: (1) a syn-rift phase from the Late Cretaceous/Paleogene to the Early Oligocene, subdivided into three episodes [8], and (2) a post-rift phase from the Late Oligocene to the present (Figure 1B). The Dongsha Event (~10.5 Ma, associated with the Taiwan Orogeny) represents the youngest significant tectonic episode, while Pliocene–Pleistocene tectonic quiescence facilitated optimal gas hydrate accumulation and preservation [40]. This regime is characterized by differential subsidence and sediment loading between depocenters and flanks, with submarine fans and gravity-flow sands constituting the primary hydrate reservoirs [8]. Importantly, multiple faults transect the gas hydrate stability zone [3], influencing fluid migration pathways.

The sediments hosting gas hydrates in the study area comprise four distinct lithotypes: siliceous–calcareous clayey silt, calcareous–siliceous clayey silt, calcareous silt, and calcareous clayey silt. Free gas accumulates predominantly as dispersed phases within the gas hydrate stability zone (GHSZ), with limited migration along fault conduits [2,3]. Notably, gas chimneys identified beneath the stability zone lack overlying fault development (Figure 1B), promoting the formation of dispersed hydrate morphologies. Both shallow biogenic and deep thermogenic gases migrate to the base of the GHSZ, diffusing into sediment pore spaces. The absence of fracture networks facilitates heterogeneous hydrate accumulation. Core sampling at site W17—targeting intervals from the seafloor to 50–100 m below seismically imaged bottom-simulating reflectors (BSRs)—was guided by log response anomalies from pilot wells [10]. Fifteen additional sites exhibited hydrate indicators in logging-while-drilling (LWD) data. At Station W17 (water depth: 1465–1475 m), analyses of core samples and infrared thermography (1490–1510 m; Figure 2) revealed that calcareous clayey medium-to-coarse silts host finely disseminated hydrate particles. Key sedimentary structures include lenticular, soup-like, and porridge-type morphologies, with subcooling anomalies (ΔT = 10–12 °C) confirming in situ dissociation. Importantly, the hydrates show homogeneous pore-space distribution, indicating uniform saturation regimes [11].

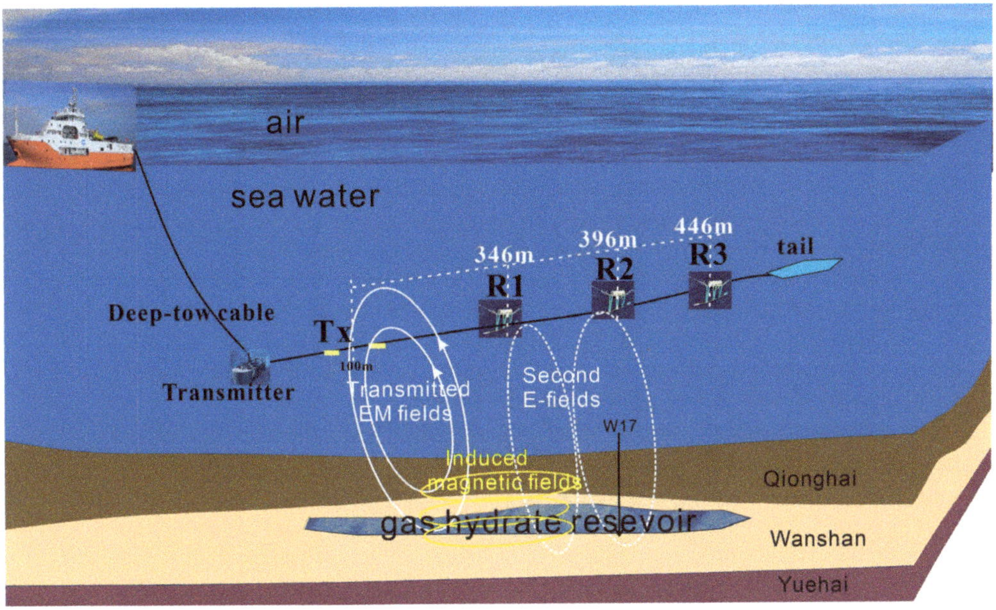

Figure 2. Schematic of the deep-towed marine controlled-source electromagnetic (TTR-MCSEM) system.

3. Data Acquisition

From September 9 to 17, 2024, GMGS conducted a TTR-MCSEM trial in the Shenhu hydrate zone aboard the R/V Haiyang Dizhi SiHao. This survey represents the first deep-towed CSEM study in this area. The 13.0 km line (Figure 1B), oriented at 58.5°, crossed the validated hydrate-rich Well W17 within the gas hydrate production test site [4]. The TTR-MCSEM system was jointly developed by GMGS and the China University of Geosciences (Beijing), building upon existing research equipment [41–44]. As shown in Figure 2, the system consists of a deck unit (step-up transformer, communication module, and server), a towed transmitter (high-voltage delivery system, chopping circuit, and 500 A output), and towed receivers (axial electric field sensors). The transmitter is equipped with a depressor, transformer, control unit, low-impedance antenna, altimeter, and ultra-short baseline (USBL) beacon.

The transmitting system is also equipped with remote telemetry, transmitted current acquisition, power supply management, attitude and orientation acquisition, altitude monitoring above the seabed, acoustic communication with the towfish tail marker, transmitted voltage acquisition, and auxiliary data acquisition [41]. The towed receiver system comprises a deck control unit, a towed master acquisition station, multiple slave electromagnetic acquisition stations, a neutrally buoyant cable, and a tail buoy. It supports connectivity with up to five slave nodes. Each slave node measures and records the axial electric field component along with motion attitude data. The deck control unit facilitates power control, GPS timing, and data monitoring. The master acquisition station is responsible for command transmission, time synchronization, data storage and uplink, and power management. Each slave acquisition station integrates a low-noise electric field sensor, a data acquisition module, an altimeter, and a motion sensor. The acquisition stations are interconnected via the neutrally buoyant cable, enabling real-time bidirectional communication of commands and data [42]. The master node aggregates data packets from all slave stations and transmits them to the deck unit via a fiber-optic cable. A PC can access the deck unit through an Ethernet connection. The deck unit and the master acquisition station are linked by an electro-optical deep-tow cable, which integrates three optical fibers and 220 VAC power transmission. This cable supplies 220 VAC power and Ethernet connectivity to the subsea acquisition stations. The master acquisition station primarily performs AC/DC and electro-optical conversion, providing Ethernet communication and 48 VDC power to the downstream slave stations while simultaneously recording attitude, altitude, and depth information [43,44].

The TTR-MCSEM acquisition system was designed with reference to the Vulcan system [31]. The transmitter broadcast a 500 A "waveform-D" [45] at a fundamental frequency of 0.5 Hz using a 100 m antenna, yielding a dipole moment of 50 kA·m. A stable tow altitude of 50 ± 10 m above the seafloor was maintained (Figure 3a). Three axial electric field receiver nodes were towed at fixed offsets of 346 m (R1), 396 m (R2), and 446 m (R3) behind the transmitter. Each EM receiver recorded the inline horizontal electric field on a 2 m dipole at a sampling rate of 1000 Hz to capture full-spectrum CSEM responses (Figure 3b). After completing the towed-source transmission, all deployed electromagnetic receivers were recovered, and time-series data of the artificially generated electromagnetic fields were successfully acquired from 15 survey stations.

Figure 3. TTR-MCSEM system components: (**a**) towed transmitter; (**b**) towed receiver.

The system maintained a mean towing velocity of 1.0 m/s (2 knots), with variations within ±0.1 knot. Precise transmitter positioning was achieved using inverted long-baseline (iLBL) acoustic navigation, ensuring spatial uncertainties of less than 1 m, which is critical for preserving long-offset amplitude integrity. Attitude sensors on the receivers (Figure 4) recorded roll, pitch, and yaw for motion correction, while pressure sensors monitored depth. These data were used in the inversion process to improve accuracy. A segment of electric field time-series data was extracted from the raw dataset, and the corresponding curve is plotted in Figure 5. As shown, the artificially generated electric field exhibits a smooth sinusoidal waveform, indicating high data quality.

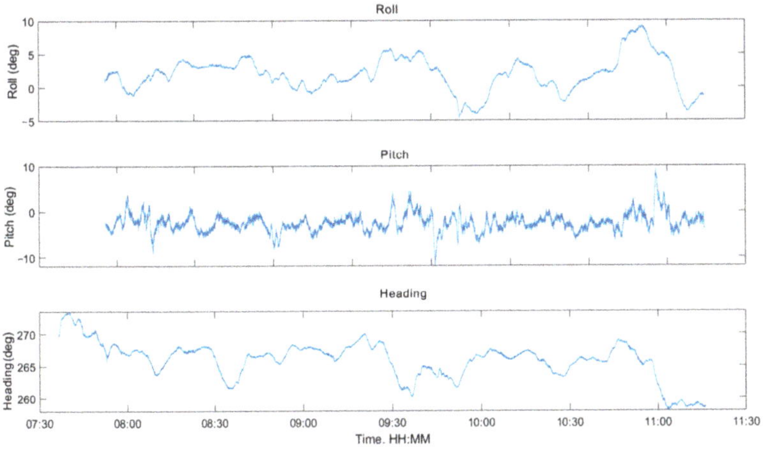

Figure 4. Attitude angles (roll, pitch, and yaw) of receiver R2 during towing.

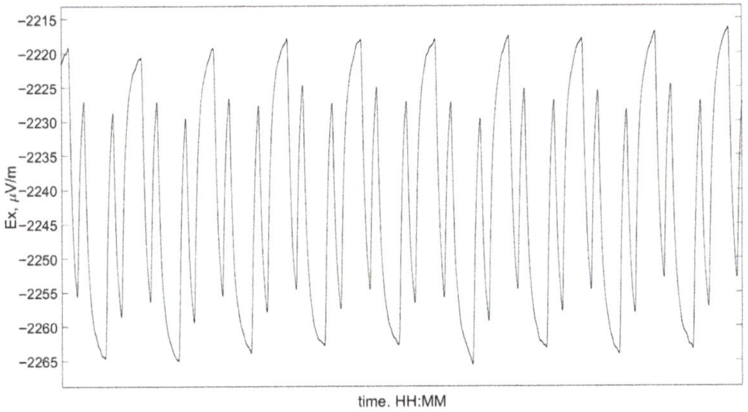

Figure 5. Raw electric field time series recorded by receiver R2 at 0.5 Hz.

4. Data Processing and Inversion

For preprocessing the marine CSEM data, we adopted a method based on the Fast Fourier Transform (FFT) as introduced by Myer et al. (2011) [45]. The data processing workflow is summarized in Figure 6 [46]. First, the Fourier spectrum of the electromagnetic field time series was computed via FFT to obtain the amplitude and phase. In the frequency domain, the frequency response function of the acquisition circuitry was determined through calibration. This calibration information was then used to correct the amplitude and phase of the electric field after FFT processing, yielding the true electric field data. The same method was applied to compute the amplitude and phase of the current data. The electromagnetic field data were then normalized by the electric dipole moment, and finally, navigation data were merged. Amplitude spectra at 0.5, 1.5, and 2.5 Hz showed robust signals (Figure 7). However, due to clock drift issues, the phase data from this survey could not be utilized.

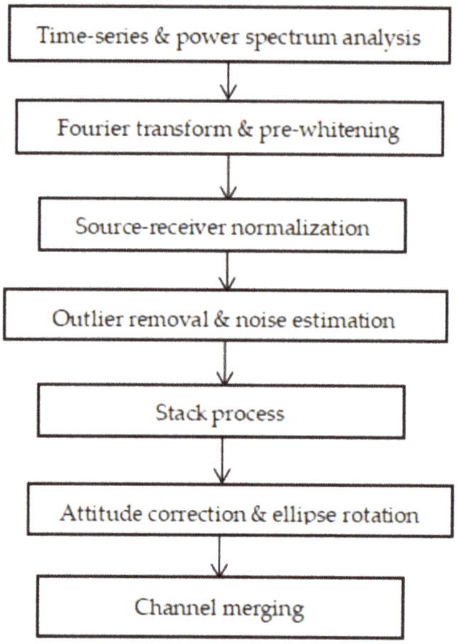

Figure 6. CSEM data preprocessing flow.

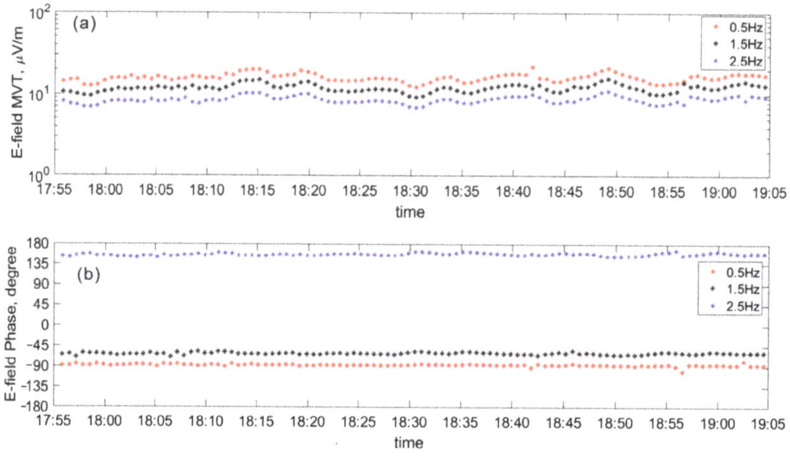

Figure 7. (**a**) Amplitude and (**b**) phase spectra at R2 for 0.5 Hz.

Two-dimensional inversion was performed using MARE2DEM [47,48] with Occam regularization. This method seeks the simplest or smoothest model that fits the data within a given tolerance, thereby avoiding overfitting. Applying Occam's inversion algorithm to marine CSEM data enables the recovery of smooth geoelectrical models while preserving major structural features, thus preventing the generation of spurious resistivity anomalies. The mesh combined quadrilaterals (incorporating bathymetry) and triangles (Figure 8), with cell sizes of 20 × 20 m in the shallow hydrate zone.

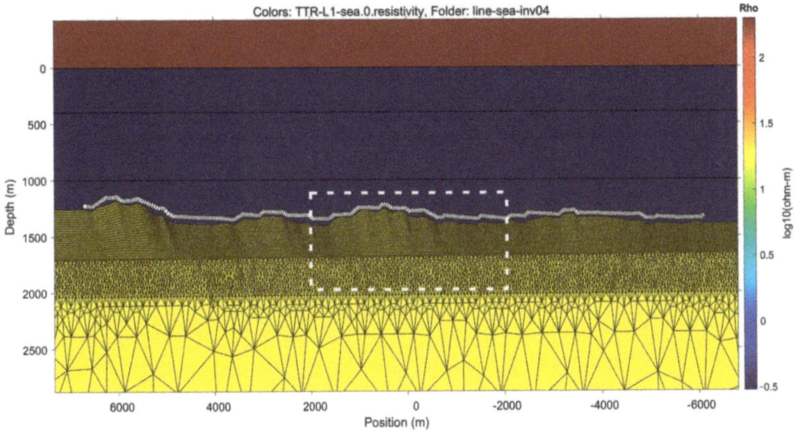

Figure 8. Initial resistivity model for 2D inversion of MCSEM data. The dashed box in the figure indicates the key interpretation area.

Above sea level, an air layer with a thickness of 100 km was included, and assigned a constant resistivity of 10^{13} Ω·m. To optimize the resolution of subsea resistivity, seawater resistivity stratification was determined during inversion [49]. Here, stratified resistivity values for the seawater layer were derived from conductivity–temperature–depth (CTD) measurements and set as fixed parameters in subsequent inversions. The starting model assumed seawater conductivity overlying sediments with a resistivity of 5 Ω·m. The vertical stratification of the seawater into three layers (0.2 Ω·m, 0.24 Ω·m, and 0.3 Ω·m) based on CT-derived resistivity profiles (Figure 9) significantly improved inversion accuracy. Data errors were set to 5% of the amplitudes or calculated standard deviations. After 10 iterations, the root mean square (RMS) misfit decreased from 2.9 to 1.05. Figure 10 shows the final fit results and normalized residuals, indicating good agreement between the model response and the measured data. As shown in Figure 10a, the RMS error stabilized, indicating convergence of the inversion process.

Figure 11 shows the resistivity profile, which correlates well with the seismic BSR. The penetration depth is limited to approximately 200 m below the seafloor due to the maximum offset of 446 m. Inversion results from Line EM (Figure 1B) provide compelling evidence for gas hydrate occurrence within the survey area. The resistivity anomalies show good agreement with seismic reflection features, particularly corresponding to the bottom-simulating reflector (BSR), confirming the robustness and reliability of the inversion results (Figure 11b). The highest resistivity value (~10 Ω·m) was observed at Hydrate Ridge (around meter 300–900). A comparative analysis with logging data from W17 along the electromagnetic profile (Figure 11a) shows that the resistivity logs increase at approximately 80 m below the seafloor, corresponding to a gas hydrate-bearing interval about 80 m thick. The inverted resistivity also exhibits a high-resistivity anomaly in this interval, which is moderately higher than the resistivity log measurements (~5 Ω·m) reported in [3]. Laterally continuous resistive anomalies extend from the base of the hydrate stability zone (HSZ) to the seafloor (Figure 11a). In contrast, between meters 1000 and 1500, weak and discontinuous resistors are observed across all survey lines, with the most prominent anomalies concentrated on the southeastern flank. At meter −600 and 0, a narrow vertical resistor extends to the seafloor from a broad tabular resistor just above the base of the HSZ, interpreted as a methane seep.

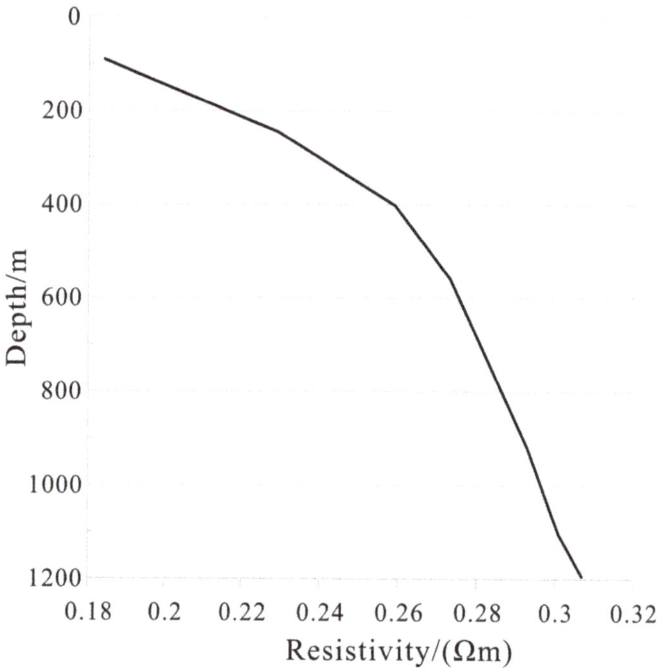

Figure 9. Measured seawater resistivity profile (CTD).

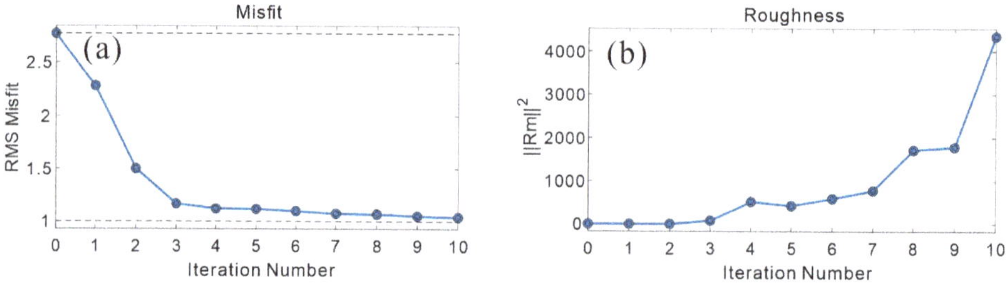

Figure 10. Inversion convergence: (**a**) RMS misfit and (**b**) model roughness.

High-resistivity conduits (approximately 10–30 Ω·m) observed beneath the gas hydrate stability zone correspond to seismically chaotic zones (gas chimneys), confirming the coexistence of gas hydrate and free gas within the reservoir in the study area. The resistivity profiles reveal several large listric faults flanking the gas hydrate reservoir. These faults are interpreted as pathways for the migration of thermogenic gas and are inferred to significantly control the distribution of both gas hydrate and free gas in the Shenhu area [50]. Moreover, these faults extend vertically upward to the seafloor, potentially facilitating the seepage of dissociated methane gas into the water column (Figure 11c).

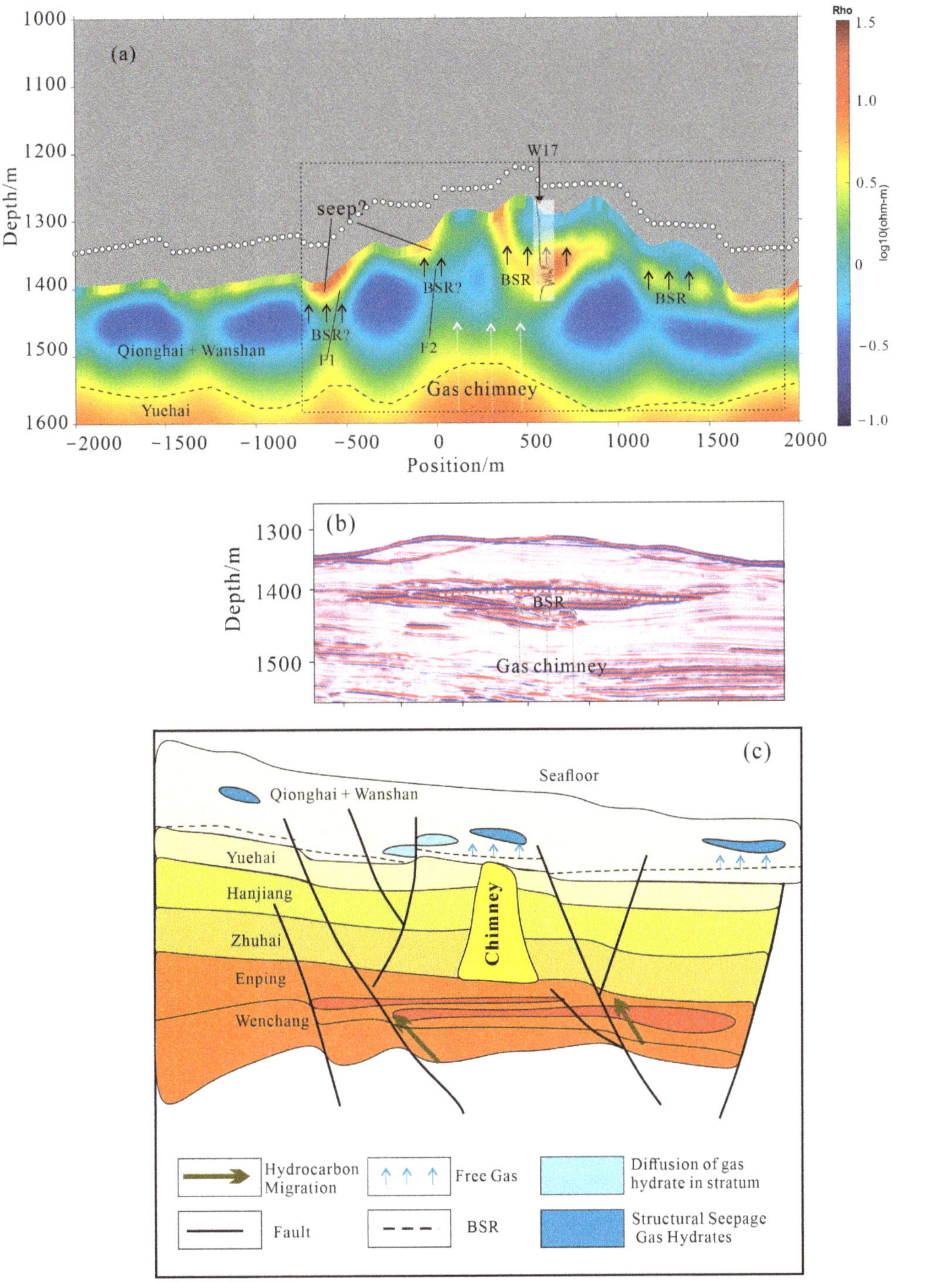

Figure 11. (**a**) Two-dimensional resistivity inversion; (**b**) corresponding seismic section; (**c**) formation model of gas hydrates in Shenhu area, South China Sea (Modified from Yang et al., 2017 [51]). The '?' indicates that the inferred BSR is relatively weak and has greater uncertainty.

5. Discussion

A primary objective in natural gas hydrate resource assessment is to quantify the saturations of both free gas and gas hydrate using pore-water chloride anomalies, resistivity measurements, and seismic data [6]. Hydrate-bearing sediments are characterized by high compressional-wave velocity and impedance, elevated resistivity, and slight density anomalies [51]. Among these, resistivity enhancement is the most prominent identifiable physical property in hydrate-bearing formations [52]. In this study, saturation estimates target large-scale sediment volumes, covering tens to hundreds of meters vertically. Archie's equation [53] was reformulated to determine the pore fraction of hydrate as follows:

$$S_h = 1 - \left(\frac{aR_w}{\phi^m R_t}\right)^{\frac{1}{n}} \tag{1}$$

where S_h is the pore space hydrate saturation, R_w is the pore fluid resistivity, ϕ is the formation porosity, and R_t is the measured formation resistivity. As shown in Table 1, the saturation index (n) is 2.00 in the Shenhu study area and close to 2.00 in adjacent areas. Therefore, we take the saturation exponent n to be 2.0, the cementation exponent m to be 2.5 [54], and the tortuosity factor a is assumed to be 1. Uncertainties in Archie parameters, especially the cementation exponent m, propagate into hydrate saturation S_h estimates. However, the dominant error source originates from inverted formation resistivity. The Occam inversion scheme constrains solutions to the most conservative resistivity model compatible with the observed data and error statistics. This approach ensures that all resistive features in the output model are data-driven. Importantly, the smoothing regularization inherent in Occam's method generates resistivity minima, meaning that S_h calculated via Archie's equation represents a lower-bound estimate for any given parameter set (m, n, a).

Table 1. Archie's empirical saturation index in the Shenhu study region [29].

Region	n	Data Source
Shenhu (SH2)	2.00	Wang et al. (2011) [52]
W11-17 and SHSC-4	1.94	Kang et al. (2020) [55]
Pearl River Mouth Basin	2.00	Liu et al. (2020) [56]

Based on a petrophysical analysis of legacy well logs, we adopted a porosity of 32% for subsequent modeling [10]. Using Archie's equation and the parameters specified above, we calculated a gas hydrate saturation (S_h) of 0.45 based on the maximum inverted reservoir resistivity (10 Ω·m). This value closely matches the well-log-derived saturation of 0.40 [8].

The inverted resistivity profile along the electromagnetic line crossing Well W17 was extracted and compared with the logged resistivity curve, as shown in Figure 12a. The logged resistivity within the gas hydrate reservoir ranges from 2.5 to 5 Ω·m. Although the inverted resistivity values are slightly higher than those from the logs, the overall trends are consistent. The saturation of the gas hydrate interval was estimated using Equation (1) (Figure 12b). The saturation derived from the inverted resistivity profile ranged between 20% and 50%, aligning well with saturation estimates of gas hydrate and associated free gas obtained from well logs. Specifically, the log-derived gas hydrate saturation in W17 varies from 0 to 76%, with an average value of 33.0% [3]. The resistivity-based saturation shows strong agreement with the log calculations, deviating by less than 5% from the average log-derived saturation.

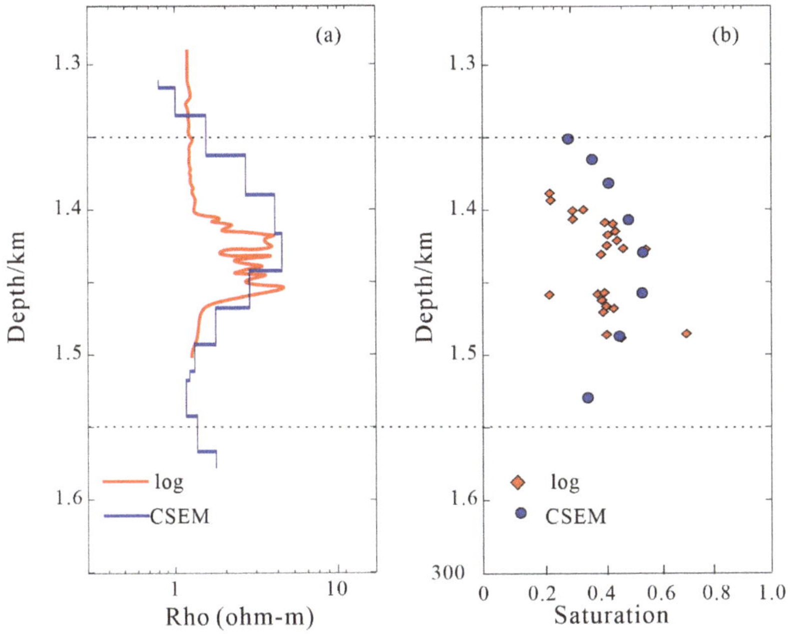

Figure 12. (**a**) Comparison between the resistivity log from Well W02 and the CSEM inverted resistivity; (**b**) vertical variations in the gas hydrate saturation in sites W17 (modified from Liang, 2022 [3]).

6. Conclusions

Our study demonstrates that the deep-towed transmitter–receiver marine controlled-source electromagnetic (TTR-MCSEM) system enables efficient basin-scale characterization of methane hydrate saturation in marine sediments. When integrated with seismic structural constraints, TTR-MCSEM data reveal critical gas migration pathways in hydrate-bearing systems. In the Shenhu area, fault zones provide effective conduits that supply hydrate reservoirs and lead to seabed methane seepage along basin flanks, whereas unfaulted horizontal sediments lack sufficient migration capacity for significant hydrate formation.

The continuous acquisition capability enhances operational efficiency, and the unprecedented near-seafloor resolution (20–100 m depth) allows the precise delineation of hydrate-bearing caprock. A comparative analysis between the resistivity profile and drilling results demonstrates a clear correlation: intervals where gas hydrates were encountered during drilling correspond to high-resistivity anomalies (~5 Ω·m) in the TTR-MCSEM inversion. In contrast, no distinct resistive anomalies are observed in zones without hydrate occurrences. The saturation estimated from the inverted resistivity ranged between 0.2 and 0.5, showing excellent agreement with the average log-derived saturation of 0.33. This consistency significantly enhances the reliability of the interpretation.

Resistivity profiles correlate strongly with seismic BSR distribution and reveal heterogeneous hydrate saturation. Localized anomalies adjacent to seabed seepage sites help map fluid migration networks. These findings confirm the transformative potential of TTR-MCSEM for the quantitative assessment of hydrate systems, seafloor massive sulfides, and marine geohazards. However, the limited transmitter–receiver offset in the current configuration resulted in insufficient resolution at depth. To enhance deep imaging capability, future surveys should incorporate more receiver nodes and employ longer offsets tailored to specific exploration objectives.

Author Contributions: Conceptualization, Z.W. and M.W. (Meng Wang); methodology, P.Y., X.L. and P.S.; investigation, Y.G., Y.Z. and X.C.; writing—original draft preparation, J.L.; writing—review and editing, K.C.; supervision, J.J.; project administration, Z.W.; funding acquisition, M.W. (Mingming Wen). All authors have read and agreed to the published version of the manuscript.

Funding: This research was funded by the Dedicated Fund for Marine Economic Development in Guangdong Province (Grant No. GDNRC [2023]40), NSFC Project Topic: Integration and Engineering Application of Intelligent Coordinated Control for Deep-Sea Resource Exploration (Grant No. 62333019), Hainan Province Key Research and Development Project (Grant No. ZDYF2024GXJS002), the National Key Research and Development Program of China (Grant No. 2022YFC2807901) and the China Geological Survey (DD20221912).

Data Availability Statement: Research data is available from the corresponding author upon reasonable request.

Acknowledgments: The authors thank Kerry Key for providing the inversion codes MARE2DEM: https://bitbucket.org/mare2dem/ (accessed on 20 August 2024).

Conflicts of Interest: The authors declare no conflicts of interest.

References

1. Kvenvolden, K.A. Gas Hydrates—Geological Perspective and Global Change. *Rev. Geophys.* **1993**, *31*, 173–187. [CrossRef]
2. Fang, Y.; Lu, J.; Liang, J.; Kuang, Z.; Cao, Y.; Chen, D. Numerical Studies of Gas Hydrate Evolution Time in Shenhu Area in the Northern South China Sea. *China Geol.* **2019**, *2*, 49–55. [CrossRef]
3. Liang, J.; Meng, M.; Liang, J.; Ren, J.; He, Y.; Li, T.; Xu, M.; Wang, X. Drilling Cores and Geophysical Characteristics of Gas Hydrate-Bearing Sediments in the Production Test Region in the Shenhu Sea, South China Sea. *Front. Earth Sci.* **2022**, *10*, 911123 [CrossRef]
4. Li, J.; Ye, J.; Qin, X.; Qiu, H.; Wu, N.; Lu, H.; Xie, W.; Lu, J.; Peng, F.; Xu, Z.; et al. The First Offshore Natural Gas Hydrate Production Test in South China Sea. *China Geol.* **2018**, *1*, 5–16. [CrossRef]
5. Waite, W.F.; Santamarina, J.C.; Cortes, D.D.; Dugan, B.; Espinoza, D.N.; Germaine, J.; Jang, J.; Jung, J.W.; Kneafsey, T.J.; Shin, H.; et al. Physical Properties of Hydrate-bearing Sediments. *Rev. Geophys.* **2009**, *47*, 2008RG000279. [CrossRef]
6. Lu, S.; McMechan, G.A. Estimation of Gas Hydrate and Free Gas Saturation, Concentration, and Distribution from Seismic Data *Geophysics* **2002**, *67*, 582–593. [CrossRef]
7. Sloan, E.D. Fundamental Principles and Applications of Natural Gas Hydrates. *Nature* **2003**, *426*, 353–359. [CrossRef] [PubMed]
8. Su, P.; Liang, J.; Peng, J.; Zhang, W.; Xu, J. Petroleum Systems Modeling on Gas Hydrate of the First Experimental Exploitation Region in the Shenhu Area, Northern South China Sea. *J. Asian Earth Sci.* **2018**, *168*, 57–76. [CrossRef]
9. Zhang, W.; Liang, J.; Wei, J.; Lu, J.; Su, P.; Lin, L.; Huang, W. Origin of Natural Gases and Associated Gas Hydrates in the Shenhu Area, Northern South China Sea: Results from the China Gas Hydrate Drilling Expeditions. *J. Asian Earth Sci.* **2019**, *183*, 103953 [CrossRef]
10. Zhang, W.; Liang, J.; Wei, J.; Lu, J.; Su, P.; Lin, L.; Huang, W.; Guo, Y.; Deng, W.; Yang, X.; et al. Geological and Geophysical Features of and Controls on Occurrence and Accumulation of Gas Hydrates in the First Offshore Gas-Hydrate Production Test Region in the Shenhu Area, Northern South China Sea. *Mar. Pet. Geol.* **2020**, *114*, 104191. [CrossRef]
11. Zhang, W.; Liang, J.; Wan, Z.; Su, P.; Huang, W.; Lin, L. Dynamic Accumulation of Gas Hydrates Associated with the Channel-Levee System in the Shenhu Area, Northern South China Sea. *Mar. Pet. Geol.* **2020**, *117*, 104354. [CrossRef]
12. Constable, S. Ten Years of Marine CSEM for Hydrocarbon Exploration. *Geophysics* **2010**, *75*, 75A67–75A81. [CrossRef]
13. Constable, S. Perspectives on Marine Electromagnetic Methods. *Perspect. Earth Space Sci.* **2020**, *1*, e2019CN000123. [CrossRef]
14. Constable, S.; Lu, R.; Stern, L.A.; Du Frane, W.L.; Roberts, J.J. Laboratory Electrical Conductivity of Marine Gas Hydrate. *Geophys. Res. Lett.* **2020**, *47*, e2020GL087645. [CrossRef]
15. Constable, S.; Orange, A.; Myer, D. Marine Controlled-Source Electromagnetic of the Scarborough Gas Field—Part 3: Multicomponent 2D Magnetotelluric/Controlled-Source Electromagnetic Inversions. *Geophysics* **2019**, *84*, B387–B401. [CrossRef]
16. Constable, S.; Srnka, L.J. An Introduction to Marine Controlled-Source Electromagnetic Methods for Hydrocarbon Exploration *Geophysics* **2007**, *72*, WA3–WA12. [CrossRef]
17. Goto, T.; Kasaya, T.; Machiyama, H.; Takagi, R.; Matsumoto, R.; Okuda, Y.; Satoh, M.; Watanabe, T.; Seama, N.; Mikada, H.; et al. A Marine Deep-Towed DC Resistivity Survey in a Methane Hydrate Area, Japan Sea. *Explor. Geophys.* **2008**, *39*, 52–59. [CrossRef]
18. Evans, R.L. Using CSEM Techniques to Map the Shallow Section of Seafloor: From the Coastline to the Edges of the Continental Slope. *Geophysics* **2007**, *72*, WA105–WA116. [CrossRef]

19. Gehrmann, R.A.S.; Schnabel, C.; Engels, M.; Schnabel, M.; Schwalenberg, K. Combined Interpretation of Marine Controlled Source Electromagnetic and Reflection Seismic Data in the German North Sea: A Case Study. *Geophys. J. Int.* **2019**, *216*, 218–230. [CrossRef]
20. Jing, J.; Chen, K.; Deng, M.; Zhao, Q.; Luo, X.; Tu, G.; Wang, M. A Marine Controlled-Source Electromagnetic Survey to Detect Gas Hydrates in the Qiongdongnan Basin, South China Sea. *J. Asian Earth Sci.* **2019**, *171*, 201–212. [CrossRef]
21. Schwalenberg, K.; Gehrmann, R.A.S.; Bialas, J.; Rippe, D. Analysis of Marine Controlled Source Electromagnetic Data for the Assessment of Gas Hydrates in the Danube Deep-Sea Fan, Black Sea. *Mar. Pet. Geol.* **2020**, *122*, 104650. [CrossRef]
22. Schwalenberg, K.; Rippe, D.; Koch, S.; Scholl, C. Marine-controlled Source Electromagnetic Study of Methane Seeps and Gas Hydrates at Opouawe Bank, Hikurangi Margin, New Zealand. *J. Geophys. Res. Solid Earth* **2017**, *122*, 3334–3350. [CrossRef]
23. Weitemeyer, K.A.; Constable, S.C.; Key, K.W.; Behrens, J.P. First Results from a Marine Controlled-source Electromagnetic Survey to Detect Gas Hydrates Offshore Oregon. *Geophys. Res. Lett.* **2006**, *33*, 2005GL024896. [CrossRef]
24. Weitemeyer, K.A.; Constable, S.; Tréhu, A.M. A Marine Electromagnetic Survey to Detect Gas Hydrate at Hydrate Ridge, Oregon: EM at Hydrate Ridge. *Geophys. J. Int.* **2011**, *187*, 45–62. [CrossRef]
25. Jing, J.E.; Wu, Z.L.; Deng, M.; Zhao, Q.X.; Luo, X.H.; Tu, G.H.; Chen, K.; Wang, M. Experiment of marine controlled source electromagnetic detection in a gas hydrate prospective region of the South China Sea. *Chin. J. Geophys.* **2016**, *59*, 2564–2572. (In Chinese) [CrossRef]
26. Jing, J.E.; Zhao, Q.X.; Deng, M.; Luo, X.H.; Chen, K.; Wang, M.; Tu, G.H. A study on natural gas hydrates and their forming model using marine controlled source electromagnetic survey in the Qiongdongnan Basin. *Chin. J. Geophys.* **2018**, *61*, 4677–4689. (In Chinese) [CrossRef]
27. Jing, J.E.; Zhao, Q.X.; Luo, X.H.; Liu, C.G.; Chen, K.; Wang, M.; Deng, M.; Wu, Z.L. A marine controlled-source electromagnetic imaging of gas hydrates and gas sources in the Shenhu area, South China Sea. *Acta Geol. Sin.* **2024**, *98*, 2694–2708. (In Chinese)
28. Duan, S.M.; Li, Y.G.; Wu, Z.Q.; Pei, J.X. Structural imaging of marine carbonatite strata in the South Yellow Sea Basin based on marine CSEM technology. *Chin. J. Geophys.* **2025**, *68*, 1555–1566. (In Chinese) [CrossRef]
29. Liu, C.; Jing, J.; Zhao, Q.; Luo, X.; Chen, K.; Wang, M.; Deng, M. High-Resolution Resistivity Imaging of a Transversely Uneven Gas Hydrate Reservoir: A Case in the Qiongdongnan Basin, South China Sea. *Remote Sens.* **2023**, *15*, 2000. [CrossRef]
30. Key, K.; Constable, S. Inverted Long-Baseline Acoustic Navigation of Deep-Towed CSEM Transmitters and Receivers. *Mar. Geophys. Res.* **2021**, *42*, 6. [CrossRef]
31. Constable, S.; Kannberg, P.K.; Weitemeyer, K. Vulcan: A Deep-towed CSEM Receiver. *Geochem. Geophys. Geosyst.* **2016**, *17*, 1042–1064. [CrossRef]
32. Kannberg, P.K.; Constable, S. Characterization and Quantification of Gas Hydrates in the California Borderlands. *Geophys. Res. Lett.* **2020**, *47*, e2019GL084703. [CrossRef]
33. King, R.B.; Constable, S. How Low Can You Go: An Investigation of Depth Sensitivity and Resolution Using Towed Marine Controlled-source Electromagnetic Systems. *Geophys. Prospect.* **2023**, *71*, 722–733. [CrossRef]
34. Zhdanov, M.S.; Anderson, C.; Endo, M.; Cox, L.H.; Čuma, M.; Wilson, G.A.; Black, N.; Gribenko, A.V. 3D Inversion of Towed Streamer EM Data: A Model Study of the Harding Field with Comparison to CSEM. *First Break.* **2012**, *30*, 71–74. [CrossRef]
35. King, R.B.; Constable, S.; Maloney, J.M. A Case Study in Controlled Source Electromagnetism: Near Seabed Hydrocarbon Seep Systems of Coal Oil Point, California, USA. *Mar. Pet. Geol.* **2022**, *139*, 105636. [CrossRef]
36. Attias, E.; Weitemeyer, K.; Hölz, S.; Naif, S.; Minshull, T.A.; Best, A.I.; Haroon, A.; Jegen-Kulcsar, M.; Berndt, C. High-Resolution Resistivity Imaging of Marine Gas Hydrate Structures by Combined Inversion of CSEM Towed and Ocean-Bottom Receiver Data. *Geophys. J. Int.* **2018**, *214*, 1701–1714. [CrossRef]
37. Sherman, D.; Kannberg, P.; Constable, S. Surface Towed Electromagnetic System for Mapping of Subsea Arctic Permafrost. *Earth Planet. Sci. Lett.* **2017**, *460*, 97–104. [CrossRef]
38. Attias, E.; Constable, S.; Sherman, D.; Ismail, K.; Shuler, C.; Dulai, H. Marine Electromagnetic Imaging and Volumetric Estimation of Freshwater Plumes Offshore Hawai'i. *Geophys. Res. Lett.* **2021**, *48*, e2020GL091249. [CrossRef]
39. Micallef, A.; Person, M.; Haroon, A.; Weymer, B.A.; Jegen, M.; Schwalenberg, K.; Faghih, Z.; Duan, S.; Cohen, D.; Mountjoy, J.J.; et al. 3D Characterisation and Quantification of an Offshore Freshened Groundwater System in the Canterbury Bight. *Nat. Commun.* **2020**, *11*, 1372. [CrossRef]
40. Liang, J.; Zhang, Z.; Su, P.; Sha, Z.; Yang, S. Evaluation of Gas Hydrate-Bearing Sediments below the Conventional Bottom-Simulating Reflection on the Northern Slope of the South China Sea. *Interpretation* **2017**, *5*, SM61–SM74. [CrossRef]
41. Wang, M.; Deng, M.; Wu, Z.; Luo, X.; Jing, J.; Chen, K. The Deep-Tow Marine Controlled-Source Electromagnetic Transmitter System for Gas Hydrate Exploration. *J. Appl. Geophys.* **2017**, *137*, 138–144. [CrossRef]
42. Wang, M.; Wu, Z.L.; Yu, P.; Deng, M.; Ren, Z.B.; Wang, C.T. 2000A class marine controlled source electromagnetic transmitter system. *Chin. J. Geophys.* **2024**, *67*, 4482–4492. (In Chinese) [CrossRef]
43. Wang, M.; Deng, M.; Yu, P.; Yin, C.C.; Chen, K.; Luo, X.H. High-power time-frequency transmission and multi-chain cable multi-componentelectromagnetic system for deeprwater exploration. *Chin. J. Geophys.* **2022**, *65*, 3664–3673. [CrossRef]

44. Chen, K.; Jing, J.E.; Zhao, Q.X.; Luo, X.H.; Tu, G.H.; Wang, M. Ocean bottom EM receiver and application for gas hydrate detection. *Chin. J. Geophys.* **2017**, *60*, 4262–4272. (In Chinese) [CrossRef]
45. Myer, D.; Constable, S.; Key, K. Broad-Band Waveforms and Robust Processing for Marine CSEM Surveys: CSEM Waveforms and Processing. *Geophys. J. Int.* **2011**, *184*, 689–698. [CrossRef]
46. Li, Y.G.; Duan, S.M. Data Preprocessing of Marine Control led—Source Electromagnetic Data. *Period. Ocean Univ. China* **2014**, *44*, 106–112. (In Chinese) [CrossRef]
47. Key, K. MARE2DEM: A 2-D Inversion Code for Controlled-Source Electromagnetic and Magnetotelluric Data. *Geophys. J. Int.* **2016**, *207*, 571–588. [CrossRef]
48. Li, Y.; Key, K. 2D Marine Controlled-Source Electromagnetic Modeling: Part 1—An Adaptive Finite-Element Algorithm. *Geophysics* **2007**, *72*, WA51–WA62. [CrossRef]
49. Key, K. 1D Inversion of Multicomponent, Multifrequency Marine CSEM Data: Methodology and Synthetic Studies for Resolving Thin Resistive Layers. *Geophysics* **2009**, *74*, F9–F20. [CrossRef]
50. Tinivella, U.; Carcione, J.M. Estimation of Gas-Hydrate Concentration and Free-Gas Saturation from Log and Seismic Data. *Lead. Edge* **2001**, *20*, 200–203. [CrossRef]
51. Yang, S.X.; Liang, J.Q.; Lu, J.A.; Qu, C.W.; Liu, B. New understandings on the characteristics and controlling factors of gas hydrate reservoirs in the Shenhu area on the northern slope of the South China Sea. *Earth Sci. Front.* **2017**, *24*, 1–14. (In Chinese)
52. Wang, X.; Wu, S.; Guo, Y.; Yang, S.; Gong, Y. Geophysical Indicators of Gas Hydrate in the Northern Continental Margin, South China Sea. *J. Geol. Res.* **2011**, *2011*, 359597. [CrossRef]
53. Archie, G.E. The Electrical Resistivity Log as an Aid in Determining Some Reservoir Characteristics. *Trans. AIME* **1942**, *146*, 54–62. [CrossRef]
54. Collett, T.S.; Ladd, J. 19. Detection of gas hydrate with downhole logs and assessment of gas hydrate concentrations (saturations) and gas volumes on the Blake Ridge with electrically resistivity log data. In Proceedings of the Ocean Drilling Program, Scientific Results, College Station, TX, USA, 12 June 2000; Volume 164.
55. Kang, D.; Lu, J.; Zhang, Z.; Liang, J.; Kuang, Z.; Lu, C.; Kou, B.; Lu, Q.; Wang, J. Fine-Grained Gas Hydrate Reservoir Properties Estimated from Well Logs and Lab Measurements at the Shenhu Gas Hydrate Production Test Site, the Northern Slope of the South China Sea. *Mar. Pet. Geol.* **2020**, *122*, 104676. [CrossRef]
56. Liu, T.; Liu, X.; Zhu, T. Joint Analysis of P-Wave Velocity and Resistivity for Morphology Identification and Quantification of Gas Hydrate. *Mar. Pet. Geol.* **2020**, *112*, 104036. [CrossRef]

Journal of
Marine Science and Engineering

MDPI

Article

Sedimentary Processes of Gas Hydrate-Bearing Layers in the Dongsha Area, South China Sea: Implications for Hydrate Accumulation

Yuhan Wang [1], Chenyang Bai [1,2,3,*], Zhe Wang [4], Wenlin Chen [1,2], Xiaolei Xu [1,2], Hongyuan Xu [1,2] and Hongbin Wang [5,*]

1 School of Ocean Sciences, China University of Geosciences (Beijing), Beijing 100083, China; m17733972073@163.com (Y.W.); 2011230001@email.cugb.edu.cn (W.C.); littlexu0830@163.com (X.X.); m18833376729@163.com (H.X.)
2 Key Laboratory of Polar Geology and Marine Mineral Resources, China University of Geosciences (Beijing), Ministry of Education, Beijing 100083, China
3 Hainan Institute, China University of Geosciences (Beijing), Sanya 572025, China
4 Bureau of Geophysical Prospecting INC., CNPC, Zhuozhou 072751, China; wangzhe02@cnpc.com.cn
5 Qingdao Institute of Marine Geology, China Geological Survey, Qingdao 266237, China
* Correspondence: baicy@cugb.edu.cn (C.B.); oceanwang7106@163.com (H.W.)

Abstract

The methane flux in the Dongsha area in the northern South China Sea is relatively high. The results indicate the presence of both shallow and deep gas hydrate reservoirs at the Site DS-W08. The gas hydrate reservoir in this area is mainly composed of fine-grained sediments, and high-saturation gas hydrates are present. The shallow-GHR (8–24 mbsf) exhibits a maximum hydrate saturation of 14% (pore volume). The deep-GHR (below 65 mbsf) shows a maximum hydrate saturation of 33% The suspended sedimentation process on the banks of turbidity currents and the deep-water traction current sedimentation process play potentially important roles in the enrichment of gas hydrates. To investigate the influence of sedimentary processes on gas hydrate accumulation, this study analyzed gas hydrate saturation, sediment grain size, grain compositions, biological components, and geochemical characteristics of hydrate-bearing and adjacent layers at Site DS-W08. Sediment grain size analysis suggests that the studied layer was formed through the interaction of turbidity current-induced overbank suspended deposition and traction current deposition. By comprehensively analyzing the comparison of sediment Sr/Ba ratios and the data of foraminifera and calcareous nannofossils, it is found that the bank deposits and traction current deposits triggered by turbidity currents correspond to glacial periods and interglacial periods, respectively. Analysis of biological components shows that layers with high foraminifera content and traction current-modified sediments are more favorable for gas hydrate accumulation. Hydrate reservoirs are all composed of traction current deposits, and the cap rock rich in foraminifera fossils at the top promotes hydrate formation; while the fine-grained turbidites formed during the turbidite deposition process inhibit hydrate accumulation. This study aims to deepen the understanding of the enrichment mechanism of natural gas hydrates and support the commercial development of fine-grained sediments in the northern South China Sea.

Keywords: gas hydrate; grain size; microfossil content; South China Sea; Dongsha area

1. Introduction

Natural gas hydrate (hereinafter referred to as hydrate) is a crystalline compound that forms under specific conditions of pressure, temperature, and water salinity [1–3]. It is an ice-like clathrate in which low-molecular-weight gases—primarily methane, but also ethane and carbon dioxide—are typically found in a marine environment at depths greater than 500 m or in terrestrial permafrost regions [2,4]. The decomposition of hydrate alters the mechanical properties and pore pressure of seabed sediments, potentially triggering geological hazards such as continental shelf collapse and submarine landslides [5–8]. The instability of hydrates releases methane, a potent greenhouse gas, which not only impacts the ocean carbon cycle but may also exacerbate global climate change [9–12]. Gas hydrates represent a potential and enormous energy treasure trove and are regarded as one of the important energy sources that may replace traditional fossil fuels in the future [13–17].

Currently, 98% of the explored global hydrate resources are located in submarine sediments, while only 2% are found on land [4,18]. In offshore hydrate accumulation systems, sediment grain size plays a critical role in the formation and enrichment of hydrates [19–21]. In sandy environments, gas hydrates preferentially accumulate in coarser sediments due to their higher porosity and permeability. For instance, highly saturated hydrates have been identified in sandy sediments within turbidity current channels in the Nankai Trough [22], Alaminos Canyon in the Gulf of Mexico [23], Krishna-Godavari Basin [24], and the Qiongdongnan area of the South China Sea [25]. Fine-grained sediments in offshore areas are one of the main carriers for gas hydrates, and such sediments usually have the characteristics of low porosity and low permeability [26]. Drilling results from the northern South China Sea [27,28] and the Ulleung Basin [29] confirm the potential for highly saturated hydrate reservoirs in fine-grained sediments. Similar findings are reported in fossil seep carbonates associated with gas hydrate destabilization, enclosed in hemipelagic and turbiditic fine-grained mudstones and siltstones [30]. The control mechanisms for hydrate accumulation in fine-grained sediments are not fully understood in some respects at present. Simulations of experimental samples confirmed that in fine-grained sediments, hydrates can form via a particle displacement model, leading to the development of massive or vein-like hydrates [31]. However, significant differences remain between experimental samples and actual field conditions. Field-based studies indicate that in fine-grained sediments, there is no linear correlation between sediment particle size and hydrate saturation [32]. Compared to non-hydrate layers, hydrate-bearing sediments may contain higher amounts of felsic detrital minerals (primarily quartz and feldspar) and lower concentrations of clay minerals, particularly illite–smectite mixed layers [33,34]. The impact of various mineral types on hydrate formation differs markedly. Molecular dynamics simulations confirm that hydrophilic quartz surfaces enhance the ordered arrangement of water molecules, significantly reducing the nucleation energy barrier of methane hydrate and effectively promoting its nucleation rate and crystal growth. In sediments, a high proportion of felsic minerals results in a looser overall structure and larger pores between sedimentary particles, providing essential spatial conditions for hydrate formation [35,36]. Clay minerals, particularly expansive clays like illite–smectite mixed layers, reduce porosity by filling pore spaces due to their high specific surface area and strong water absorption and expansion properties, thus inhibiting gas diffusion. Additionally, the complex microstructure formed hinders hydrate aggregation [37,38]. Sediments rich in illite–smectite mixed layers strongly bind methane gas and fluids, hindering their migration. Some studies have also suggested a correlation between foraminiferal abundance and the occurrence of high-quality hydrates [39]. However, within the hydrate stability zone, not all intervals with high foraminiferal content led to hydrate enrichment [26]. The sedimentary process is influenced by paleo-marine environments and paleoclimatic conditions, such as sea-level fluctuations, and affects the microstructure and composition of sediments [40–43]. Therefore, understanding sedimentary processes is essential for revealing hydrate accumulation in fine-grained sediments [44–46]. Turbidites are considered favorable for hydrate enrichment. In the Dongsha and Shenhu areas of the northern SCS, water depths are large, ranging from 800 to 1100 m, with a relatively low sediment source supply flux [47]. The sediments are mainly composed of silty clay or clay, with grain sizes all below silt grade (64 μm) [48]. Fine-grained turbidites in the submarine canyons are primarily formed by low-density turbidity currents [49–52]. The grain size of these sediments may not differ significantly from that of suspended deposits near the overflow margin. Therefore, the extent to which sedimentary processes influence hydrate accumulation in fine-grained layers requires further investigation. Sedimentary processes provide a link between the macroscopic accumulation of hydrates and the microscopic sediment structure, as well as mechanisms of hydrate formation.

The Dongsha area contains abundant gas hydrate resources. During the second China National Gas Hydrate Program Expedition (GMGS2) conducted by the Guangzhou Marine Geological Survey in 2013, both shallow and deep gas hydrate reservoirs were identified in the Dongsha area, offering valuable insights into the sedimentary processes associated with hydrate formation. We performed grain-size, geochemical, and biogenic analyses on sediment samples from hydrate reservoirs and adjacent strata in the Dongsha area. These analyses revealed differences in sedimentary features between hydrate-bearing and adjacent layers, provided insights into the evolution of the paleo-marine environment, and highlighted the role of sedimentary processes in controlling hydrate reservoir formation.

2. Geological Setting

The South China Sea is bordered to the west by a fossil subduction zone and to the east by a transform fault zone. Its tectonic framework shows notable similarities to other marginal seas in the western Pacific, such as the Komandor Basin in the Sea of Okhotsk, the Sulu Sea, and the Sulawesi Sea [53–55]. The northern slope of the South China Sea represents a typical passive continental margin, where numerous oil and gas basins have developed along a southwest-northeast direction. Notably, the eastern part of the northern South China Sea exhibits features of both passive and active continental margins [56]. The study site is situated in the Dongsha area, on the northeastern slope of the South China Sea (Figure 1a). The Dongsha area lies in the eastern part of the Pearl River Mouth Basin, near the central uplift zone of Taixinan Basin. Water depth in this area varies significantly, increasing gradually from 200 m to over 2400 m (Figure 1a). The geothermal gradient ranges from 3.3 to 4.5 °C/km, which favors the formation of hydrates. The seafloor morphology is complex, featuring numerous seamounts, canyons, and hills. A prominent submarine canyon extends from the northwest to the southeast, spanning from the upper shelf to the lower slope, and transports a large amount of sediment to the slope and rises [57]. The Dongsha area also contains numerous mud diapirs and faults. Some faults penetrate the basin floor, while others extend to the seafloor, facilitating the migration of methane and hydrocarbons, and promoting the formation of natural gas hydrates [58,59]. Gas chimneys are widely observed in the Dongsha area, particularly along submarine ridges [57,60]. Faults and gas chimneys serve as key pathways for the upward migration of fluids in the Dongsha area [21,57]. In the Dongsha area, natural gas hydrates primarily occur in massive, layered, veined, nodular, and scattered forms. However, multiple episodes of methane leakage have occurred within the hydrate-bearing layers, accompanied by anaerobic oxidation of methane (AOM), leading to the formation of abundant authigenic minerals, including carbonates, pyrite, and gypsum.

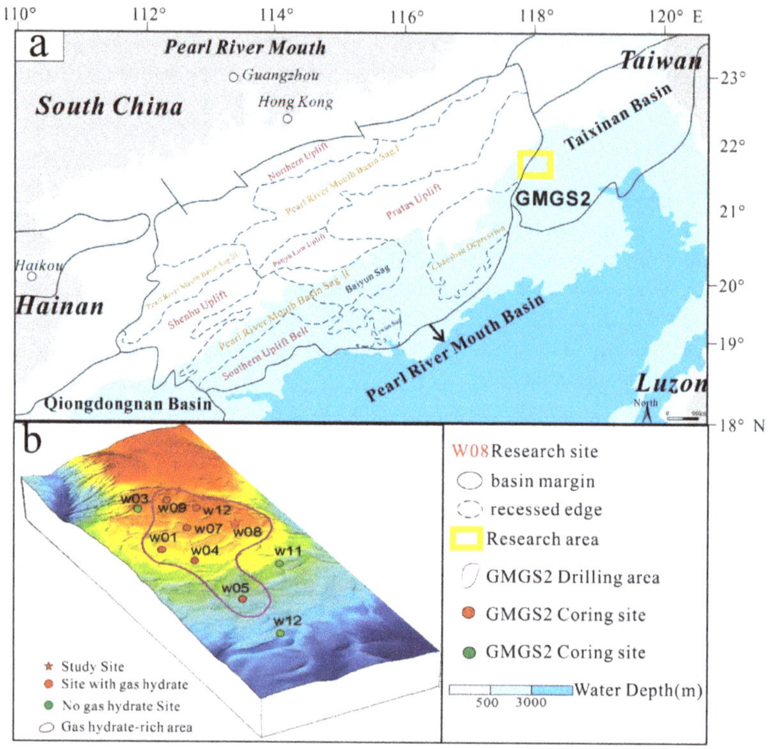

Figure 1. (**a**) Study area and marine currents (modified from [57]); (**b**) Enlarged view of the red box in (**a**) showing the drilling area of GMGS2 (modified from [21]). Multibeam-reconstructed seafloor topography, where water depth shallows from blue to red.

In 2013, the GMGS carried out the China Notions Gas Hydrate Program Expedition 2 (GMGS2), which marked a significant breakthrough by obtaining high-purity gas hydrate samples for the first time in China. At the Site W08 (Figure 1b), located at a water depth of 801 m and a drilling depth of 95 m, the lower boundary of the gas hydrate stability zone is approximately 120 m below the seafloor (mbsf). Two hydrate-bearing layers were identified at the depths of 8–23 mbsf and 66–92 mbsf.

3. Materials and Methods

3.1. Site and Samples

Core drilling at Site DS-W08 reached a depth of 95 mbsf, with samples primarily obtained from two intervals: the upper (0–28 mbsf) and lower (58–94 mbsf) intervals. Shallow gas hydrate reservoirs at Site W08 are found at 8–24 mbsf, primarily in massive, layered, veined, and nodular forms. Since the primary exploration target at this site is the hydrate-bearing zone, and the interval from 28 to 58 mbsf did not show clear hydrate occurrence characteristics in seismic exploration and well logging analysis, coring operations were not performed (Figure 2). Deep gas hydrate reservoirs occur at 66–92 m, predominantly in massive and dispersed forms. Additionally, authigenic carbonate minerals were identified near the seafloor and around the 60 mbsf sedimentary layer, leading to increased resistivity [61]. Sediment samples were extensively collected from both hydrate-bearing and adjacent non-hydrate layers. A total of 78 samples were analyzed for grain size, major and trace elements, and biological contents.

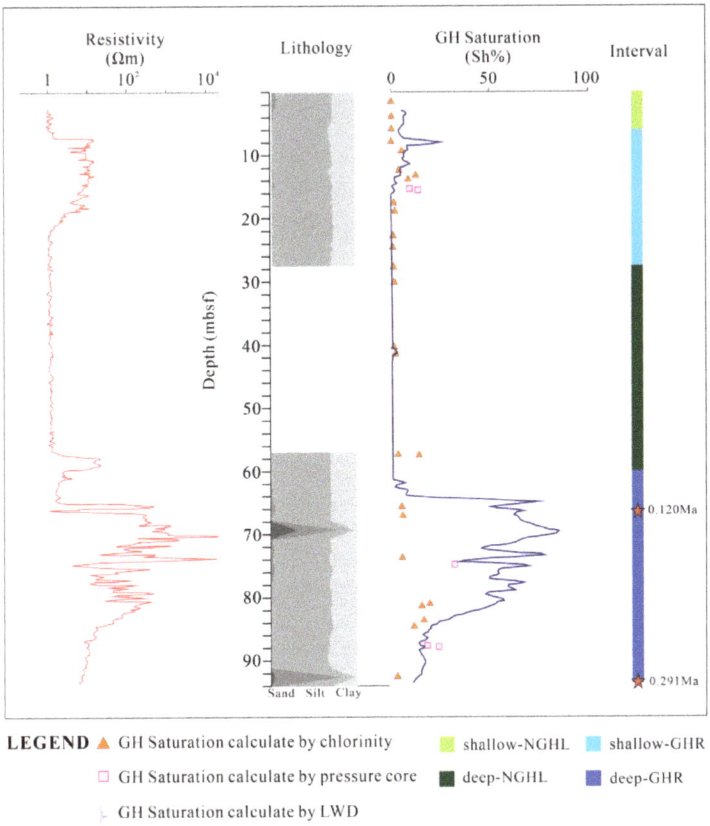

Figure 2. Logs of gas hydrate saturation, resistivity, lithology, and grain-size parameters at Site DS-W08. Resistivity data is obtained through logging-while-drilling. Interval refers to the gas hydrate reservoirs and the non-gas hydrate layers (modified from [57,60]).

3.2. Grain-Size

This study employed laser diffraction for particle size analysis, requiring sample pretreatment prior to experimentation. Before sampling, the samples were thoroughly mixed to ensure representativeness. A gram of sediment sample was placed in a glass cup, with 5 ml of purified water and 0.5 mol/L sodium hexametaphosphate ($[NaPO_3]_6$) added. The sample was soaked for 24 h, with gentle stirring every 8 h to ensure complete dispersion. All soaked samples were transferred to the laser sample chamber, where ultrasonic vibration and high-speed centrifugation were applied to achieve thorough dispersion again. The testing program was then immediately initiated.

Laser diffraction was used for particle size analysis. The samples were previously mixed thoroughly to ensure representativeness. A Mastersizer 2000 laser particle size analyzer, manufactured by Malvern Instruments Ltd. (Malvern, Britain), was used for the measurements. This instrument has a measuring range of 0.02~2200 μm and a relative error of less than 3% for repeatability measurements [62]. The obtained data were converted using the Φ grain size scale of the Wudun–Wendehua system, with the formula $\Phi = -\log2D$ (where D is the particle diameter in millimeters). The McManus' method was employed to calculate grain size parameters, including mean size (Mz), sorting values (δi), skewness (Ski), kurtosis (Kg), and median size.

$$\mathrm{Mz} = \frac{\varphi_{16} + \varphi_{50} + \varphi_{84}}{3} \tag{1}$$

$$\delta\mathrm{i} = \frac{\varphi_{84} - \varphi_{16}}{4} + \frac{\varphi_{95} - \varphi_{5}}{6.6} \tag{2}$$

$$\mathrm{Ski} = \frac{\varphi_{16} + \varphi_{84} - 2\varphi_{50}}{2(\varphi_{84} - \varphi_{16})} + \frac{\varphi_{5} + \varphi_{95} - 2\varphi_{50}}{2(\varphi_{95} - \varphi_{5})} \tag{3}$$

$$\mathrm{Kg} = \frac{\varphi_{95} - \varphi_{5}}{2.44(\varphi_{75} - \varphi_{25})} \tag{4}$$

where φ16, φ84, etc., are the particle size φ corresponding to the 16th, 84th, etc., percentiles on the probability cumulative curve, in units of mm.

3.3. Geochemistry

Major element analyses were performed at the GMGS Laboratory using a PerkinElmer Optima 830 ICP-OES instrument (PerkinElmer, Waltham, MA, USA). After cleaning, drying, and grinding, the samples were completely digested into solutions using a hydrofluoric acid, nitric acid, and perchloric acid system. Main elements (such as Si, Al, Fe, Mg) are directly determined by ICP-OES with relatively high precision. This method can accurately determine samples within the range of copper content (345–907 ppm) and zinc content (130–216 ppm). Trace element concentrations measured in this experiment were all in the ppm range, with copper content from 6.9 to 38 ppm and zinc content from 25.3 to 127 ppm. Inductively coupled plasma mass spectrometry (ICP-MS) was used for elements present below the quantification limit of ICP-OES. The sample was decomposed in a closed vessel with nitric, hydrofluoric, and perchloric acids. Hydrofluoric acid was subsequently evaporated in an open system. The resulting salts were dissolved by heating with hydrochloric acid and then converted to a nitric acid medium. Forty-four trace elements were analyzed using ICP-MS. Interference coefficients were determined following sample analysis. A single-element standard solution was used for interference correction, treated as a sample solution, and the concentration of each element was measured. The interference coefficient for each interfering element (barium, manganese, titanium, lanthanum, cerium, neodymium) was calculated as the ratio of its measured concentration to its concentration in the standard solution.

3.4. Microfossils

All samples were prepared using the direct smear method. Coarse-grained sediments were removed, and the fine-grained suspension was evenly spread across the surface of the glass slide. The slides were examined under a Zeiss Axio Imager A1 and A2m polarizing microscope at 1000 times magnification. A total of 200 fields of view were observed, and fossils in 10 randomly selected fields were counted to estimate the relative abundance of microfossil content. Fossil preservation was categorized into four grades: very good, good, moderate, and poor. Calcareous nannofossils were analyzed at the Key Laboratory of Submarine Mineral Resources, Guangzhou Marine Geological Survey.

A total of 16–100 g of dried sample was soaked thoroughly in water, dispersed, and then sieved using a 0.063 mm standard copper sieve. The retained fraction on the sieve was dried and examined under a Zeiss Discovery. V20/LEICA M165C binocular solid microscope (Zeiss, Oberkochen, Germany). Foraminiferal specimens larger than 0.15 mm were counted, and their abundance was quantified based on this size fraction. Foraminifera identification and relative abundance statistics were conducted at the Key Laboratory of Submarine Mineral Resources, Guangzhou Marine Geological Survey.

4. Results

4.1. Hydrate Saturation

Hydrate saturation data are primarily derived from chloride ion anomalies in pore water and gas release measurements from pressure-retaining cores. In addition, high resistivity anomalies are used as an indicator of hydrate-bearing reservoirs [61].

As shown in Figure 2, based on the GMGS2 cruise report and the logging-while-drilling (LWD) resistivity data at Site DS-W08, abnormally high resistivity values are observed between 8–24 mbsf and below 58 mbsf. Additionally, resistivity fluctuations are present within these high-resistivity intervals. In the 8–24 mbsf interval, hydrate saturation estimated from anomalous chloride concentrations in pore water ranges from 0% to 14%. Saturation derived from gas release volumes of pressure-retained cores also ranges from 10% to 14%, with a peak value at 17 mbsf.In the deeper layer (below 65 mbsf), hydrate saturation estimated from chloride anomalies ranges from 30% to 50%, with an average of 33%. Hydrate saturation derived from gas release volumes of pressure-retained cores ranges from 17% to 33%.

4.2. Particle Size

The grain-size parameters of 78 sediment samples from station W08 are presented in Figure 3. These parameters include average grain size, sorting, skewness, kurtosis, and median grain size. The average grain size ranges from 5.94 to 7.65 φ. Sorting values range from 1.44 to 2.07. Skewness ranges from 0 to 0.28, while kurtosis ranges from 0.87 to 1.14. The median grain size ranges from 6.81 to 7.78 φ.

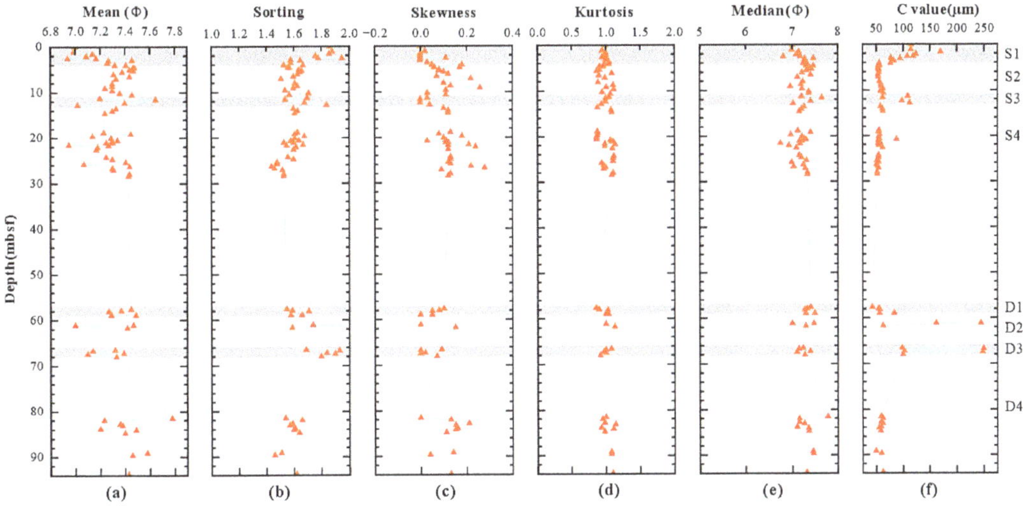

Figure 3. Logs of grain-size parameters at Site DS-W08. (**a**) Mean grain size, (**b**) Sorting coefficient, (**c**) Skewness, (**d**) Kurtosis, (**e**) Median grain size, and (**f**) C value.

Grain size data indicate minimal lithological variation within the sediment succession at Site DS-W08. Silt dominates the sediment, with clayey silt being common. Silt is the most abundant component, followed by clay, whereas the sand content is minimal. Both shallow and deep hydrate reservoirs are closely associated with clayey silt.

Previous reports on Site DS-W08 indicate that sediment grains primarily consist of quartz, feldspar, carbonate minerals, and clay minerals, with minimal amounts of other heavy and metallic minerals (Figure 4b) [63]. Quartz and feldspar dominate the silt fraction, while the remaining fine-grained sediments are primarily clay minerals. Carbonate minerals mainly comprise calcareous microfossils and authigenic carbonates.

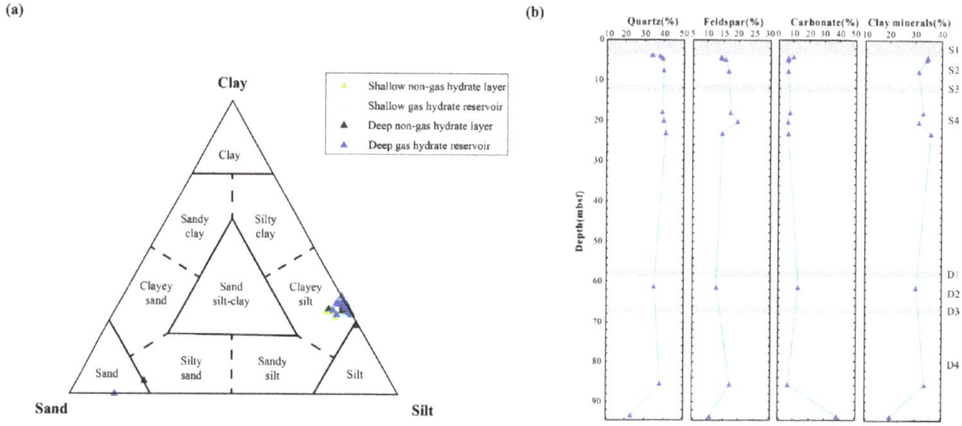

Figure 4. Ternary plot of the lithology (**a**) and logs of mineral contents (**b**) of sediment from Site DS-W08.

4.3. Geochemical Data

As shown in Figure 5, the main elements analyzed included Al_2O_3, CaO, K_2O, and Al_2O_3, and ranged from 1.46% to 16.28%. The range of CaO was 3.58% to 50.29%. The range of K_2O is 0.87% to 3.38%. The Sr/Ba, a trace element indicator used to infer ancient sea-level fluctuations, was also analyzed. The Sr/Ba ratio ranged from 0.33 to 23.90.

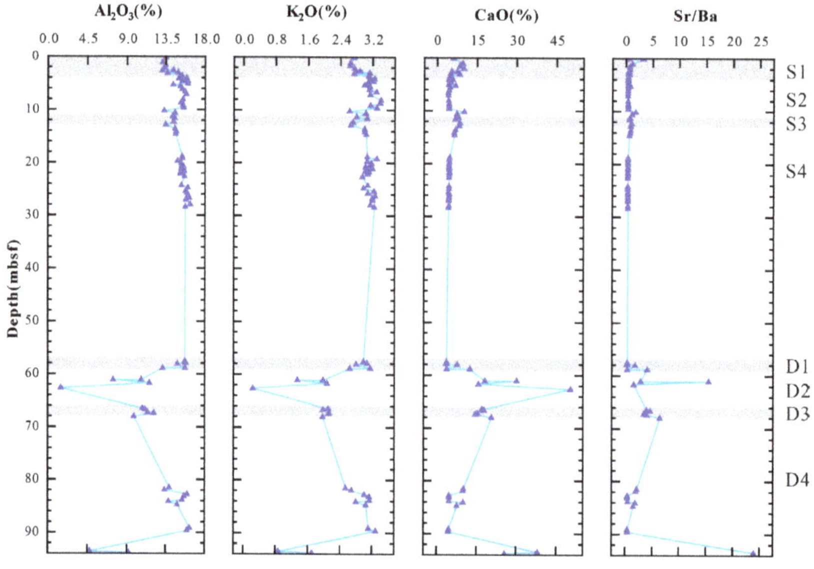

Figure 5. Depth profiles of major element oxide contents (expressed as mass percentages of the whole rock) and the Sr/Ba ratio.

4.4. Microfossil Content

Figure 6 presents the abundance of foraminifera and calcareous nannofossils in the sediment samples from Site DS-W08. According to the cruise report, the samples are dominated by planktonic foraminifera, with *Globigerinoides ruber*, *Neogloboquadrina dutertrei*, and *Globigerina bulloides* as the dominant species. Benthic foraminifera have high abundance and diversity, mainly endobenthic species, with relatively high contents of *Bolivina spathulata*, *Bolivinita quadrilatera*, *Cassidulina carinata*, *Cassidulinoides bradyi*, *Uvigerina peregrina*, *Bolivina subaenariensis*, *Bulimina mexicana*, and *Globobulimina affinis*. The total foraminifera abundance ranged from 7.98 to 4352 individuals per gram (ind/g), with an average of 377.36 ind/g. Planktonic foraminifera ranged from 7.5 to 4108.8 ind/g, with an average of 359.18 ind/g. Benthic foraminifera ranged from 2.4 to 243.2 ind/g, with an average of 18.18 ind/g. The ratio of planktonic to benthic foraminifera abundance varied between 1.75 to 53.43, with an average ratio of 14.85. According to the cruise report, the calcareous nannofossils are dominated by *Florisphaera profunda*, *Gephyrocapsa* spp. (small), *Emiliania huxleyi*, and *G. oceanica*. Obvious recrystallization of calcareous nannofossils occurs at the interval of 59–62 mbsf. The calcareous nannofossils abundance ranged from 0 to 91.3 individuals per field of view (ind/F).

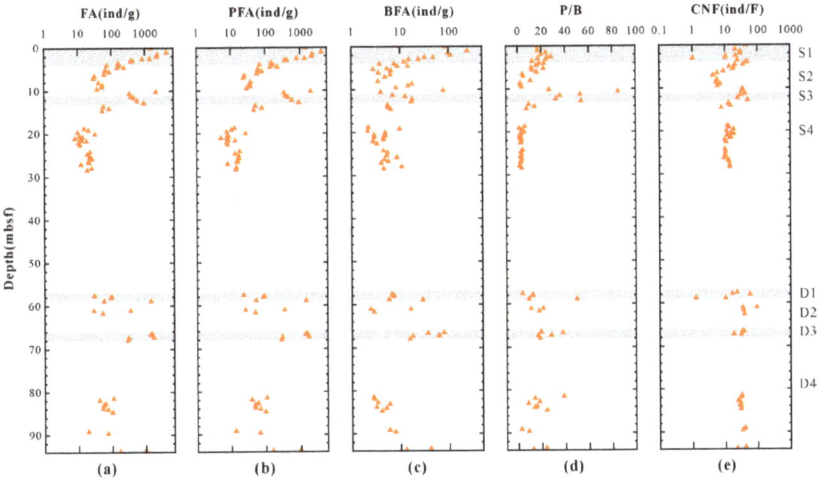

Figure 6. Variation in biogenic content with depth. (**a**) total foraminiferal abundance, (**b**) planktonic foraminiferal abundance, (**c**) benthic foraminiferal abundance, (**d**) ratio of planktonic to benthic foraminiferal abundance, and (**e**) calcareous nannofossil abundance.

5. Discussion

5.1. Variation Pattern of Hydrate Saturation

Hydrate saturation is a key indicator for assessing hydrate enrichment. It is primarily determined by quantifying chloride ion anomalies in pore water or gas release from pressure-retaining cores [63]. Alternatively, hydrate saturation can be derived from LWD resistivity calculations [64,65]. Among these methods, saturation estimates based on gas release from pressure-retaining cores offer the highest accuracy. However, due to the complexities and high costs associated with pressure-retaining cores, data acquired using this approach are typically limited. Furthermore, the method based on chloride ion anomalies in pore water is not applicable to sediment intervals lacking core sampling. Although saturation estimates derived from LWD resistivity data exhibit good vertical continuity, they may be subject to multiple interpretations. Therefore, this study adopts the mutual verification and comparative analysis of these three methods to accurately determine the hydrate enrichment.

The hydrate saturation data exhibit nearly identical vertical variation trends across the three methods. However, within the 59–62 mbsf interval, resistivity values are markedly elevated. This phenomenon is attributed to the elevated CaO content present in this interval (Figure 5). In the study area, CaO primarily originates from authigenic carbonate minerals and calcareous fossils (specifically foraminifera and calcareous nannofossils) [66]. Figures 5 and 6 indicate that the abundance of calcareous microfossil fossils in the 59–62 mbsf interval does not significantly increase compared with the underlying layers. Thus, these observations suggest a relative enrichment of authigenic carbonate minerals in this interval, which aligns with previous studies and the drilling results presented in the cruise report [67,68].

Based on variations in hydrate saturation, the research layer at Site DS-W08 can be divided into four intervals: shallow non-gas hydrate layer (shallow-NGHL), shallow gas hydrate reservoir (shallow-GHR), deep non-gas hydrate layer (deep-NGHL), and deep gas hydrate reservoir (deep-GHR). The shallow-GHR exhibits a maximum hydrate saturation of 14%. The deep-GHR shows a maximum hydrate saturation of 33% [63]. Clearly, the deep-GHR exhibits higher maximum hydrate saturation than the shallow-GHR.

5.2. Sedimentary Process Identification

Sedimentary processes may significantly control and indicate the accumulation of gas hydrates. Numerous studies on sandy reservoirs have long demonstrated that turbidite deposits facilitate hydrate accumulation. This observation is attributed to the coarse particle size and poor sorting of turbidite deposits in sandy sediments, which create greater reservoir space [69]. However, in fine-grained sediments, the theory may be less applicable because their particle sizes are more uniform. At Site DS-W08, the dominant lithology is silty clay. In the vertical direction, conventional particle size parameters exhibit minimal variation. Multiple peaks are observed in the C values—which represent the coarsest one percentile of sediments—in the study layer (Figure 3f). The C value is a robust indicator of hydrodynamic changes induced by various sedimentary processes and may also reveal major transformative events in these processes [31,70]. Based on these peaks of C values, the study depth was further subdivided: shallow coring layers were designated as sections S1 (0–4 mbsf), S2 (4–11 mbsf), S3 (11–13 mbsf), and S4 (13–28 mbsf), whereas deep coring layers were labeled sections D1 (57–59 mbsf), D2 (59–62 mbsf), D3 (66–68 mbsf), and D4 (80–94 mbsf) (Figure 3f).

Determining the primary hydrodynamic conditions of various sedimentary intervals depends on sediment grain size data. The C-M pattern diagram typically illustrates the relationship between the coarsest and median grain sizes (Figures 7 and 8). Points representing deep-water fine-grained sediment grain size parameters on the C-M diagram can determine if the sediment was formed by gravitational flow deposition, suspension deposition, or other non-gravitational processes [70]. Numerous studies indicate the presence of widespread traction currents, such as contour currents or internal waves, in the deep-water environment of the study area [71,72]. Consequently, the optimal fitting curve for the contourite current, derived from particle size parameters (average size and sorting) of modern contourite sediments (Figures 7 and 8), effectively characterizes the dynamic conditions of traction flow deposition [73]. The sedimentary hydrodynamic analysis for the study's layered section was performed using the two aforementioned diagrams (Figures 7 and 8).

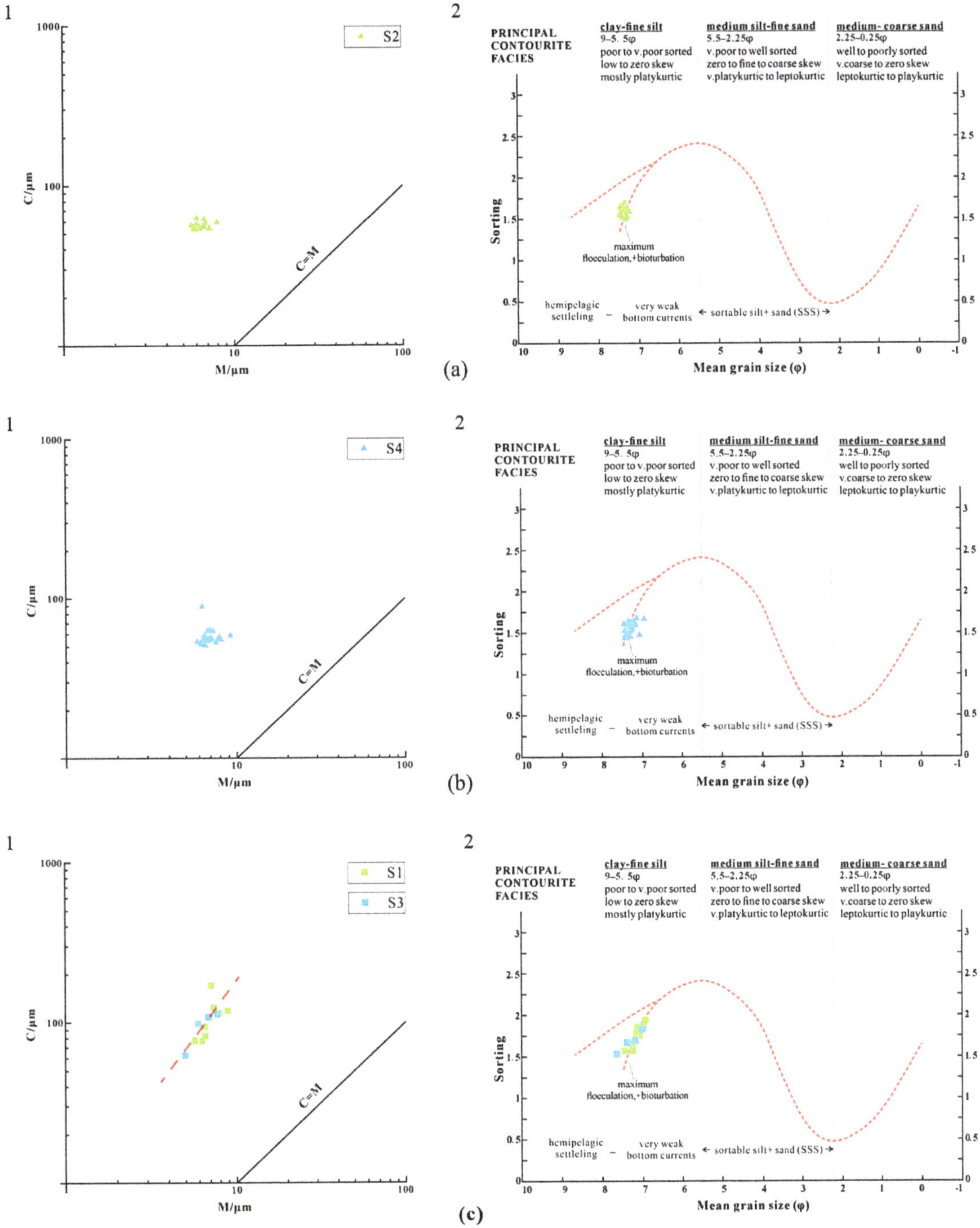

Figure 7. Identification of sedimentary processes in shallow core interval. C-M diagrams of S2 (**a1**), S4 (**b1**) and S1 and 3 (**c1**); traction current identification diagrams of S2 (**a2**), S4 (**b2**) and S1 and 3 (**c2**).

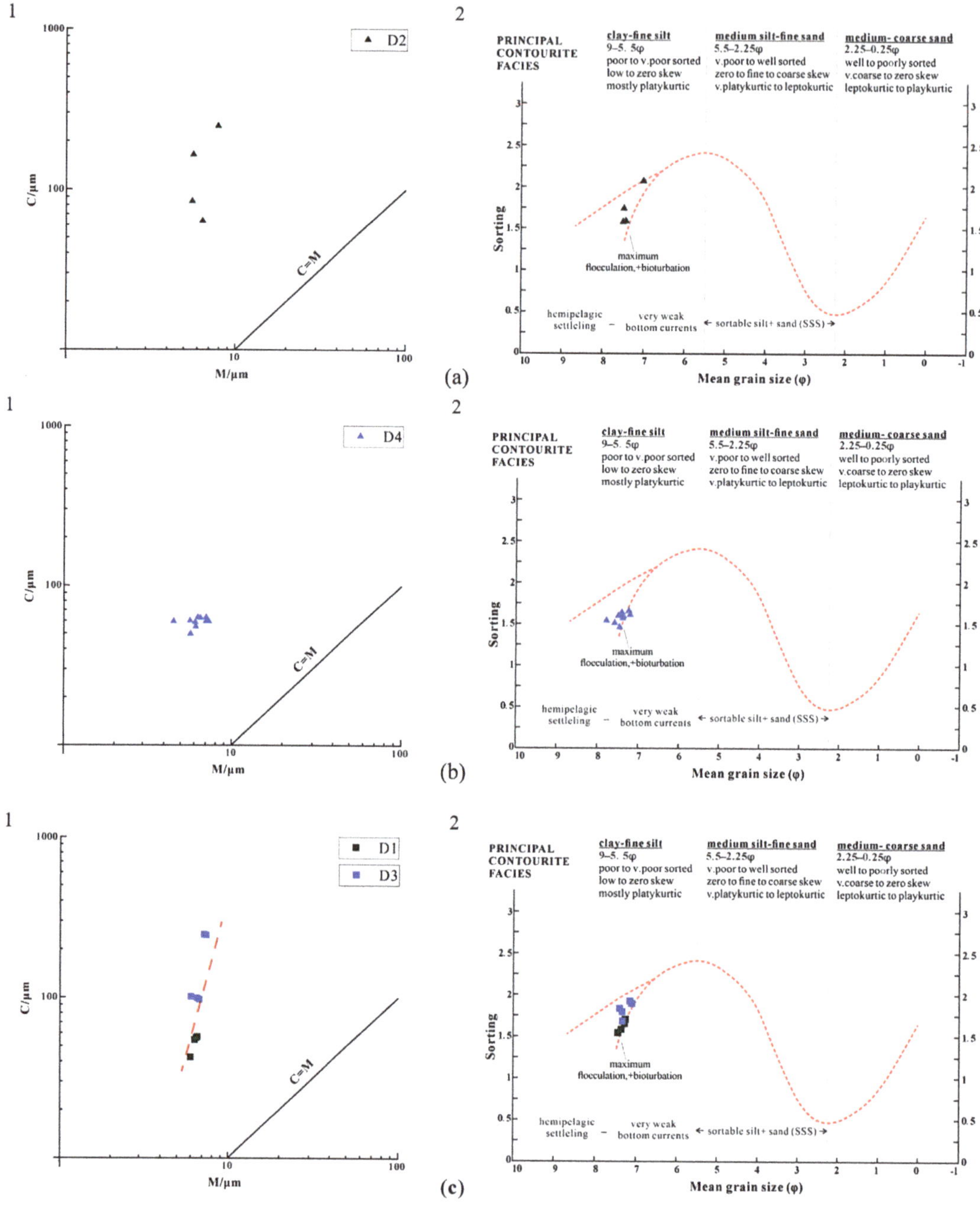

Figure 8. Identification of sedimentary processes in shallow core interval. C-M diagrams of D2 (**a1**), D4 (**b1**) and D1 and 3 (**c1**); traction current identification diagrams of D2 (**a2**), D4 (**b2**) and D1 and 3 (**c2**).

Notably, the increase in the C value in Section D2 may not be attributed to hydrodynamic factors. This is evidenced by the significant increase in carbonate minerals and CaO content in Section D2 (Figures 4b and 5), while the abundance of calcareous microfossils remains unchanged (Figure 6), suggesting a rise in authigenic carbonates. Furthermore, cruise reports and previous studies indicate that hydrate decomposition throughout geological history has led to the accumulation of authigenic carbonates in Section D2 [74]. These authigenic carbonates form large aggregates, with particle sizes surpassing fine sediment particles, reaching up to the millimeter scale [75]. These authigenic carbonates may influence the grain size parameters of Section D2, thus excluding this section from the hydrodynamic sedimentation analysis.

In the C-M diagram, trend lines for sediment samples in Section S1 and S3 are inclined relative to both the C-M baseline and the *X*-axis (Figure 7c). Grain size parameters exhibit strong agreement with the fitting curve associated with weak bottom current traction flow identified in previous studies, indicating that traction currents predominantly control sedimentary dynamics [70]. Sediment samples from Section S2 align parallel to the *X*-axis and primarily consist of fine-grained sediments (Figure 7a). Furthermore, grain-size parameters do not align well with the fitting curve associated with contourite currents—a feature of suspension deposition typically observed at the end of gravity flows or during normal deep-water sedimentation [76]. On the C-M diagram, the trend line for sediment samples from Section S4 runs nearly parallel to the *X*-axis and does not align well with the fitting curve associated with contourite currents (Figure 7b). This pattern is characteristic of suspension deposition [77].

The trend line of sediment samples from Section D1 and D3 forms an angle with the C-M baseline (Figure 8c). The grain size distribution aligns with the fitting curve that characterizes weak bottom-current traction flow [73]. Consequently, this section exhibits pronounced traction flow characteristics. The sediment sample from Section D2 contains a substantial amount of authigenic carbonate and does not accurately reflect sediment dynamics (Figure 8a). The sediment samples from Section D4 are parallel to the *X*-axis and do not align well with the fitting curve associated with contourite currents (Figure 8b). This observation is indicative of suspension deposition [77].

5.3. Sea-Level Change

Due to sample limitations, Site DS-W08 lacks AMS radiocarbon dating and oxygen isotope data to support its chronostratigraphic framework. Biostratigraphic analysis indicates that the last occurrence of *Globigerinoides ruber* (pink) at 67.90 mbsf in Site DS-W08 corresponds to an estimated age of approximately 120 kyr [21]. The first appearance of the calcareous nannofossil *Emiliania huxleyi* at 94 mbsf marks an estimated age of ~291 kyr [78] (Figure 2). Integrating biostratigraphic dating data, the core interval at Site DS-W08 reflects sedimentary records dating since 291 kyr, whereas the interval above 65 mbsf corresponds to records since the last glacial period (120 kyr) [22]. Previous studies on the ancient marine environment of the northern SCS indicate significant sea level fluctuations during this period [22–24]. Sea level fluctuations are closely linked to the formation and decomposition of hydrates and also influence sediment properties [79,80]. Variations in sediment properties significantly impact the dynamic adjustment of hydrate reservoirs [60,81,82].

The Sr/Ba ratio is commonly used as a proxy for paleosalinity. It also serves as an important indicator of sea-level fluctuations [25]. Due to their high abundance and widespread distribution, foraminifera are extensively used in marine paleoclimate reconstruction. The P/B ratio, defined as the ratio of planktonic to benthic foraminifera, generally increases with water depth [26,29]. Therefore, both Sr/Ba and P/B ratios exhibit similar trends: they increase during sea-level rise and decrease during sea-level fall. Al_2O_3 and K_2O typically exhibit higher concentrations in terrigenous sediments, serving as indicators of variations in terrigenous sediment input and indirectly reflecting sea level changes. Analysis of microfossil abundance and geochemical data revealed that sections S1, S3, D1, and D3 exhibit high-value intervals of P/B and Sr/Ba (Figures 5 and 6), while the Al_2O_3 and K_2O contents are relatively low in these sections (Figure 5). This suggests that sections S1, S3, D1, and D3 correspond to periods of high sea level. The remaining sections likely correspond to periods of low sea level. Sections D2 and the bottom of Section D4 (below 90 mbsf) contain outliers, with abnormally high CaO and carbonate mineral contents (Figure 4b). As previously mentioned, this anomaly may result from authigenic carbonates, indicating that the high Sr/Ba ratios in these intervals do not represent high sea-level periods.

Sedimentary processes exhibit a clear coupling with sea-level fluctuations. At Site DS-W08, sediments deposited during high sea level periods are primarily influenced by coastal bottom traction currents, including internal waves and contour currents. During low sea-level periods, sedimentation is dominated by overbank suspended deposits resulting from turbidity current processes.

5.4. Coupling Relationship Between Sedimentation Processes and Gas Hydrate Enrichment

A comparison of hydrate reservoirs at Site DS-W08, along with sedimentary process evolution and sea-level changes, reveals that the tops of both shallow and deep hydrate reservoirs are located within sedimentary sections influenced by traction currents, rather than in overbank deposits formed by turbidity currents. The shallow hydrate reservoir corresponds to sedimentary layers deposited during Sections S3 and S4 (Figure 2). The deep hydrate reservoir is associated with sedimentary layers from sections D3 and D4. Additionally, both shallow and deep hydrate reservoirs exhibit high foraminiferal abundance in the sediments (Figure 6a). Due to their relatively large particle size and internal chamber structures, foraminifera can provide additional pore space for hydrate accumulation. However, not all layers with high foraminiferal content correspond to hydrate reservoirs across all hydrate exploration sites in the SCS [83,84]. At Site DS-W08, layers with high foraminiferal content do not consistently correspond to zones of high hydrate saturation. Some studies suggest that clay minerals in fine-grained sediments significantly reduce effective reservoir space by occluding the chambers of foraminifera [85,86].

In shallow hydrate reservoirs, the high abundance of foraminifera provides favorable pore structures for hydrate accumulation [19,85]. Meanwhile, traction flow during Section S3 effectively removed fine-grained sediments, enhancing reservoir quality (Figure 9d). In the overlying section of shallow hydrate reservoirs, although foraminifera are also abundant, sedimentation is dominated by overbank suspension deposits from turbidity currents (Figure 9a). These deposits contain abundant fine-grained materials and lack effective transport dynamics to remove the fine-grained sediments (Figure 9b). As a result, the infilling of foraminiferal chambers with fine particles hinders hydrate enrichment. The deep hydrate reservoir exhibits similar conditions to the shallow reservoir. In the D1 section, although traction flow dominates, foraminiferal content is low and pore space is limited, which restricts hydrate accumulation. The D2 section may represent a layer enriched in authigenic carbonate minerals, potentially formed by hydrate decomposition. The D3 section also contains abundant foraminifera and is influenced by traction flow dynamics, which facilitates hydrate formation (Figure 9). In both shallow and deep hydrate reservoirs, hydrate formation significantly reduces porosity and permeability, forming a cap layer that restricts upward gas and fluid migration and promotes downward hydrate growth into underlying sediments. Consequently, hydrate saturation increases sharply from the top downward, followed by a gradual decline (Figure 2).

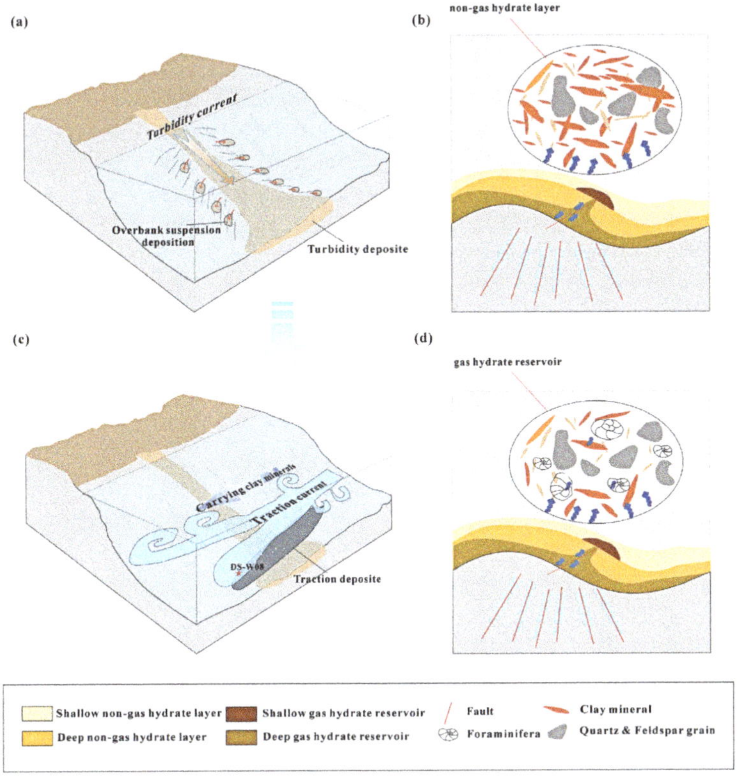

Figure 9. The model of the coupling relationship between sedimentary process and gas hydrate accumulation. The macroscopic (**a**) and microscopic (**b**) patterns of the constraints on gas hydrate enrichment imposed by the overbank suspended deposits formed by gravity flow; the macroscopic (c) and microscopic (**d**) controlling effects of traction current deposition on the enrichment of hydrates.

6. Conclusions

(1) The sedimentary process of gas hydrate reservoirs and adjacent layers in the Dongsha Sea is complex. At Site DS-W08, overbank suspended sediments from turbidity currents are superimposed with deep-water traction current deposits. Traction currents dominate during the high sea level periods, while turbidity currents prevail during low sea level periods.

(2) In the Dongsha area, both shallow and deep gas hydrate reservoirs are present. A strong coupling exists between the gas hydrate reservoirs and sedimentary processes. The tops of both shallow and deep gas hydrate reservoirs are associated with sedimentary layers influenced by traction currents.

(3) Sedimentary layers formed by traction currents, rich in foraminifera fossils, promote the formation of high-saturation hydrate reservoirs. Fine-grained turbidites, resulting from turbidity flows, lack effective storage space, are not conducive to hydrate formation.

(4) In foraminifera-rich sedimentary layers affected by traction flow transformation, gas hydrate accumulation leads to a sudden reduction in porosity and permeability. This results in the formation of a cap that inhibits the upward migration of gases and fluids, promoting hydrate growth into the overbank suspended sedimentary layers deposited by gravity-driven flows.

Author Contributions: Conceptualization, C.B. and H.W.; methodology, Y.W. and Z.W.; validation, C.B., W.C. and X.X.; formal analysis, X.X.; investigation, Y.W. and H.X.; resources, C.B. and H.W.; data curation, Y.W.; writing—original draft preparation, Y.W.; writing—review and editing, C.B., Z.W., and H.W.; visualization, Y.W. and W.C.; supervision, C.B.; project administration, H.W. and C.B.; funding acquisition, H.W. and C.B. All authors have read and agreed to the published version of the manuscript.

Funding: This research was funded by the Fundamental Research Funds for the Central Universities of China, grant number 3-7-10-2025-03, National Natural Science Foundation of China, grant number 42376217, and National Key Research and Development Program of China, grant number 2024YFC2814702.

Data Availability Statement: The datasets supporting the results of the study and findings are accessible through Figshare (DOI: http://doi.org/10.6084/m9.figshare.29493305).

Acknowledgments: We would like to thank the Guangzhou Marine Geological Survey (GMGS) for allowing the data of this research to be shared. We wish to thank all those who contributed to the China Notions Gas Hydrate Program Expedition 2.

Conflicts of Interest: Author Zhe Wang was employed by the company Bureau of Geophysical Prospecting INC., CNPC. The remaining authors declare that the research was conducted in the absence of any commercial or financial relationships that could be construed as a potential conflict of interest.

References

1. Sloan, E.D. Fundamental Principles and Applications of Natural Gas Hydrates. *Nature* **2003**, *426*, 353–359. [CrossRef]
2. Kvenvolden, K.A. Methane Hydrate—A Major Reservoir of Carbon in the Shallow Geosphere? *Chem. Geol.* **1988**, *71*, 41–51. [CrossRef]
3. Sloan, E.D., Jr.; Koh, C.A. *Clathrate Hydrates of Natural Gases*; CRC Press: Boca Raton, FL, USA, 2007; ISBN 0-429-12914-9.
4. Ruppel, C. Permafrost-Associated Gas Hydrate: Is It Really Approximately 1% of the Global System? *J. Chem. Eng. Data* **2015**, *60*, 429–436. [CrossRef]
5. Paull, C.K.; Buelow, W.J.; Ussler, W., III; Borowski, W.S. Increased Continental-Margin Slumping Frequency during Sea-Level Lowstands above Gas Hydrate–Bearing Sediments. *Geology* **1996**, *24*, 143–146. [CrossRef]
6. Paull, C.K.; Ussler, W., III; Dillon, W.P. Potential Role of Gas Hydrate Decomposition in Generating Submarine Slope Failures. *Nat. Gas Hydrate Ocean. Permafr. Environ.* **2000**, *12*, 149–156. [CrossRef]
7. Maslin, M.; Owen, M.; Betts, R.; Day, S.; Dunkley Jones, T.; Ridgwell, A. Gas Hydrates: Past and Future Geohazard? *Philos. Trans. R. Soc. Math. Phys. Eng. Sci.* **2010**, *368*, 2369–2393. [CrossRef]
8. Liu, J.; Gupta, S.; Rutqvist, J.; Ma, Y.; Wang, S.; Wan, K.; Fan, C.; Yan, W. Can Glacial Sea-Level Drop-Induced Gas Hydrate Dissociation Cause Submarine Landslides? *Geophys. Res. Lett.* **2024**, *51*, e2023GL106772. [CrossRef]
9. Kvenvolden, K.A. Methane Hydrate in the Global Organic Carbon Cycle. *Terra Nova* **2002**, *14*, 302–306. [CrossRef]
10. Ruppel, C.D.; Kessler, J.D. The Interaction of Climate Change and Methane Hydrates. *Rev. Geophys.* **2017**, *55*, 126–168. [CrossRef]
11. Burton, Z.F.; Kroeger, K.F.; Hosford Scheirer, A.; Seol, Y.; Burgreen-Chan, B.; Graham, S.A. Tectonic Uplift Destabilizes Subsea Gas Hydrate: A Model Example from Hikurangi Margin, New Zealand. *Geophys. Res. Lett.* **2020**, *47*, e2020GL087150. [CrossRef]
12. Ruffine, L.; Tang, A.M.; O'neill, N.; Toffin, L.; Paris, J.-D.; Yang, J.; Georgiev, V.; Fietzek, P.; Giustiniani, M.; Tinivella, U. Environmental Challenges Related to Methane Hydrate Decomposition from Climate Change Scenario and Anthropic Activities: State of the Art, Potential Consequences and Monitoring Solutions. *Earth-Sci. Rev.* **2023**, *246*, 104578. [CrossRef]
13. Collett, T.S. Energy Resource Potential of Natural Gas Hydrates. *AAPG Bull.* **2002**, *86*, 1971–1992. [CrossRef]
14. Boswell, R. Is Gas Hydrate Energy within Reach? *Science* **2009**, *325*, 957–958. [CrossRef] [PubMed]
15. Makogon, Y.F.; Holditch, S.A.; Makogon, T.Y. Natural Gas-Hydrates—A Potential Energy Source for the 21st Century. *J. Pet. Sci. Eng.* **2007**, *56*, 14–31. [CrossRef]
16. Boswell, R.; Collett, T.S. Current Perspectives on Gas Hydrate Resources. *Energy Environ. Sci.* **2011**, *4*, 1206–1215. [CrossRef]
17. Chong, Z.R.; Yang, S.H.B.; Babu, P.; Linga, P.; Li, X.-S. Review of Natural Gas Hydrates as an Energy Resource: Prospects and Challenges. *Appl. Energy* **2016**, *162*, 1633–1652. [CrossRef]
18. Dai, J.; Ni, Y.; Huang, S.; Peng, W.; Han, W.; Gong, D.; Wei, W. Genetic Types of Gas Hydrates in China. *Pet. Explor. Dev.* **2017**, *44*, 887–898. [CrossRef]

19. Handwerger, A.L.; Rempel, A.W.; Skarbek, R.M. Submarine Landslides Triggered by Destabilization of High-saturation Hydrate Anomalies. *Geochem. Geophys. Geosyst.* **2017**, *18*, 2429–2445. [CrossRef]
20. Dai, S.; Santamarina, J.C.; Waite, W.F.; Kneafsey, T.J. Hydrate Morphology: Physical Properties of Sands with Patchy Hydrate Saturation. *J. Geophys. Res. Solid Earth* **2012**, *117*, 2012JB009667. [CrossRef]
21. Zhang, G.; Liang, J.; Lu, J.; Yang, S.; Zhang, M.; Holland, M.; Schultheiss, P.; Su, X.; Sha, Z.; Xu, H.; et al. Geological Features, Controlling Factors and Potential Prospects of the Gas Hydrate Occurrence in the East Part of the Pearl River Mouth Basin, South China Sea. *Mar. Pet. Geol.* **2015**, *67*, 356–367. [CrossRef]
22. Cao, X.; Sun, J.; Gu, Y.; Qin, F.; Li, Y.; Yu, Y.; Ning, F. Revised Inflow Performance Relationship for Productivity Prediction from Marine Sandy Hydrate Reservoirs in Nankai Trough. *Geoenergy Sci. Eng.* **2024**, *238*, 212845. [CrossRef]
23. Lee, M.W.; Collett, T.S.; Lewis, K.A. Anisotropic Models to Account for Large Borehole Washouts to Estimate Gas Hydrate Saturations in the Gulf of Mexico Gas Hydrate Joint Industry Project Leg II Alaminos Canyon 21 B Well. *Mar. Pet. Geol.* **2012**, *34*, 85–95. [CrossRef]
24. Lee, M.W.; Collett, T.S. Gas Hydrate Saturations Estimated from Fractured Reservoir at Site NGHP-01-10, Krishna-Godavari Basin, India. *J. Geophys. Res. Solid Earth* **2009**, *114*, 2008JB006237. [CrossRef]
25. Ren, J.; Cheng, C.; Xiong, P.; Kuang, Z.; Liang, J.; Lai, H.; Chen, Z.; Chen, Y.; Li, T.; Jiang, T. Sand-Rich Gas Hydrate and Shallow Gas Systems in the Qiongdongnan Basin, Northern South China Sea. *J. Pet. Sci. Eng.* **2022**, *215*, 110630. [CrossRef]
26. Su, M.; Luo, K.; Fang, Y.; Kuang, Z.; Yang, C.; Liang, J.; Liang, C.; Chen, H.; Lin, Z.; Wang, C.; et al. Grain-Size Characteristics of Fine-Grained Sediments and Association with Gas Hydrate Saturation in Shenhu Area, Northern South China Sea. *Ore Geol. Rev.* **2021**, *129*, 103889. [CrossRef]
27. Wei, J.; Fang, Y.; Lu, H.; Lu, H.; Lu, J.; Liang, J.; Yang, S. Distribution and Characteristics of Natural Gas Hydrates in the Shenhu Sea Area, South China Sea. *Mar. Pet. Geol.* **2018**, *98*, 622–628. [CrossRef]
28. Yang, S.; Zhang, M.; Liang, J.Q.; Lu, J.; Zhang, Z.J.; Holland, M.; Schultheiss, P.; Fu, S.Y.; Sha, Z.B. Preliminary Results of China's Third Gas Hydrate Drilling Expedition: A Critical Step from Discovery to Development in the South China Sea. *Cent. Nat. Gas Oil* **2015**, *412*, 386–7614.
29. Ryu, B.-J.; Collett, T.S.; Riedel, M.; Kim, G.Y.; Chun, J.-H.; Bahk, J.-J.; Lee, J.Y.; Kim, J.-H.; Yoo, D.-G. Scientific Results of the Second Gas Hydrate Drilling Expedition in the Ulleung Basin (UBGH2). Mar. *Pet. Geol.* **2013**, *47*, 1–20. [CrossRef]
30. Argentino, C.; Conti, S.; Fioroni, C.; Fontana, D. Evidences for Paleo-Gas Hydrate Occurrence: What We Can Infer for the Miocene of the Northern Apennines (Italy). *Geosciences* **2019**, *9*, 134. [CrossRef]
31. Bai, C.; Su, P.; Xu, X.; Zhang, Y.; Han, S.; Liang, J. Controlling Effect of Particle Size on Gas Hydrate Enrichment in Fine-Grained Sediments. *Earth Space Sci.* **2024**, *11*, e2024EA003594. [CrossRef]
32. Wang, X.; Hutchinson, D.R.; Wu, S.; Yang, S.; Guo, Y. Elevated Gas Hydrate Saturation within Silt and Silty Clay Sediments in the Shenhu Area, South China Sea. *J. Geophys. Res.* **2011**, *116*, B05102. [CrossRef]
33. Bian, H.; Qin, X.; Sun, J.; Luo, W.; Lu, C.; Zhu, J.; Ma, C.; Zhou, Y. The Impact of Mineral Compositions on Hydrate Morphology Evolution and Phase Transition Hysteresis in Natural Clayey Silts. *Energy* **2023**, *274*, 127303. [CrossRef]
34. Kumari, A.; Balomajumder, C.; Arora, A.; Dixit, G.; Gomari, S.R. Physio-Chemical and Mineralogical Characteristics of Gas Hydrate-Bearing Sediments of the Kerala-Konkan, Krishna-Godavari, and Mahanadi Basins. *J. Mar. Sci. Eng.* **2021**, *9*, 808. [CrossRef]
35. Wang, X.; Feng, J.; Zhang, L.; Yuan, Y.; Xu, C.; Liang, J.; Yang, J. Mineralogy and Pore Characteristics of Marine Gas Hydrate-Bearing Sediments in the Northern South China Sea. *Mar. Pet. Geol.* **2022**, *141*, 105711. [CrossRef]
36. He, Z.; Gupta, K.M.; Linga, P.; Jiang, J. Molecular Insights into the Nucleation and Growth of CH4 and CO2 Mixed Hydrates from Microsecond Simulations. *J. Phys. Chem. C* **2016**, *120*, 25225–25236. [CrossRef]
37. Zhou, S.; Li, Q.; Zhu, J.; Gu, Q.; Wang, L.; Wu, Z. Experimental Study on Permeability of Methane Hydrate Clayey Interbedded Sediments Considering Effective Stress and Hydrate Dissociation. *Energy Fuels* **2023**, *37*, 11113–11124. [CrossRef]
38. Quainoo, A.K.; Negash, B.M.; Bavoh, C.B.; Idris, A.; Shahpin, H.B.A.; Yaw, A.D. Inhibition Impact of Amino Acids on Swelling Clays: An Experimental and COSMO-RS Simulation Evaluation. *Energy Fuels* **2020**, *34*, 13985–14000. [CrossRef]
39. Fang, C. Relations between Biogenic Component (Foraminifera) and Highly Saturated Gas Hydrates Distribution from Shenhu Area, Northern South China Sea. *Earth Sci.-J. China Univ. Geosci.* **2013**, *38*, 0907. [CrossRef]
40. Wilson, M.D.; Pittman, E.D. Authigenic Clays in Sandstones; Recognition and Influence on Reservoir Properties and Paleoenvironmental Analysis. *J. Sediment. Res.* **1977**, *47*, 3–31. [CrossRef]
41. Sun, S.Q.; Esteban, M. Paleoclimatic Controls on Sedimentation, Diagenesis, and Reservoir Quality: Lessons from Miocene Carbonates. *AAPG Bull.* **1994**, *78*, 519–543. [CrossRef]
42. Burton, Z.F.; McHargue, T.; Kremer, C.H.; Bloch, R.B.; Gooley, J.T.; Jaikla, C.; Harrington, J.; Graham, S.A. Peak Cenozoic Warmth Enabled Deep-Sea Sand Deposition. *Sci. Rep.* **2023**, *13*, 1276. [CrossRef] [PubMed]
43. Burton, Z.F.; McHargue, T.R.; Graham, S.A. Global Eocene-Oligocene Unconformity in Clastic Sedimentary Basins. *Earth-Sci. Rev.* **2024**, *258*, 104912. [CrossRef]

44. Boswell, R.; Rose, K.; Collett, T.S.; Lee, M.; Winters, W.; Lewis, K.A.; Agena, W. Geologic Controls on Gas Hydrate Occurrence in the Mount Elbert Prospect, Alaska North Slope. *Mar. Pet. Geol.* **2011**, *28*, 589–607. [CrossRef]
45. You, K.; Flemings, P.B.; Malinverno, A.; Collett, T.S.; Darnell, K. Mechanisms of Methane Hydrate Formation in Geological Systems. *Rev. Geophys.* **2019**, *57*, 1146–1196. [CrossRef]
46. Dafov, L.N.; Burton, Z.F.; Haines, S.S.; Scheirer, A.H.; Masurek, N.; Boswell, R.; Frye, M.; Seol, Y.; Graham, S.A. Terrebonne Basin, Gulf of Mexico Gas Hydrate Resource Evaluation and 3-D Modeling of Basin-Scale Sedimentation, Salt Tectonics, and Hydrate System Evolution since the Early Miocene. *Mar. Pet. Geol.* **2025**, *176*, 107330. [CrossRef]
47. Liang, J.; Zhang, W.; Lu, J.A.; Wei, J.; Kuang, Z.; He, Y. Geological Occurrence and Accumulation Mechanism of Natural Gas Hydrates in the Eastern Qiongdongnan Basin of the South China Sea: Insights from Site GMGS5-W9-2018. *Mar. Geol.* **2019**, *418*, 106042. [CrossRef]
48. Shengqiang, Y.; Shiguo, W.; Thomas, L.; Genshun, Y.; Fuliang, L.; Feng, C.; Hairong, W.; Li, L. Fine-Grained Pleistocene Deepwater Turbidite Channel System on the Slope of Qiongdongnan Basin, Northern South China Sea. *Mar. Pet. Geol.* **2009**, *26*, 1441–1451. [CrossRef]
49. Lowe, D.R. Sediment Gravity Flows; II, Depositional Models with Special Reference to the Deposits of High-Density Turbidity Currents. *J. Sediment. Res.* **1982**, *52*, 279–297. [CrossRef]
50. Shanmugam, G. 50 Years of the Turbidite Paradigm (1950s–1990s): Deep-Water Processes and Facies Models—A Critical Perspective. *Mar. Pet. Geol.* **2000**, *17*, 285–342. [CrossRef]
51. Piper, D.J.; Normark, W.R. Processes That Initiate Turbidity Currents and Their Influence on Turbidites: A Marine Geology Perspective. *J. Sediment. Res.* **2009**, *79*, 347–362. [CrossRef]
52. Meiburg, E.; Kneller, B. Turbidity Currents and Their Deposits. *Annu. Rev. Fluid Mech.* **2010**, *42*, 135–156. [CrossRef]
53. Taylor, B.; Hayes, D.E. The Tectonic Evolution of the South China Basin. In *Geophysical Monograph Series*; Hayes, D.E., Ed.; American Geophysical Union: Washington, DC, USA, 1980; Volume 23, pp. 89–104. ISBN 978-0-87590-023-0.
54. Tang, W.; Zhao, Z.; Song, S.; Wang, Y.; Xie, X.; Liu, S. Differences in the Tectonic Evolution of Basins in the Central-Southern South China Sea and Their Hydrocarbon Accumulation Conditions. *Acta Geol. Sin.-Engl. Ed.* **2021**, *95*, 30–40. [CrossRef]
55. Taylor, B.; Hayes, D.E. Origin and History of the South China Sea Basin. In *Geophysical Monograph Series*; Hayes, D.E., Ed.; American Geophysical Union: Washington, DC, USA, 1983; Volume 27, pp. 23–56. ISBN 978-0-87590-053-7.
56. Li, X.; Wang, Z.; Yao, Y.; Gao, H.; Zhu, S.; Xu, Z. Formation and Evolution of the South China Sea since the Late Mesozoic: A Review. *China Geol.* **2023**, *6*, 154–167. [CrossRef]
57. Sha, Z.; Liang, J.; Zhang, G.; Yang, S.; Lu, J.; Zhang, Z.; McConnell, D.R.; Humphrey, G. A Seepage Gas Hydrate System in Northern South China Sea: Seismic and Well Log Interpretations. *Mar. Geol.* **2015**, *366*, 69–78. [CrossRef]
58. Shyu, C.-T.; Hsu, S.-K.; Liu, C.-S. Heat Flows off Southwest Taiwan: Measurements over Mud Diapirs and Estimated from Bottom Simulating Reflectors. *Terr. Atmos. Ocean. Sci.* **1998**, *9*, 795. [CrossRef]
59. Yan, P.; Deng, H.; Liu, H. The Geological Structure and Prospect of Gas Hydrate over the Dongsha Slope, South China Sea. *Terr. Atmos. Ocean. Sci.* **2006**, *17*, 645. [CrossRef]
60. Bai, C.; Zhang, G.; Lu, J.; Liang, J.; Yang, Z.; Yan, W.; Zhu, D.; Tian, Y. Deep-Water Sediment Waves as a Special Gas Hydrate Reservoirs in the Northeastern South China Sea. *Mar. Pet. Geol.* **2019**, *101*, 476–485. [CrossRef]
61. Liu, C.; Jing, J.; Zhao, Q.; Luo, X.; Chen, K.; Wang, M.; Deng, M. High-Resolution Resistivity Imaging of a Transversely Uneven Gas Hydrate Reservoir: A Case in the Qiongdongnan Basin, South China Sea. *Remote Sens.* **2023**, *15*, 2000. [CrossRef]
62. Li, G.; Du, R.; Tang, J.; Li, Z.; Xia, Q.; Shi, B.; Zhou, L.; Yang, Y.; Zhang, W. Comparison of the Graphic and Moment Methods for Analyzing Grain-Size Distributions: A Case Study for the Chinese Inner Continental Shelf Seas. *Int. J. Sediment Res.* **2022**, *37*, 729–736. [CrossRef]
63. Bai, C.; Wang, H.; Li, Q.; Zhang, Y.; Xu, X. Controls on Deep and Shallow Gas Hydrate Reservoirs in the Dongsha Area, South China Sea: Evidence from Sediment Properties. *J. Mar. Sci. Eng.* **2024**, *12*, 696. [CrossRef]
64. Li, H.; Liu, J.; Qu, C.; Song, H.; Zhuang, X. A Method for Calculating Gas Hydrate Saturation by Dual Parameters of Logging. *Front. Earth Sci.* **2022**, *10*, 986647. [CrossRef]
65. Kim, G.Y.; Yi, B.; Kang, N.K.; Yoo, D.-G.; Lee, J.Y. Comparison of Logging-While-Drilling and Wireline Logging Data from Gas Hydrate-Bearing Deep-Sea Sediments, the Ulleung Basin, East Sea. *Explor. Geophys.* **2022**, *53*, 92–103. [CrossRef]
66. Turchyn, A.V.; Bradbury, H.J.; Walker, K.; Sun, X. Controls on the Precipitation of Carbonate Minerals Within Marine Sediments. *Front. Earth Sci.* **2021**, *9*, 618311. [CrossRef]
67. Wang, S.; Yan, W.; Magalhães, V.H.; Chen, Z.; Pinheiro, L.M.; Gussone, N. Factors Influencing Methane-Derived Authigenic Carbonate Formation at Cold Seep from Southwestern Dongsha Area in the Northern South China Sea. *Environ. Earth Sci.* **2014**, *71*, 2087–2094. [CrossRef]
68. Tong, H.; Feng, D.; Cheng, H.; Yang, S.; Wang, H.; Min, A.G.; Edwards, R.L.; Chen, Z.; Chen, D. Authigenic Carbonates from Seeps on the Northern Continental Slope of the South China Sea: New Insights into Fluid Sources and Geochronology. *Mar. Pet. Geol.* **2013**, *43*, 260–271. [CrossRef]

69. Yoneda, J.; Oshima, M.; Kida, M.; Kato, A.; Konno, Y.; Jin, Y.; Jang, J.; Waite, W.F.; Kumar, P.; Tenma, N. Permeability Variation and Anisotropy of Gas Hydrate-Bearing Pressure-Core Sediments Recovered from the Krishna–Godavari Basin, Offshore India. *Mar. Pet. Geol.* **2019**, *108*, 524–536. [CrossRef]
70. Passega, R. Grain Size Representation by CM Patterns as a Geologic Tool. *J. Sediment. Res.* **1964**, *34*, 830–847. [CrossRef]
71. Wang, X.; Cai, F.; Sun, Z.; Li, Q.; Li, A.; Sun, Y.; Wang, H.; Sun, Q. Tectonic and Oceanographic Controls on the Slope-Confined Dendritic Canyon System in the Dongsha Slope, South China Sea. *Geomorphology* **2022**, *410*, 108285. [CrossRef]
72. Wang, X.; Kneller, B.; Sun, Q. Sediment Waves Control Origins of Submarine Canyons. *Geology* **2023**, *51*, 310–314. [CrossRef]
73. Stow, D.; Smillie, Z. Distinguishing between Deep-Water Sediment Facies: Turbidites, Contourites and Hemipelagites. *Geosciences* **2020**, *10*, 68. [CrossRef]
74. Zhuang, C.; Chen, F.; Cheng, S.; Lu, H.; Wu, C.; Cao, J.; Duan, X. Light Carbon Isotope Events of Foraminifera Attributed to Methane Release from Gas Hydrates on the Continental Slope, Northeastern South China Sea. *Sci. China Earth Sci.* **2016**, *59*, 1981–1995. [CrossRef]
75. Chen, F.; Hu, Y.; Feng, D.; Zhang, X.; Cheng, S.; Cao, J.; Lu, H.; Chen, D. Evidence of Intense Methane Seepages from Molybdenum Enrichments in Gas Hydrate-Bearing Sediments of the Northern South China Sea. *Chem. Geol.* **2016**, *443*, 173–181. [CrossRef]
76. Brackenridge, R.E.; Stow, D.A.V.; Hernández-Molina, F.J.; Jones, C.; Mena, A.; Alejo, I.; Ducassou, E.; Llave, E.; Ercilla, G.; Nombela, M.A.; et al. Textural Characteristics and Facies of Sand-rich Contourite Depositional Systems. *Sedimentology* **2018**, *65*, 2223–2252. [CrossRef]
77. Passega, R. Significance of CM Diagrams of Sediments Deposited by Suspensions. *Sedimentology* **1977**, *24*, 723–733. [CrossRef]
78. Fang, C.; Chang, Z.; Yang, Z.; Xin, S.; Xiao, D.; Guanghu, L.; Cong, W.; Xia, J. Calcareous Nannofossils and Foraminifera Biostratigraphy on the Northeastern Slope of the South China Sea and Variation in Sedimentation Rates. *Earth Sci.-J. China Univ. Geosci.* **2016**, *41*, 416. [CrossRef]
79. Liu, F.; Huang, E.; Jin, X.; Tian, J. Glacial Methane Hydrate Dissociation in the South China Sea Margin During the Oligocene. *Geophys. Res. Lett.* **2025**, *52*, e2024GL114439. [CrossRef]
80. Song, Y.; Lei, Y.; Zhang, L.; Cheng, M.; Li, C.; Liu, N. Spatial-Temporal Evolution of the Gas Hydrate Stability Zone and Accumulation Patterns of Double BSRs Formation in the Shenhu Area. *Front. Earth Sci.* **2022**, *10*, 880933. [CrossRef]
81. Ren, J.; Yin, Z.; Lu, H.; Xu, C.; Kuang, Z.; Deng, W.; Liu, Y.; Linga, P. Effects of South China Sea Clayey-Silty Sediments on the Kinetics and Morphology of CH4 Hydrate: Implication on Energy Recovery. *Appl. Energy* **2024**, *367*, 123399. [CrossRef]
82. Li, Y.; Xu, C.; Zhu, J.; Lu, H.; Liu, Y.; Gu, Y.; Pan, Z.; Linga, P.; Yin, Z. Comprehensive Characterizations of Core Sediments Recovered from Shenhu W17 Well in South China Sea and Its Impact on Methane Hydrate Kinetics. *Gas Sci. Eng.* **2024**, *131*, 205482. [CrossRef]
83. Huang, Y.; Cheng, J.; Wang, M.; Wang, S.; Yan, W. Gas Hydrate Dissociation Events During LGM and Their Potential Trigger of Submarine Landslides: Foraminifera and Geochemical Records From Two Cores in the Northern South China Sea. *Front. Earth Sci.* **2022**, *10*, 876913. [CrossRef]
84. Lu, Y.; Yang, H.; Huang, B.; Liu, Y.; Lu, H. Foraminifera Associated with Cold Seeps in Marine Sediments. *Front. Mar. Sci.* **2023**, *10*, 1157879. [CrossRef]
85. Li, C.; Hu, G.; Zhang, W.; Ye, Y.; Liu, C.; Li, Q.; Sun, J. Influence of Foraminifera on Formation and Occurrence Characteristics of Natural Gas Hydrates in Fine-Grained Sediments from Shenhu Area, South China Sea. *Sci. China Earth Sci.* **2016**, *59*, 2223–2230. [CrossRef]
86. Choi, K.; Jo, J. Morphodynamics and Stratigraphic Architecture of Compound Dunes on the Open-Coast Macrotidal Flat in the Northern Gyeonggi Bay, West Coast of Korea. *Mar. Geol.* **2015**, *366*, 34–48. [CrossRef]

Journal of
*Marine Science
and Engineering*

MDPI

Article

Research on Reservoir Identification of Gas Hydrates with Well Logging Data Based on Machine Learning in Marine Areas: A Case Study from IODP Expedition 311

Xudong Hu [1,2,3,4,*], Wangfeng Leng [1], Kun Xiao [1], Guo Song [1], Yiming Wei [1] and Changchun Zou [5,*]

1 Nanchang Key Laboratory of Intelligent Sensing Technology and Instruments for Geological Hazards, East China University of Technology, Nanchang 330013, China
2 Shaanxi Key Laboratory of Petroleum Accumulation Geology, Xi'an Shiyou University, Xi'an 710065, China
3 Hubei Key Laboratory of Intelligent Geo-Information Processing, China University of Geosciences, Wuhan 430078, China
4 Key Laboratory of Gas Hydrate, Guangzhou Institute of Energy Conversion, Chinese Academy of Sciences, Guangzhou 510640, China
5 School of Geophysics and Information Technology, China University of Geosciences, Beijing 100083, China
* Correspondence: huxudong@foxmail.com (X.H.); zoucc@cugb.edu.cn (C.Z.)

Abstract

Natural gas hydrates, with their efficient and clean energy characteristics, are deemed a significant pillar within the future energy sector, and their resource quantification and development have a profound impact on the transformation of global energy structure. However, how to accurately identify gas hydrate reservoirs (GHRs) is currently a hot research topic. This study explores the logging identification method of marine GHRs based on machine learning (ML) according to the logging data of the International Ocean Drilling Program (IODP) Expedition 311. This article selects six ML methods, including Gaussian process classification (GPC), support vector machine (SVM), multilayer perceptron (MLP), random forest (RF), extreme gradient boosting (XGBoost), and logistic regression (LR). The internal relationship between logging data and hydrate reservoir is analyzed through six ML algorithms. The results show that the constructed ML model performs well in gas hydrate reservoir identification. Among them, RF has the highest accuracy, precision, recall, and harmonic mean of precision and recall (F1 score), all of which are above 0.90. With an area under curve (AUC) of nearly 1 for RF, it is confirmed that ML technology is effective in this area. Research has shown that ML provides an alternative method for quickly and efficiently identifying GHRs based on well logging data and also offers a scientific foundation and technical backup for the future prospecting and mining of natural gas hydrates.

Keywords: reservoir identification; gas hydrates; well logging; machine learning; IODP Expedition 311

1. Introduction

Natural gas hydrate is a kind of crystalline object formed by the combination of water molecules and gas molecules under low temperature and high pressure. It has an appearance reminiscent of ice and is predominantly found within the sedimentary deposits of permafrost-bearing regions and along the continental margins [1–3]. It is a new type of energy resource with astonishing reserves and widespread distribution. Owing to its singular physical and chemical attributes, it has emerged as a central point of interest in the international energy community and the field of geosciences in recent years [4–10]. The precise identification of gas hydrate reservoirs (GHRs) holds significant importance as it serves as a crucial prerequisite for both the assessment and exploitation of hydrate reserves. The interplay among natural gas hydrates, the surrounding sediments, and fluids is an intricate occurrence. As a consequence, it often makes the process of identifying GHRs more challenging [11–13]. In order to develop natural gas hydrate resources in a reasonable and feasible manner in the future, it is necessary to find methods that can effectively identify GHRs.

Over the past few decades, many experts have proposed a variety of natural gas hydrate identification methods through the comprehensive use of geology, geochemistry, geophysics, and other multidisciplinary knowledge and technical means. In the early days, scientists mainly obtained core samples through drilling and judged the existence of hydrate through naked-eye observation or low-temperature anomaly obtained by an infrared thermal imager [14]. As technology advances steadily, the research findings indicated a decline in the chlorine concentration or salinity of pore water within natural gas hydrate deposits. Therefore, it is viable to identify natural gas hydrate by analyzing the chloride ion anomaly present in pore water [15]. Although these identification methods are very accurate, drilling and coring are inefficient and costly [16,17]. In addition, due to the easy decomposition of hydrate, it is difficult to collect complete hydrate samples. Geophysical methods have relatively high cost-effectiveness and are suitable for preliminary large-scale exploration. Among many geophysical methods, seismic and logging are considered to be the two most effective methods to identify gas hydrates. Within seismic exploration, bottom-simulating reflectors (BSRs) are regarded as a crucial geophysical indicator for confirming the presence of marine gas hydrates. But natural gas hydrate and BSRs are not in one-to-one correspondence [18]. Geophysical logging techniques present the advantage of directly, continuously, and economically assessing the attributes of GHRs at the actual site. This holds substantial importance for the qualitative characterization of GHRs as well as the quantitative computation of reservoir porosity and hydrate saturation parameters. Tian and Liu [19] put forward an innovative attribute that integrates velocity and density to identify hydrates, which has achieved notable success in the Dongsha area of the South China Sea and the Hydrate Ridge along the Oregon continental margin. Wu et al. [20] proposed a gas hydrate reservoir identification method based on a petrophysical model and sensitive elastic parameters. The method was applied to the Mallik permafrost region in Canada and the Shenhu sea area in the South China Sea, and the effectiveness of the method in GHR recognition was verified. These methods are usually based on a rock physical volume model. The presence of multiple micro-existence modes of natural gas hydrate gives rise to numerous challenges in the micro-response mechanism of natural GHR logging. This, in turn, results in the fact that many theories and methods of conventional reservoir logging evaluation cannot be fully applied in hydrate exploration. Therefore, these model-driven hydrate reservoir identification methods have the problems of difficult identification, low accuracy, and low processing efficiency. At the same time, they are also vulnerable to human factors and a complex geological environment.

With the continuous progress of technology, artificial intelligence has penetrated into various fields and has had a very positive and far-reaching impact. As a segment in artificial intelligence, machine learning (ML) is widely used in oil and gas reservoir identification [21–24], fluid identification [25,26], lithology identification [27–32], and other fields and has yielded favorable outcomes. This approach can be used in the area of natural gas hydrates as well. Lee et al. [33] proposed the K-means clustering method, which used a variety of attributes to identify the distribution of potential GHRs. Zhu et al. [34] proposed an effective classification of GHRs based on K-means clustering and the AdaBoost method.

Tian et al. [35] used seven ML algorithms to identify hydrates in the Oregon Hydrate Ridge by P-wave and S-wave velocities. In summary, there are relatively few studies on hydrate reservoir identification using multiple conventional logging curves based on the ML method, and the identification methods still have shortcomings. It is evident that further research is required in order to develop high-precision hydrate reservoir identification methods.

In this study, well logging data from four boreholes located within the accretionary prism of the Cascadia subduction zone will be utilized to select well logging curves that exhibit heightened sensitivity to gas hydrates. Subsequently, six ML algorithms, namely support vector machine (SVM), Gaussian process classification (GPC), multilayer perceptron (MLP), random forest (RF), extreme gradient boosting (XGBoost), and logistic regression (LR), will be employed to identify GHRs within the study area.

2. Geological Setting

The specific objective of Expedition 311 of the International Ocean Drilling Program (IODP) is to study the enrichment and formation mechanism of natural gas hydrates in aggradation complexes [36]. The study area is located in the accretionary prism of the Cascadia subduction zone. Its tectonic setting is on the accretionary wedge formed by the subduction of the Juan de Fuca plate to the North American plate, and the Juan de Fuca plate subducts downward at a speed of ~45mm/y [37,38]. Tectonic plate subduction conveys considerable sediment volumes from deep-sea basins to the outermost edge of the subduction–accretion prism, where the sediment gradually builds up. Favorable temperature and pressure conditions, coupled with an abundant supply of hydrocarbons, enable this region to host natural gas hydrates, making it a quintessential example of an active continental margin hydrate research area.

The IODP Expedition 311 was conducted from September to October 2005. Throughout the entire voyage, five stations were established at the edge of the Cascadia subduction zone, namely sites U1325, U1326, U1327, U1328, and U1329. As depicted in Figure 1, the five stations are placed in the front, middle, and back parts of the accretionary wedge, representing the different stages of gas hydrate evolution in the context of active continental margins [39]. The location information (longitude and latitude) of each borehole is shown in Table 1. Among them, the depths of U1326A, U1327A, and U1328A are 300 m below the seafloor (mbsf), while the depths of U1325A and U1329A are 350 mbsf and 220 mbsf, respectively [39].

Table 1. Information on the drilling locations of boreholes U1325A, U1326A, U1327A, U1328A, and U1329A during the IODP Expedition 311 [39].

Hole	Latitude	Longitude	Total Penetration
U1325A	48°38.691′ N	126°58.991′ W	350 mbsf
U1326A	48°37.635′ N	127°3.029′ W	300 mbsf
U1327A	48°41.887′ N	126°51.921′ W	300 mbsf
U1328A	48°40.072′ N	126°51.022′ W	300 mbsf
U1329A	48°47.369′ N	126°40.713′ W	220 mbsf

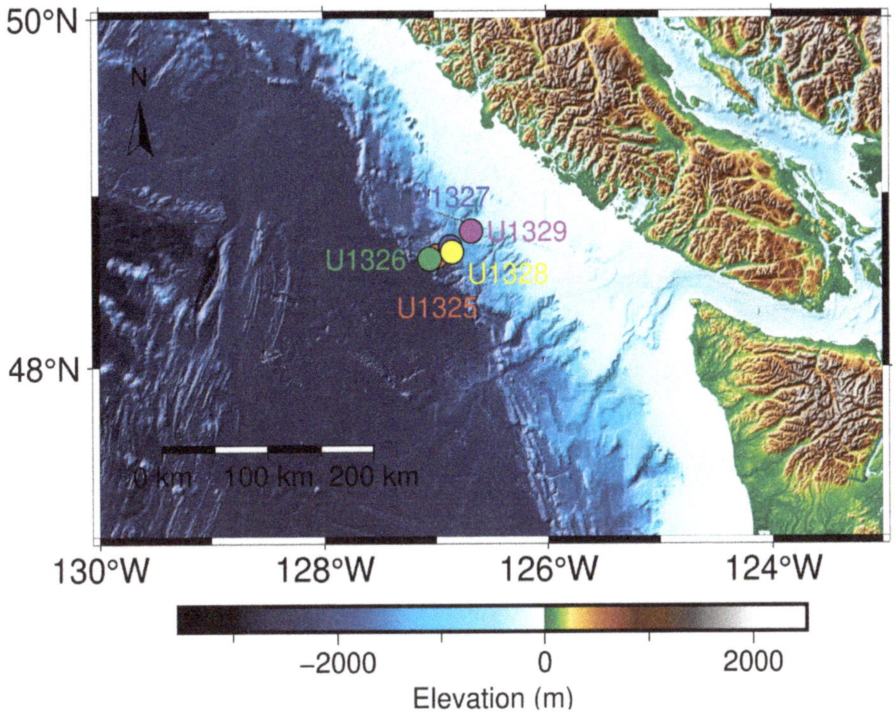

Figure 1. The geographical location of the five sites (U1325, U1326, U1327, U1328, and U1329) during the IODP Expedition 311 [39].

3. Methodology

In this study, six algorithms are proposed to be used: GPC, SVM, MLP, RF, XGBoost, and LR. For model training and testing, we used logging data from boreholes in the study area and the identification of reservoir segments for gas hydrates. The underlying theoretical principles and corresponding model evaluation metrics of these algorithms are concisely summarized below.

3.1. Algorithm Introduction

3.1.1. Gaussian Classification Process

As an ML paradigm grounded in the Gaussian process, GPC is specifically formulated to tackle classification problems. The Gaussian process is a random process in which each point x corresponds to a random variable f. The joint distribution of these variables exhibits a Gaussian nature, defined via the mean and covariance functions [40]. In GPC, since the target value y of the classification problem is discrete, it is no longer assumed that $y \mid x$ obeys the Gaussian distribution, but other suitable distributions, such as the Bernoulli distribution, are selected. In order to solve this problem, a latent function f (implicit variable) is introduced, and its normal prior is given. Then, the observed data (x, y) is used to calculate the potential function f and its output after passing through the response function (such as the sigmoid or probit function). These outputs are projected onto the [0, 1] range to derive the classification probability [41]. The implementation of GPC includes selecting appropriate kernel functions (such as RBF kernel, etc.) to define the covariance function of the Gaussian process, using training datasets (x, y) to train the model and obtain the posterior distribution of potential function f, and to finally predict the new input x^* to obtain the classification label.

GPC performs well in dealing with nonlinear classification problems and has a strong generalization ability. However, due to its high computational complexity, approximation methods (such as Laplace approximation, expected propagation, etc.) are often used to reduce the computational complexity [42].

3.1.2. Support Vector Machine

SVM stands as a potent instrument within the domain of supervised learning, showcasing remarkable proficiency in handling both classification and regression analysis assignments [43]. The core principle involves identifying the most suitable hyperplane that maximizes the margin separating distinct categories of sample points, thereby enabling precise classification. For linear separable data, SVM constructs a decision boundary; for linear non-separable data, it is solved by mapping kernel function to high-dimensional space [44]. In addition, SVM also introduces the concept of a soft interval to allow for certain classification errors to deal with noise and outliers, and the penalty coefficient C controls the penalty intensity. SVM demonstrates robust performance in handling high-dimensional and noisy datasets, making them highly adaptable to diverse classification tasks. Additionally, SVM enables nonlinear classification via kernel-induced feature mapping, thereby extending its applicability to complex decision boundaries. Its training process involves solving convex quadratic programming problems, and its performance is highly dependent on parameter selection. It is often optimized by grid search, cross-validation, and other methods.

With regard to nonlinear functions, the SVM algorithm unfolds its general operational procedure as detailed below:

(I) First select the kernel function: $K(x_i, x_j)$.

(II) In the feature space, we construct the following optimization problem to find the optimal separation hyperplane:

$$\min_\alpha \frac{1}{2}\sum_{i=1}^{N}\sum_{j=1}^{N}\alpha_i\alpha_j y_i y_j K(x_i, x_j) - \sum_{i=1}^{N}\alpha_i. \tag{1}$$

(III) Constraints:

$$\begin{cases} \sum\limits_{i=1}^{N}\alpha_i y_i = 0 \\ 0 \le \alpha_i \le C,\ \forall i = 1, 2, \ldots, N \end{cases}, \tag{2}$$

where α_i is the Lagrange multiplier, C is the regularization parameter, which is used to control the punishment for classification errors, and y_i is the label of the data point x_i (value $\pm$ 1).

(IV) For the purpose of resolving the above-described constrained optimization problem, the sequential minimal optimization (SMO) method, along with other convex optimization algorithms, is typically employed. After solving, we obtain the optimal Lagrange multiplier vector:

$$\alpha^* = (\alpha_1^*, \alpha_2^*, \ldots, \alpha_N^*)^T. \tag{3}$$

(V) Calculate the support vector in the decision function:

$$\sum_{i=1}^{N}\alpha_i^* y_i K(x, x_i). \tag{4}$$

(VI) Calculate the bias term and solve this equation. We can obtain a possible value of the bias term b^*, which is usually calculated by using the average value of all support vectors:

$$y_s\left(\sum_{i=1}^{N}\alpha_i^* y_i K(x_s, x_i) + b^*\right) = 1. \quad (5)$$

(VII) Finally, we build a decision function to classify the following:

$$f(x) = sign\left(\sum_{i=1}^{N} a_i^* y_i K(x, x_i) + b^*\right). \quad (6)$$

3.1.3. Multilayer Perceptron

MLP serves as the foundational architecture within the realm of deep learning, encompassing an input layer, one or more hidden layers, and an output layer. The input layer assumes the responsibility of capturing raw data information. The hidden layer is dedicated to conducting feature extraction and applying nonlinear mapping to the data, and the output layer is employed to yield the final prediction results [45]. The operations of linear transformation and activation function processing are performed on the data during forward propagation. Operations are performed step-by-step through the layers, leading to the data being conveyed to the output layer [46]. The loss function is employed to gauge the discrepancy between the anticipated outcomes and the true values in the dataset. Consequently, the choice of the loss function holds significant importance. Following this, the gradient is calculated through the backpropagation mechanism, and subsequently, the model's parameters are adjusted utilizing the gradient descent approach to attain model optimization. MLP possesses a powerful ability to represent complex relationships, making it appropriate for nonlinear issues and high-dimensional datasets. Nevertheless, it entails a substantial training time and has high requirements for data preprocessing.

Figure 2 shows the MLP with only one hidden layer, where $f(x)$ represents the activation function, which is a nonlinear function. We determine whether neurons should be activated by calculating the weighted sum and adding bias, mainly including Sigmoid, Tanh, and ReLu.

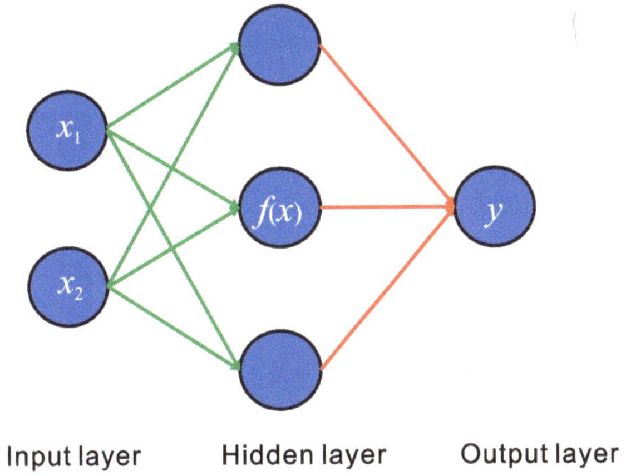

Figure 2. MLP structure diagram.

In a multilayer perceptron, every neural network layer comprises numerous neurons. The architectural details of an individual neuron are presented in Figure 3, which encompasses four components:

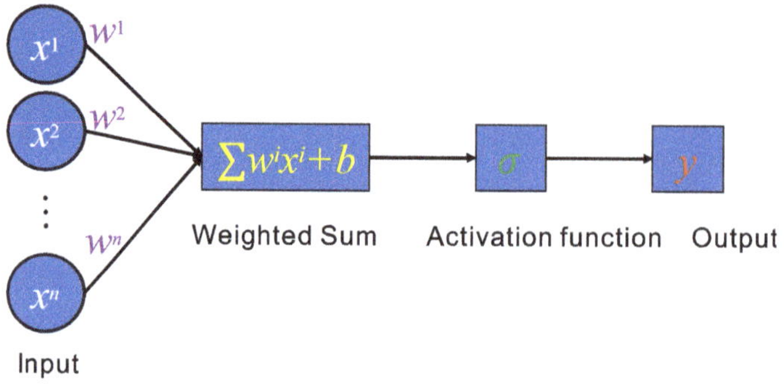

Figure 3. Structure of single neuron.

(I) Input: The neuron receives multiple input signals from the previous layer or external sources. These input signals are usually weighted values; that is, each input value is multiplied by a corresponding weight, and its vector expression is as follows:

$$X = [x^1, x^2, x^3, \ldots, x^n]^T. \tag{7}$$

(II) Weighted summation: $w_1, w_2, \ldots, w_n$ is the weight of neurons, and its vector expression is as follows:

$$W = [w^1, w^2, w^3, \ldots, w^n]. \tag{8}$$

The neuron multiplies all input signals by their corresponding weights and adds these products to obtain a weighted sum. However, in some cases, especially in neural networks, the formula of weighted summation may be extended to include a constant term b; that is, the bias term. The relevant formula is as follows:

$$z = \sum_{i=1}^{n} w^i x^i + b. \tag{9}$$

(III) Activation Function: A nonlinear activation function is employed to transform the weighted sum, thereby resulting in the output of the neuron.

(IV) Output: The outcomes derived from the processing of values by the nonlinear activation mechanism constitute the outputs produced by neurons, and its value is as follows:

$$y = \sigma(z) = \sigma(WX + b). \tag{10}$$

3.1.4. Logistic Regression

In the statistical learning methodology system, LR plays a crucial role and has been widely applied in various classification tasks, especially in the field of binary classification, demonstrating excellent applicability [47]. Grounded in linear models, LR employs the Sigmoid function to project input features onto a probability scale spanning from 0 to 1. This allows for a quantitative assessment of the probability that samples fall into positive classes. In particular, the Sigmoid function is employed to map the output generated by a linear model onto a probabilistic scale, and the decision function sets a threshold (such as 0.5) based on this probability value to achieve sample classification judgment. In LR, the log-likelihood loss serves to assess the deviation between the forecasted outcomes and the actual labels, thereby enabling the evaluation of prediction accuracy. When it comes to the pursuit of optimal parameters for LR, gradient descent stands as a broadly utilized optimization algorithm. It computes the gradient of the loss function relative to the parameters and iteratively modifies the parameters in the direction opposite to the gradient, thus progressively attaining the minimization of the loss function.

LR is constructed upon a linear modeling framework. Within this framework, the model's predictions of the output are achieved through a linear combination of input features. To be more specific, linear models can be characterized by the following formula:

$$z = \beta_0 + \beta_1 x_1 + \beta_2 x_2 + \cdots + \beta_n x_n, \tag{11}$$

where β_0 is the intercept term, $\beta_1, \beta_2, \ldots, \beta_n$ represents the coefficient associated with the feature, and $x_1, x_2, \ldots, x_n$ is the input feature.

The Sigmoid function converts the output z of the linear model into the probability value p, and the formula is as follows:

$$p = \frac{1}{1 + e^{-z}}, \tag{12}$$

where the probability value p represents the possibility that the sample belongs to a positive class. LR uses a decision function to convert the probability value p into a classification result. Generally, when $p \geq 0.5$, the samples are classified as positive classes; when $p < 0.5$, the sample are classified as negative. This threshold (0.5) can be adjusted according to the actual demand.

In the LR model system, logarithmic likelihood loss is used to accurately measure the degree of difference between predicted probabilities and true labels. For the common and important application scenario of binary classification, logarithmic likelihood loss has a specific mathematical expression, and its formula is presented as follows:

$$L = -\frac{1}{m} \sum_{i=1}^{m} [y_i \log(p_i) + (1 - y_i) \log(1 - p_i)], \tag{13}$$

where m denotes the quantity of samples, y_i represents the actual label (either 0 or 1) of the i-th sample, and p_i signifies the predicted probability for the i-th sample.

3.1.5. Random Forest

RF is composed of multiple decision trees, which is an integrated learning method. The calculation process is a complex and delicate process, which combines the prediction results of multiple decision trees to improve the accuracy and stability of the model.

RF uses the self-help sampling method to randomly extract multiple sample book subsets $D_1, D_2, \ldots, D_n$ from the original training dataset D by means of put-back sampling, where n is the number of decision trees. Each sample subset D_i ($i = 1, 2, \ldots, n$) contains the same number of samples as the original dataset, but the samples in D_i may be duplicated due to the put-back sampling. Each sample subset D_i is used to train a decision tree T_i independently. In the process of constructing each decision tree T_i, the RF algorithm will not use all features to split nodes. Let all available features be set to F, and the RF algorithm will randomly select a subset of features $F_i \subseteq F$ from F. Then select the best feature to split nodes. This random feature selection method further increases the diversity of the model, which helps to reduce overfitting and improves the generalization ability of the model. When the new sample x is predicted, each decision tree T_i gives the prediction result y_i.

For the classification problem, the majority voting method is used to determine the final category y. If C is the set of category labels, the final category y can be determined by the following formula:

$$y = \underset{c \in C}{\operatorname{argmax}} \sum_{i=1}^{n} I(y_i = c), \tag{14}$$

where $I(\cdot)$ is the indicating function, which is 1 when $y_i = c$; otherwise, it is 0.

The RF algorithm has high accuracy and robustness and reduces error by integrating the prediction results of multiple decision trees. It has a strong anti-overfitting ability and uses self-help sampling method and random feature selection to increase the diversity of the models. It can process high-dimensional data and automatically select important features.

3.1.6. Extreme Gradient Boosting

XGBoost is an efficient ensemble learning algorithm that is based on the gradient lifting decision tree framework and gradually optimizes the prediction results by iteratively training multiple decision trees. Each new tree will focus on correcting the prediction error of the previous tree and finally the weighted sum of the prediction results of all trees to obtain a more accurate final prediction. In order to prevent the model from overfitting, XGBoost introduces regularization technology to control the complexity of the model. At the same time, it also supports parallel computing, which can significantly accelerate the training process. In addition, XGBoost can automatically process the missing values in the data without the need for cumbersome manual preprocessing. Because of its excellent performance and ease of use, XGBoost has been widely used in ML tasks such as classification and regression, especially for processing structured data and high-dimensional sparse data.

3.2. *Performance Evaluation of Machine Learning Model*

For this research endeavor, in order to obtain a comprehensive and accurate evaluation of the classification model's performance, we adopted multiple evaluation indicators, including accuracy, precision, recall, the harmonic mean of precision and recall (F1 score), the receiver operating characteristic (ROC) curve, and the confusion matrix. These evaluation indexes play an important role in quantifying model accuracy, evaluating model robustness, and identifying model problems.

Accuracy, serving as a pivotal metric for assessing the overall efficacy of a model, holds significant importance in classification tasks. It offers a straightforward reflection of the model's precision in forecasting sample categories, namely the proportion of samples correctly predicted by the model (including samples correctly identified as hydrate and non-hydrate) in the total samples:

$$\text{Accuracy} = \frac{\text{TP} + \text{TN}}{\text{TP} + \text{TN} + \text{FP} + \text{FN}}. \quad (15)$$

In the realm of model evaluation, precision signifies the fraction of samples genuinely belonging to the positive category within the set of model-predicted positive samples; that is, the proportion of samples predicted to be hydrates that are actually hydrates. Its essential significance resides in its capacity to significantly mitigate the risk of false positives in the model, subsequently enhancing the reliability of the prediction outcomes:

$$\text{Precision} = \frac{\text{TP}}{\text{TP} + \text{FP}}. \quad (16)$$

Recall rate measures the coverage rate of positive samples; that is, the proportion correctly predicted by the model in the samples that are actually hydrate to identify more positive instances:

$$\text{Recall} = \frac{\text{TP}}{\text{TP} + \text{FN}}. \quad (17)$$

As a holistic evaluation indicator, the F1 score ingeniously amalgamates the performance of precision and recall. It is not a simple arithmetic mean but a harmonic average calculation that takes these two factors into account. When the F1 score is high, it indicates that the model has a well-balanced and excellent performance regarding both accuracy and recall, showcasing its ability to accurately discern and fully cover classification tasks:

$$F1 = 2 \times \frac{\frac{TP}{TP+FP} \times \frac{TP}{TP+FN}}{\frac{TP}{TP+FP} + \frac{TP}{TP+FN}} = \frac{2 \times TP}{2 \times TP + FP + FN}. \tag{18}$$

The true positive rate (TPR), which is also referred to as sensitivity or recall within the realm of model evaluation, is employed to ascertain the ratio of samples that a model accurately predicts as positive out of all the samples that are truly positive. It serves as a significant metric for gauging the model's capability to recognize positive samples:

$$TPR = \frac{TP}{TP + FN} = \text{Recall}. \tag{19}$$

The false positive rate is the percentage of samples that are really negative but are wrongly predicted as positive by the model:

$$FPR = \frac{FP}{FP + TN}. \tag{20}$$

In the field of binary classification model evaluation, the ROC curve and the area under curve (AUC) are indispensable tools. They offer an intuitive way to reflect the model's classification performance under different threshold conditions, which is of great practical significance for optimizing model performance and making informed decisions. The ROC curve is a crucial analytical tool. This curve takes the true case rate as the vertical axis and the false positive case rate as the horizontal axis. By depicting the dynamic relationship between the two under different classification thresholds, it intuitively presents the changing trend of model performance with the adjustment of classification thresholds. This feature provides a solid theoretical foundation and analytical basis for comprehensively evaluating model performance, effectively screening key features, and reasonably evaluating model performance on imbalanced datasets. AUC, as the core quantitative indicator for measuring the classification performance of the model represented by the ROC curve, is defined as the area enclosed by the ROC curve and the coordinate axis. The range of AUC values is strictly defined between 0 and 1. When the AUC value approaches 1, it indicates that the model has a stronger ability to distinguish positive and negative samples; that is, the classification performance of the model is better. On the contrary, if the AUC value is closer to 0, it means that the classification performance of the model is worse. They are often used together to comprehensively evaluate the performance of classifiers.

The AUC, in mathematical terms, is defined as the area encompassed beneath the ROC curve. The corresponding calculation formula is as follows:

$$AUC = \int_0^1 (1 - FPR(t))dt. \tag{21}$$

The confusion matrix serves as an analytical tool within the ML sphere for gauging the performance of classification models and visually representing the classification outcomes. A differentiation exists between the rows and columns within the confusion matrix. Typically, the columns of the confusion matrix correspond to predicted instances, whereas the rows denote real instances. Data points where the number of rows and columns are equal signify the count of correct predictions, while those with unequal row and column counts indicate the number of incorrect predictions. In the confusion matrix, we can not only see the overall accuracy of the prediction but also obtain the specific error point of the prediction error.

4. Results

4.1. Data

The data of this study are mainly from the logging while drilling (LWD) data of IODP Expedition 311 during the inspection using Schlumberger measurement tools. LWD tools provide conventional logging curves and special logging data. Among them, conventional logging includes Density Caliper, Average (DCAV), Gamma Ray (GR), Bulk Density (RHOB), Best Thermal Neutron Porosity, Average (BPHI), Deep Resistivity Average (BDAV), Medium resistivity, shallow resistivity, and P-wave velocity (Vp), and the sampling interval of these logging curves is 0.1524 m. Special logging includes LWD image log-Resistivity At Bit (RAB) images, and the sampling interval is 0.0305 m. As the hydrate reservoir in borehole U1329A is not clearly defined, in order not to affect the research results, the data of U1329A are not used in this study. The logging data of the other four wells and the hydrate reservoir location are shown in Table 2. According to statistics, in borehole U1325A, the interval where natural gas hydrate exists is within the depth range of 195–230 mbsf [48–51]. For borehole U1326A, scientific analysis reveals the existence of gas hydrate within the depth intervals spanning from 73 to 94 mbsf and from 252 to 261 mbsf [49,52]. In borehole U1327A, the interval containing gas hydrate is in the depth range of 120–138 mbsf [39,49,53]. In borehole U1328A, gas hydrate is distributed in the depth intervals of 20–35 mbsf and 215–222 mbsf [39,48,51,54]. We do not use sonic velocity logging data because sonic velocity logging curves are missing at some depths. Pursuant to the correlation analysis delineated in Figure 4, among the conventional logging data in the study area, five types of logging parameters, namely DCAV, GR, BPHI, RHOB, and BDAV, are chosen due to their strong correlation and high sensitivity to gas hydrate. Therefore, in this study, these five logging curves from four boreholes (U1325A, U1326A, U1327A, and U1328A) are selected as the input characteristic parameters, as shown in Figure 5.

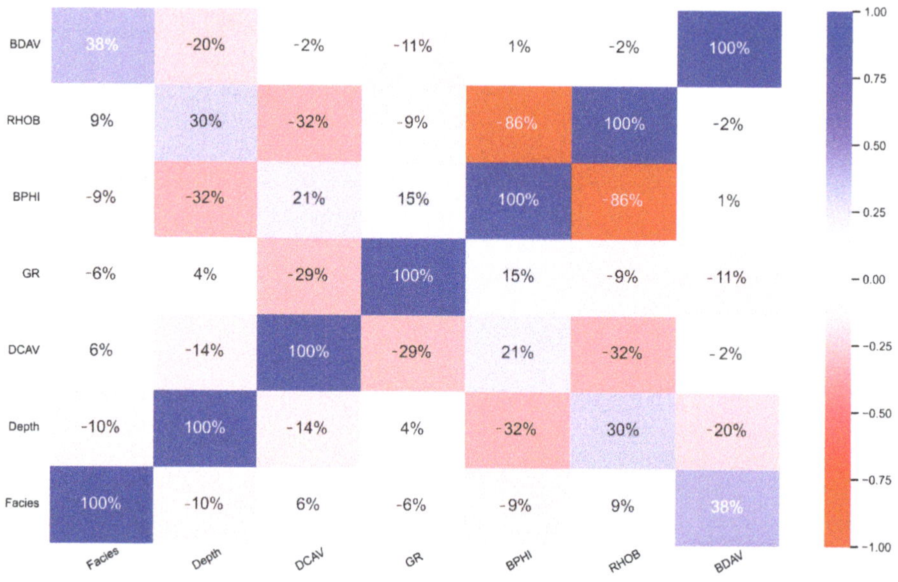

Figure 4. Correlation heatmap of seven parameters: a comprehensive visual analysis.

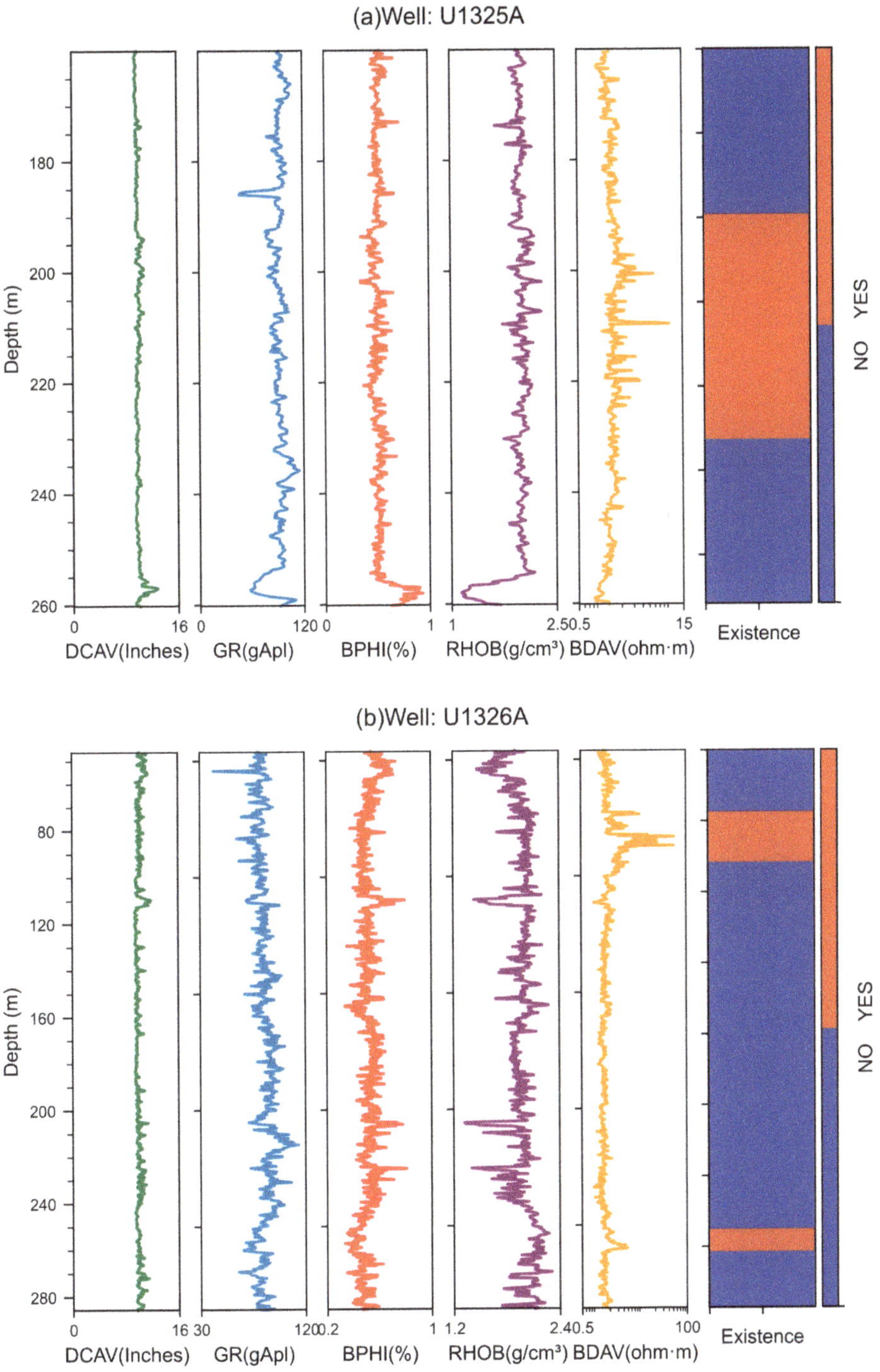

Figure 5. *Cont.*

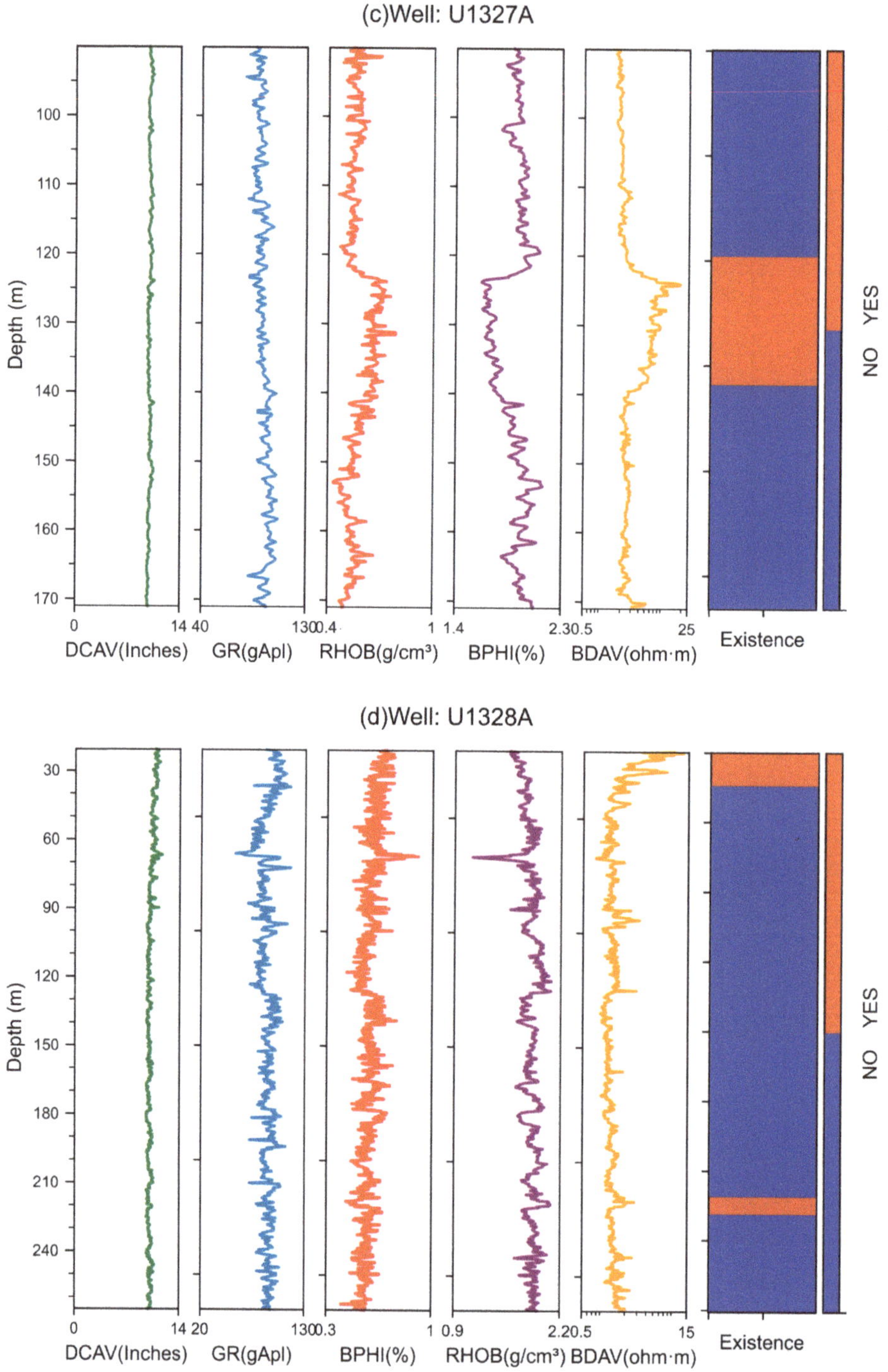

Figure 5. Logging curves of four boreholes, including DCAV, GR, BPHI, RHOB, and BDAV. Intervals containing natural gas hydrate are marked with red (YES), and intervals without natural gas hydrate are marked with blue (NO). (**a**) U1325A, (**b**) U1326A, (**c**) U1327A, and (**d**) U1328A.

Table 2. Logging data information and hydrate reservoir depth range [39,50,52–54].

Well ID	Logging Curves	Sampling Interval (m)	Gas Hydrate Depth Range (mbsf)
U1325A	DCAV, GR, RHOB, BPHI, BDAV, Vp	0.1524	195–230
	RAB	0.0305	
U1326A	DCAV, GR, RHOB, BPHI, BDAV, Vp	0.1524	73–94, 252–261
	RAB	0.0305	
U1327A	DCAV, GR, RHOB, BPHI, BDAV, Vp	0.1524	120–138
	RAB	0.0305	
U1328A	DCAV, GR, RHOB, BPHI, BDAV, Vp	0.1524	20–35, 215–222
	RAB	0.0305	

Analysis of Log Curve Cross Plot

We selected five logging data corresponding to hydrate-bearing and hydrate-free intervals in the four boreholes and drew the transaction diagram. The results are shown in Figure 6. Among them, the blue dot represents no hydrate, and the red dot represents hydrate. It can be seen from Figure 6 that the cross plot is composed of four curves of well diameter, density, natural gamma, and porosity. The two types of data points are mixed, and the points containing hydrate and those without hydrate can hardly be distinguished. However, when the resistivity curves interact with the other four kinds of logging curves, most hydrate-bearing intervals and hydrate-free intervals are well distinguished. Nevertheless, some of the two types of data points are mixed together, which makes it difficult to distinguish quickly and accurately. To accurately differentiate between intervals containing gas hydrates and those without, this research employs a series of ML algorithms geared toward the identification of GHRs.

4.2. Data Preprocessing, Model Training, and Optimization

The data need to be preprocessed before model training. We extracted the logging data of the interval needed in this study and labeled the extracted data. Among these intervals, the one with natural gas is assigned a value of 1, while the one without is given a value of 0. Finally, a total of 3753 data were selected from the data of four wells, including 743 data containing gas hydrate and 3010 data without gas hydrate, as illustrated in Figure 7. In order to improve the fairness of the model, we use the synthetic minority oversampling technique (SMOTE) to increase the number of small samples. In the XGBoost model, we increase the minority sample to the same level as the majority sample. In the other five models, we increase the minority sample to 0.3 times the majority sample for experiments. With the aim of optimizing the model's training efficiency and overall performance, scaling the data to a standardized range and carrying out standardization are indispensable procedures.

The standardized formula, commonly referred to as the Z-score standardization formula, serves to transform the original data into a standard normal distribution, where the mean equals 0 and the standard deviation is 1. The formula is presented below:

$$Z = \frac{X - \mu}{\sigma}, \tag{22}$$

where Z is the standardized value, and it indicates the degree to which data points deviate from the mean, X corresponds to the raw value of each individual data point, μ is the arithmetic mean of the entire dataset, and σ is the measure of dispersion, representing how much the data points deviate from the mean within the dataset.

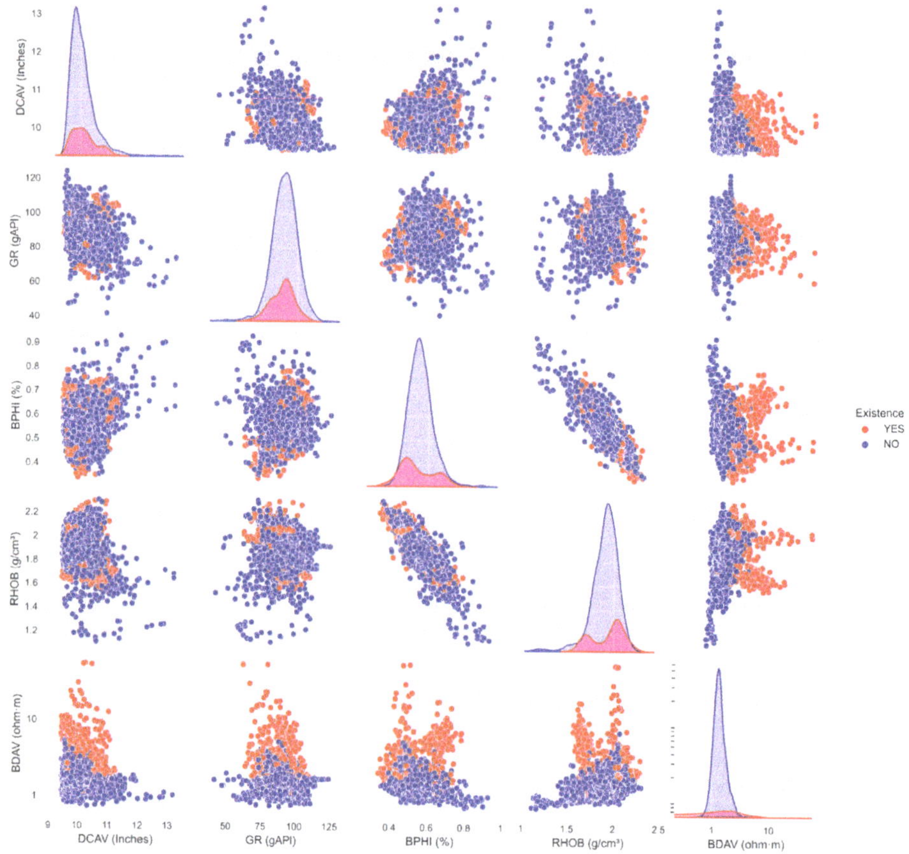

Figure 6. Cross plot of five logging curves.

Upon finishing data preprocessing, the dataset samples are arbitrarily split into a training set and a test set in a 3:1 ratio. We utilize the training set to support model learning, enabling the model to identify the characteristics and hidden rules within the data. In contrast, the test set is used to assess how well the trained model functions, testing its generalization capability and its accuracy. The categorization of dataset samples could potentially result in bias during model performance assessment, and cross-validation is thus adopted as a more reliable evaluation approach. This approach trains and assesses the model through the process of splitting the dataset into several subsets. Each subset is used in rotation as a validation set, while the others are employed as training sets. By adopting this approach, every subset has a chance to act as a validation set, enabling a more all-encompassing assessment of the model's capabilities across various data subsets. In the pursuit of model performance optimization, hyperparameter optimization is an indispensable link [55]. Grid search functions as a method for optimizing hyperparameters. It can systematically traverse the combination of multiple parameters and determine the best effect parameters through cross-validation. Its general principle is to first specify a value range for each hyperparameter. Then, the grid search algorithm will traverse all the combinations in the grid and use cross-validation for each group of parameters to evaluate the effect. Finally, the grid search will output the best combination of parameters based on the results of cross-validation; that is, those parameters that perform best in all cross-validation rounds. For this investigation, the grid search strategy is employed to optimize the hyperparameters of the four classifiers, and the 5-fold cross-validation approach is adopted to assess the model's performance across every parameter configuration. This strategy not only enhances the efficiency while elevating the accuracy of model optimization but also offers us a more comprehensive and robust model evaluation framework, laying a solid foundation for subsequent research and application.

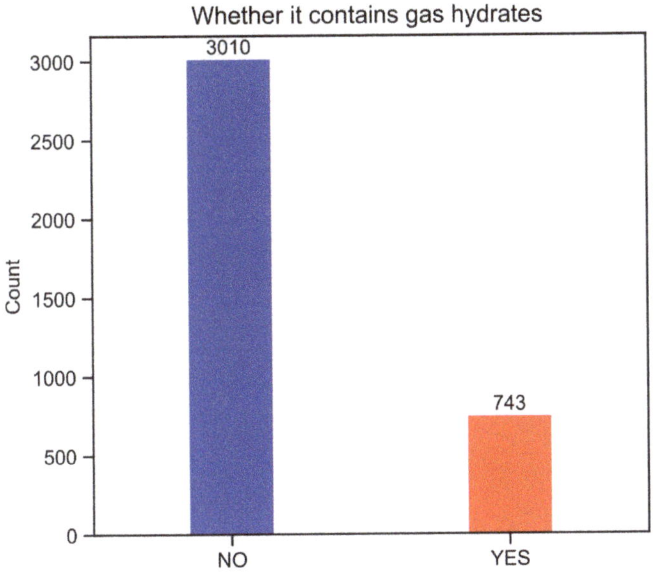

Figure 7. Histogram of sample size distribution of logging dataset. NO represents the sample data without hydrate, and YES represents the data with hydrate.

4.3. Identification Results

In this study, we use six different models to identify marine GHRs. The results of each model are obtained from the fine-tuning optimal model using the grid search method. The best parameter selection is shown in Table 3. Table 4 presents the numerical values of the determined evaluation indicators. Figure 8 displays the ROC curves along with the corresponding AUC values for the six models.

Table 3. Optimal hyperparameter selection of six models.

Model	Hyperparameter (Symbol)	Parameter Value
GPC	Default setting	Default setting
SVM	Kernel	RBF
	Penalty coefficient (c)	10
	Gamma	1
MLP	Regularization parameters (alpha)	0.0001
	Maximum number of iterations (max iter)	100
	Initialization learning rate (learning rate init)	0.01
	Optimization algorithms (solver)	Adam
	Batch size (batch size)	100
	Hidden layer configuration (hidden layer sizes)	100

Table 3. *Cont.*

Model	Hyperparameter (Symbol)	Parameter Value
RF	Random selection of features (max features)	3
	Minimum sample size for node splitting (min samples split)	6
	The maximum depth of a tree (max depth)	15
	The number of trees (n estimators)	30
XGBoost	Learning rate	0.1
	The number of trees (n estimators)	200
	The maximum depth of a tree (max depth)	7
	Minimum sample size for node splitting (min child weight)	1
	Gamma	0.1
	Subsample	0.6
	Colsample bytree	0.8
LR	Default setting	Default setting

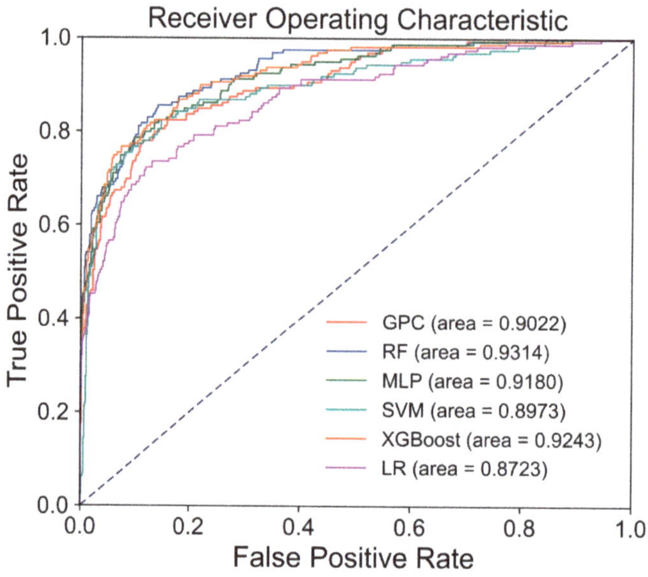

Figure 8. ROC curves and AUC values of six models.

Table 4. Accuracy, precision, recall, and F1 score of six models.

Model Type	Accuracy	Precision	Recall	F1 Score
SVM	0.9105	0.9062	0.9105	0.9070
GPC	0.8988	0.8925	0.8988	0.8926
MLP	0.9095	0.9052	0.9095	0.9060
RF	0.9191	0.9154	0.9191	0.9153
XGBoost	0.8860	0.6340	0.7736	0.6969
LR	0.8871	0.8791	0.8871	0.8746

4.3.1. GPC Identification Results

We use logging data as input characteristics to construct a Gaussian process to simulate the potential distribution of data based on the GPC algorithm and then classify and predict GHRs. During the construction of the GPC and recognition model, the kernel sets the initial values of a set of hyperparameters to start the first optimization process. In the subsequent iterative optimization, the value of the hyperparameter will be randomly selected from the preset reasonable value range. As shown in Table 3, the default parameter values of the Gaussian classification algorithm in scikit learn are used in this study.

Within the test set, the GPC recognition accuracy stands at 0.8988, fully attesting to its effectiveness in recognizing natural gas hydrates. The excellent performance of GPC with respect to this index indicates its capability to precisely detect the presence of natural gas hydrate based on logging data. In terms of precision, GPC also performed well, reaching 0.8925. In the task of natural gas hydrate identification, positive samples usually refer to the logging data containing natural gas hydrate. The high performance of GPC in terms of precision shows that it can effectively identify these positive samples, thus reducing the possibility of misjudgment. At the same time, its recall rate is as high as 0.8988, which means that the algorithm can identify the vast majority of actual positive samples in the test set. This result further verifies the stability and reliability of the GPC algorithm in the natural gas hydrate recognition task. The F1 score of the algorithm is 0.8926, which is between the accuracy and recall metrics. This shows that the algorithm can balance these two indicators well. In addition, the AUC value (0.9022), a key metric for evaluating the model's classification capability, also underscores the GPC model's excellent performance in differentiating gas hydrate and non-hydrate samples. Therefore, it can be inferred that the algorithm performs well in the task of natural gas hydrate recognition.

4.3.2. SVM Identification Results

SVM separates different types of samples by constructing the optimal hyperplane. SVM achieves high-precision classification prediction on logging data. During the construction of an SVM model, the choice of parameters substantially influences the model's performance. To precisely ascertain the optimal parameters, this research employed the grid search approach integrated with 5-fold cross-validation. Finally, the parameter value that shows the best classification accuracy on the test set is determined as $c = 10$, $\gamma = 1$, and the kernel function is RBF type.

Following the optimization of model parameters and their deployment on the testing dataset, a set of crucial evaluation metrics is acquired. In the evaluation, the model yields an accuracy of 0.9105, a precision of 0.9062, a recall rate of 0.9105, and an F1 score of 0.9070. Concurrently, the area under the ROC curve of the model registers at 0.8973, with the ROC curve depicted in Figure 8. In light of the preceding metrics, we ascertain that the SVM model demonstrates robust performance in discerning natural gas hydrate, boasting high accuracy, precision, recall rate, and F1 score. At the same time, the AUC value additionally indicates that the model exhibits competent classification capabilities. This implies that the model exhibits high reliability and practicability in practical application, thereby offering robust assistance in the detection and identification of natural gas hydrate.

4.3.3. MLP Identification Results

MLP learns the nonlinear characteristics of data by constructing multiple hidden layers, and the MLP model achieves high-precision classification and prediction on logging data. In this model, we optimize the regularization parameters, maximum number of iterations, initialization learning rate, optimization algorithm, batch size, and hidden layer size. As indicated in Table 3, via the grid search approach, the regularization parameter is configured at 0.0001, while the model's maximum iteration count is 100 and the initial learning rate is set to 0.01. Opt for the 'Adam' optimization algorithm, with each training iteration utilizing 100 samples, and set the number of neurons in the hidden layer to 100. Upon deploying the meticulously tuned model on the test dataset, a comprehensive set of evaluation metrics pertaining to the model's performance is derived. The accuracy, precision, recall rate, and F1 score are 0.9095, 0.9052, 0.9095, and 0.9060, respectively. The AUC value of this model reaches 0.9180, and the ROC curve is shown in Figure 8. According to the comprehensive analysis of the above evaluation indexes, the MLP model shows excellent classification performance in the identification of natural gas hydrate. It has high-precision prediction ability, stable evaluation index, strong feature learning ability, and performs well in dealing with complex nonlinear relationships and data feature extraction.

4.3.4. LR Identification Results

As a classical linear classification algorithm, LR has also been applied to the task of gas hydrate recognition. Although its performance may not be as good as that of nonlinear models (such as GPC, SVM, and MLP), the LR model can still play a certain role in some specific cases. Following the implementation of grid search coupled with cross-validation, the parameters adopted by LR are determined to be the default parameters in the scikit learn library. After being applied to the test data, the evaluation index of the model is obtained. The model achieved an accuracy of 0.8871, a precision of 0.8791, a recall rate of 0.8871, and an F1 score of 0.8746. Additionally, the AUC metric for the model was recorded as 0.8723, with the corresponding ROC curve depicted in Figure 8.

Among the six models, the reason for the poor performance of LR may be that the algorithm has some limitations in dealing with nonlinear features, which may affect its performance in the task of natural gas hydrate recognition. It also indicates that the model exhibits relatively poor performance in differentiating between gas hydrate specimens and their non-hydrate counterparts. In summary, the LR model shows good performance in the task of gas hydrate recognition. In some specific cases, such as linear separable data, the LR model is still a choice worth considering.

4.3.5. RF Identification Results

RF builds a strong classification model by integrating multiple decision trees and using a bagging strategy. The RF model shows excellent classification performance in the task of gas hydrate recognition. In order to determine the optimal parameters, the grid search method and 5-fold cross-validation are used to find the optimal parameter combination in the process of establishing the RF model. As shown in Table 3, we set the random selection of features to 3, the minimum sample size for node splitting to 6, the maximum depth of the tree to 15, and the number of trees to 30. Finally, we apply the set parameters to the model and obtain a series of evaluation indexes of the model. The accuracy, precision, recall rate, and F1 score were 0.9191, 0.9154, 0.9191, and 0.9153, respectively, and the AUC value of the model reached 0.9314. The ROC curve is shown in Figure 7. Based on the above indicators, it can be clearly concluded that among the six comparison models, the indicators of the RF model rank first, especially in the identification of natural gas hydrate. The RF model has high accuracy, precision, recall rate, F1 score, and AUC value. This means that the RF model, with its excellent classification performance and generalization ability, shows significant application value in the actual detection scenario of natural gas hydrate and provides strong technical support for the efficient identification of potential reservoirs and optimization of exploration decisions.

4.3.6. XGBoost Identification Results

XGBoost, as an integrated learning algorithm based on a gradient lifting decision tree, shows certain feature learning ability in natural gas hydrate recognition task by iteratively optimizing residuals and introducing regularization items to prevent overfitting. Parameter settings are shown in Figure 3. However, according to the evaluation results of the current model, its comprehensive performance does not meet the expected application standards.

In the test set, the accuracy, precision, recall rate, F1 score, and AUC value are 0.8860, 0.6340, 0.7736, 0.6969, and 0.9243, respectively.

Although the AUC value (0.9243) indicates that the model has a certain classification ability, the following key indicators reveal its limitations: (1) The accuracy is insufficient. The precision (0.6340) is significantly lower than that of similar models (such as 0.9108 of SVM), indicating that the actual positive samples accounted for only 63.4% of the samples predicted as "natural gas hydrate" by the model, and the risk of false positives is high, which may increase the exploration cost. (2) The F1 score is low. The F1 value (0.6969) comprehensively reflects the balance between accuracy and recall. The index below 0.7 indicates that the model has obvious shortcomings in distinguishing positive and negative samples, which cannot meet the needs of high-precision detection.

Comprehensive evaluation conclusion: The current XGBoost model performs poorly in the task of natural gas hydrate recognition, and its lack of precision, low F1 score, and low feature overlap may lead to a high misjudgment rate in practical application. Combined with the excellent performance of SVM and other comparative models, we recommend carefully evaluating the applicability of XGBoost in this scenario and giving priority to other algorithms or further optimizing the performance through feature engineering.

4.4. SHapley Additive exPlanations Value Analysis of Six Models

SHapley Additive exPlanations (SHAP) analysis reveals differences in feature contribution patterns and model behavior of different ML models in natural gas hydrate recognition tasks. The SHAP summary diagram of the six models of SVM, RF, LR, GPC, MLP, and XGBoost is shown in Figure 9.

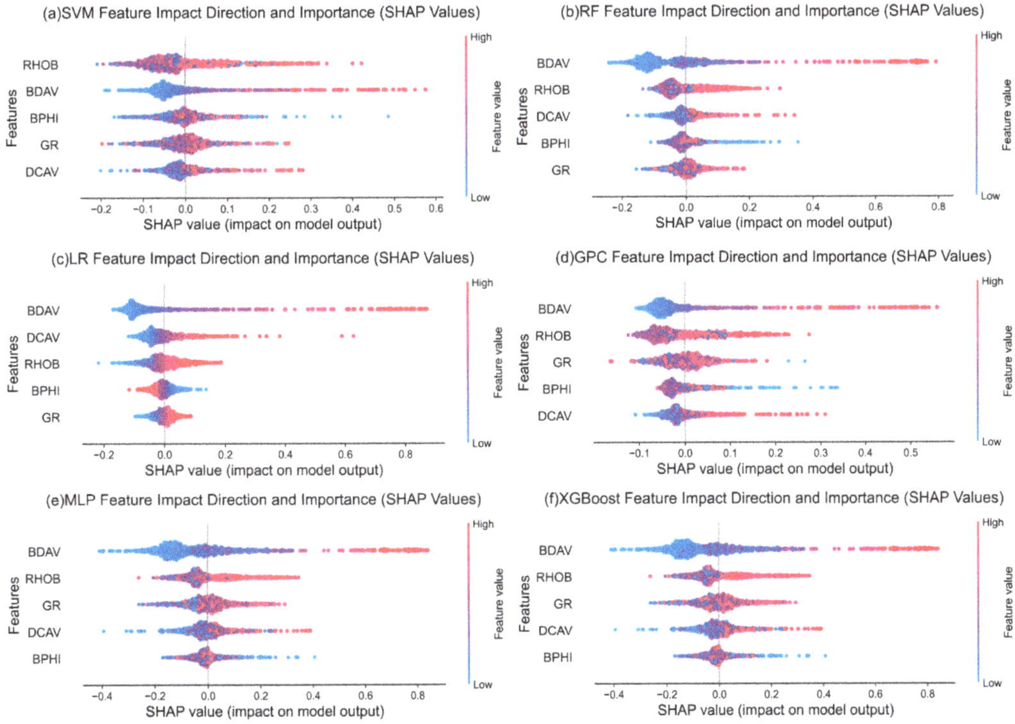

Figure 9. SHAP value plots for six models: (**a**) SVM model, (**b**) RF model, (**c**) LR model, (**d**) GPC model, (**e**) MLP model, and (**f**) XGBoost model.

As shown in Figure 9, RHOB and BDAV are key features for identifying natural gas hydrates in the SVM model, which have significant impacts on the model output and have diverse directions of influence. The RF model shows that BDAV and RHOB are much more important than other features, and these two features may have a positive or negative impact on the model output in different situations. In the LR model, BDAV dominates and mainly exhibits a positive impact, while GR mainly exhibits a negative impact, reflecting the linear regression model's ability to capture the linear relationship between features and target variables. In the GPC model, BDAV and RHOB are equally important, and all features have diverse directions of influence, demonstrating the flexibility of GPC models in handling complex nonlinear relationships. In the MLP model, the importance of BDAV and RHOB is prominent, and their influence directions are diverse, which reflects the strong capturing ability of the multilayer perceptron model as a neural network model for nonlinear relationships. Finally, in the XGBoost model, BDAV is a key predictive factor, mainly manifested as a positive impact, indicating that the XGBoost model highly relies on BDAV features when identifying natural gas hydrates.

In summary, through a simple analysis of the SHAP summary graphs of six models, we have revealed the key features, feature influence directions, and differences between models in the task of identifying natural gas hydrates for each model. These pieces of information provide an important basis for understanding model behavior, optimizing model performance, and selecting more suitable models for natural gas hydrate identification.

5. Discussion

In this study, six ML algorithms are applied to train the logging data from four wells of the IODP Expedition 311. Through data preprocessing and hyperparameter optimization of the model, the samples with and without gas hydrate layers are identified. A range of evaluation metrics, encompassing accuracy, precision, recall, the F1 score, the AUC value, and the ROC curve, are leveraged to assess the efficacy of distinct ML models in hydrate recognition tasks. Next, model performance, parameter optimization, model comparison, and research limitations will be discussed in depth based on the experimental results.

5.1. Comparison of Six Machine Learning Methods

This study demonstrates the various evaluation metrics (Table 4), ROC curves and AUC values (Figure 8), and the prediction confusion matrices (Figure 10) of each model for hydrate identification by the six ML methods. From the synthesis of these indicators, the six ML models have shown certain accuracy in the task of GHR identification. Among them, the indicators of SVM, MLP, and RF have reached more than 0.90, and the RF indicators are the highest, which is the best performance of the six models. This shows that for this binary classification problem, the RF model constructs a strong classification model by integrating multiple decision trees and using the Bagging strategy and realizes the optimal distinction between GHRs and non-reservoirs. Its accuracy, precision, recall rate, F1 score, and AUC value rank first among the six comparison models, thus achieving the effect of accurately and efficiently identifying GHRs. For the high score indicators presented by MLP and SVM, it also shows that the MLP model, as a kind of deep learning, can learn the complex mapping relationship between logging data and hydrate reservoir characteristics through the training of a multilayer neural network and can also play a good role in recognition. The SVM model can effectively distinguish GHRs from non-reservoirs by finding the optimal hyperplane, so as to achieve the effect of accurately identifying GHRs. For GPC and LR models, the evaluation indexes are close to 0.90, although they are not reached, and the overall recognition effect is good. The reason why the GPC effect is slightly inferior to the first two may be related to parameter settings or data characteristics. The reason why the

LR effect is slightly inferior may be that the LR model, as a linear classifier, is relatively weak in nonlinear problems, which has a certain impact on its ability to identify GHRs under complex geological conditions. For XGBoost, its overall performance is relatively weak compared with the other five models, which may be due to the model's insufficient ability to capture complex nonlinear relationships in logging data or its limited ability to distinguish GHR boundary characteristics.

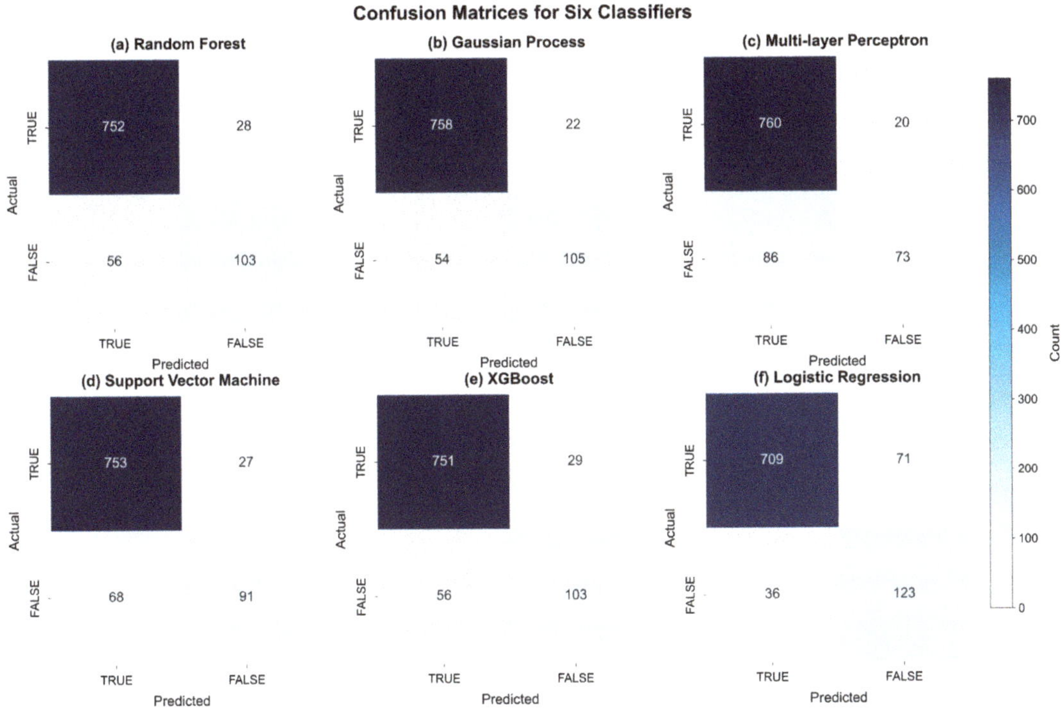

Figure 10. Prediction confusion matrix of six models: (**a**) RF model, (**b**) GPC model, (**c**) MLP model, (**d**) SVM model, (**e**) XGBoost model, and (**f**) LR model.

The AUC value is a key indicator for evaluating classifier performance. As presented in Figure 8, the AUC values of the six models are ordered as follows: RF > XGBoost > MLP > GPC > SVM > LR. Among them, the AUC value of RF is the highest, reaching 0.9314, and GPC is slightly lower than MLP, which is 0.9022. Although the AUC value of MLP is slightly higher than that of GPC, considering that the calculation of the AUC value has certain randomness and volatility, the difference between them is not significant. It can be considered that the performance of MLP and GPC models is equivalent. The AUC value of SVM is 0.8973, which is not as high as the first two, but still at a high level, which indicates that the model has a certain application value in the task of GHR identification. The AUC value of LR is 0.8723, which has reached an acceptable level, but there is still a certain gap compared with other models. Although other indicators of XGBoost are low, its AUC value is 0.9243. The high AUC of XGBoost indicates that it has the overall classification potential to be developed.

According to the analysis of various indicators, although the XGBoost model performs well in terms of its AUC value, its accuracy and F1 score are significantly lower than other models. Through further analysis of model characteristics and data characteristics, we found that the reason may be the mismatch between model complexity and data support.

Although the class distribution is balanced through smote, the limited amount of data (especially labeled samples) is still not enough to support XGBoost to fully learn complex nonlinear patterns. Unlike the linear constraint of the LR model, the main challenge of XGBoost is to avoid overfitting and optimize the interactive expression of features.

5.2. Advantages and Limitations of Machine Learning Methods

The traditional method of identifying GHRs using logging data is usually based on an empirical formula and the intersection diagram method. However, the recognition effect of these methods usually depends on the experience of logging interpretation personnel, and the recognition efficiency is low. It commonly poses a challenge to precisely define the nonlinear relationship between logging data and GHRs. This study is based on a data-driven method, using different ML algorithms to train the relationship between the input parameters (logging data) and the output results (whether there is a hydrate layer or not). Except for XGBboost, which needs to be used with caution, the accuracy of the other five models is higher than 0.88. The recognition effect is good, and the discrimination classification model constructed by them is not interfered with by logging interpretation personnel, with high efficiency.

This study leverages an extensive collection of well logging data along with labeled hydrate reservoir data to form the training samples. Given the paucity of samples, the model's training performance could be profoundly impacted, thereby leading to a reduction in its generalization capacity and prediction accuracy. Furthermore, due to the fact that the number of hydrate-bearing reservoirs is less than that of non-hydrate-bearing reservoirs in actual boreholes, the number of samples in these two categories is unbalanced, which may have a certain impact on the accuracy of ML.

5.3. Model Limitations and Transferability Across Geological Settings

The study area of this paper is located in the accretionary prism area of the Cascadia subduction zone. The geological conditions of the region (such as marine sedimentary environment and specific tectonic background) may limit the universality of the model. The limitations of the model are as follows: (1) The correlation between the response characteristics of logging curves and geological parameters may vary due to differences in sedimentary facies or lithological combinations. For example, in areas with active tectonic activity or permafrost on land, there are significant differences in the occurrence state of hydrates compared to the marine environment, leading to a decrease in the predictive ability of model input parameters. (2) Although ML models can capture nonlinear relationships, they may overlook the physical mechanisms of geological processes, leading to prediction biases in extreme geological conditions. (3) The weight or feature combination of logging curves may be optimal in the IODP Expedition 311 region, but the key parameters in other regions may be different. Consequently, the algorithm of the model developed in this research is predominantly suitable for regions with similar formation lithology as the study area. When dealing with regions that have considerable lithological differences relative to the study area, retraining the model and adjusting its parameters become necessary.

5.4. Ideas for Further Investigations

This study relies on the available logging data, whose quality and volume might impose constraints. As a recommendation for future studies, the multi-source data of geology, geophysics, geochemistry, and other disciplines can be combined to conduct a more comprehensive study on the reservoir identification method of natural gas hydrate. Moreover, the ML approach utilized in this research can be integrated with other cutting-edge techniques, including, but not limited to, deep learning and ensemble learning, to enhance the model's recognition precision and generalization capacity.

6. Conclusions

In this study, the logging information from four boreholes related to natural gas hydrates within the research zone is integrated into the training samples as feature inputs, and whether the formation contains gas hydrates is used as a label. Six ML algorithms are used to identify GHRs. The evaluation of the classification model's performance is carried out by a range of performance indicators, encompassing accuracy, precision, recall, the F1 score, the ROC, and the AUC. The following conclusions can be drawn:

(1) Except for the XGBoost model, the accuracy of the other five models is higher than 0.89, which shows that these models can effectively identify GHRs. The evaluation scores of these models are RF, SVM, MLP, GPC, LR, and XGBoost in descending order.

(2) Compared with the traditional intersection diagram method to identify GHRs, the ML method can independently establish a model to characterize the nonlinear relationship between logging data and GHRs through a data-driven method. The discriminant classification model constructed by the ML method can effectively reduce the impact of human intervention in the process of logging interpretation and significantly improve the efficiency of interpretation. Consequently, it introduces an innovative concept and approach for the identification of GHRs, which establishes a basis for the subsequent prospecting and exploitation of natural gas hydrate resources.

Author Contributions: Conceptualization, X.H. and C.Z.; methodology, X.H. and W.L.; software, W.L.; validation, K.X.; formal analysis, W.L.; investigation, W.L.; resources, K.X. and X.H.; data curation, X.H.; writing—original draft preparation, W.L.; writing—review and editing, X.H. and W.L.; visualization, G.S. and Y.W.; supervision, X.H. and C.Z.; project administration, X.H.; funding acquisition, X.H. All authors have read and agreed to the published version of the manuscript.

Funding: This research was funded by the National Natural Science Foundation of China (No. 42404150), Open Fund of Shaanxi Key Laboratory of Petroleum Accumulation Geology (No. PAG-202403), Key Laboratory of Gas Hydrate, Guangzhou Institute of Energy Conversion, Chinese Academy of Sciences (No. E229kf18), Open Research Project of the Hubei Key Laboratory of Intelligent Geo-Information Processing (No. KLIGIP-2023-A03), the Academic and Technical Leader Training Program of Jiangxi Province (No. 20204BCJ23027), Science and Technology Research Project of Jiangxi Provincial Department of Education (No. GJJ2200747), and the Postgraduate Innovation Fund from the East China University of Technology (No. YC2025-S406 and No. YC2025-S408).

Data Availability Statement: The data underlying this article can be accessed at https://iodp.tamu.edu (accessed on 1 October 2024).

Conflicts of Interest: The authors declare no conflicts of interest.

Abbreviations

The following abbreviations are used in this manuscript:

AUC	Area under curve
BDAV	Deep Resistivity Average
BPHI	Best Thermal Neutron Porosity, Average
BSRs	Bottom-simulating reflectors
DCAV	Density Caliper, Average
F1 score	The harmonic mean of precision and recall
GHRs	Gas hydrate reservoirs
GPC	Gaussian process classification
GR	Gamma Ray
IODP	International Ocean Drilling Program
LR	Logistic regression
LWD	Logging While Drilling

Mbsf	Meters below seafloor
ML	Machine learning
MLP	Multilayer perceptron
MSE	Mean-square error
RAB	Resistivity At Bit
RF	Random forest
RHOB	Bulk Density
ROC	Receiver operating characteristic
SHAP	SHapley Additive exPlanations
SMOTE	Synthetic minority oversampling technique
SVM	Support vector machine
Vp	P-wave velocity
XGBoost	Extreme gradient boosting

References

1. Collett, T.S.; Lee, M.W.; Agena, W.F.; Miller, J.J.; Lewis, K.A.; Zyrianova, M.V.; Boswell, R.; Inks, T.L. Permafrost-associated natural gas hydrate occurrences on the Alaska North Slope. *Mar. Pet. Geol.* **2011**, *28*, 279–294. [CrossRef]
2. Sloan, E.D., Jr.; Koh, C.A. *Clathrate Hydrates of Natural Gases*, 3rd ed.; CRC Press: Boca Raton, FL, USA, 2007; p. 752. [CrossRef]
3. Waite, W.F.; Santamarina, J.C.; Cortes, D.D.; Dugan, B.; Espinoza, D.N.; Germaine, J.; Jang, J.; Jung, J.W.; Kneafsey, T.J.; Shin, H.; et al. Physical properties of hydrate-bearing sediments. *Rev. Geophys.* **2009**, *47*, RG4003. [CrossRef]
4. Gajanayake, S.M.; Gamage, R.P.; Li, X.S.; Huppert, H. Natural gas hydrates–Insights into a paradigm-shifting energy resource. *Energy Rev.* **2023**, *2*, 100013. [CrossRef]
5. Hu, X.D.; Zou, C.C.; Lu, Z.Q.; Yu, C.Q.; Peng, C.; Li, W.; Tang, Y.Y.; Liu, A.Q.; Kouamelan, K.S. Evaluation of gas hydrate saturation by effective medium theory in shaly sands: A case study from the Qilian Mountain permafrost, China. *J. Geophys. Eng.* **2019**, *16*, 215–228. [CrossRef]
6. Hu, X.D.; Zou, C.C.; Qin, Z.; Yuan, H.; Song, G.; Xiao, K. Numerical simulation of resistivity and saturation estimation of pore-type gas hydrate reservoirs in the permafrost region of the Qilian Mountains. *J. Geophys. Eng.* **2024**, *21*, 599–613. [CrossRef]
7. Kianoush, P.; Mesgari, F.; Jamshidi, E.; Gomar, M.; Kadkhodaie, A.; Varkouhi, S. Investigating the effect of hole size, bottom hole temperature, and composition on cement bonding quality of exploratory wells in Iran. *Sci. Rep.* **2024**, *14*, 29653. [CrossRef]
8. Wei, N.; Pei, J.; Li, H.; Sun, W.; Xue, J. Application of in-situ heat generation plugging removal agents in removing gas hydrate: A numerical study. *Fuel* **2022**, *323*, 124397. [CrossRef]
9. Xing, L.; Gao, L.; Ma, Z.; Lao, L.; Wei, W.; Han, W.; Ge, X. A permittivity-conductivity joint model for hydrate saturation quantification in clayey sediments based on measurements of time domain reflectometry. *Geoenergy Sci. Eng.* **2024**, *237*, 212798. [CrossRef]
10. Zhu, X.; Liu, T.; Ma, S.; Liu, X.; Li, A. Morphology identification of gas hydrate based on a machine learning method and its applications on saturation estimation. *Geophys. J. Int.* **2023**, *234*, 1307–1325. [CrossRef]
11. Wu, S.Y.; Liu, J.; Xu, H.N.; Liu, C.L.; Ning, F.L.; Chu, H.X.; Wu, H.R.; Wang, K. Application of frequency division inversion in the prediction of heterogeneous natural gas hydrates reservoirs in the Shenhu Area, South China Sea. *China Geol.* **2022**, *5*, 251–266. [CrossRef]
12. Wang, Y.; Wang, Y.F. Quantitative evaluation of gas hydrate reservoir by AVO attributes analysis based on the Brekhovskikh equation. *Pet. Sci.* **2023**, *20*, 2045–2059. [CrossRef]
13. Zhan, L.; Matsushima, J.; Liu, H.; Lu, H. Evaluation and modeling of velocity dispersion and frequency-dependent attenuation in gas hydrate-bearing sediments. *Mar. Pet. Geol.* **2025**, *171*, 107204. [CrossRef]
14. Yamamoto, K. Overview and introduction: Pressure core-sampling and analyses in the 2012–2013 MH21 offshore test of gas production from methane hydrates in the eastern Nankai Trough. *Mar. Pet. Geol.* **2015**, *66*, 296–309. [CrossRef]
15. Lu, S.; McMechan, G.A. Estimation of gas hydrate and free gas saturation, concentration, and distribution from seismic data. *Geophysics* **2002**, *67*, 582–593. [CrossRef]
16. Liu, C.; Meng, Q.; He, X.; Li, C.; Ye, Y.; Zhang, G.; Liang, J. Characterization of natural gas hydrate recovered from Pearl River Mouth basin in South China Sea. *Mar. Pet. Geol.* **2015**, *61*, 14–21. [CrossRef]
17. Wang, X.; Collett, T.S.; Lee, M.W.; Yang, S.; Guo, Y.; Wu, S. Geological controls on the occurrence of gas hydrate from core, downhole log, and seismic data in the Shenhu area, South China Sea. *Mar. Geol.* **2014**, *357*, 272–292. [CrossRef]
18. Yuan, J.; Edwards, R.N. The assessment of marine gas hydrates through electrical remote sounding: Hydrate without a BSR? *Geophys. Res. Lett.* **2000**, *27*, 2397–2400. [CrossRef]

19. Tian, D.; Liu, X. A new approach for the identification of gas hydrate in marine sediments. *Mar. Geophys. Res.* **2020**, *41*, 13. [CrossRef]
20. Wu, C.; Han, L.; Zhang, F.; Liu, J.; Chen, H.; Di, B. Gas hydrate reservoir identification based on rock physics modeling and sensitive elastic parameters. *J. Geophys. Eng.* **2023**, *20*, 117–127. [CrossRef]
21. Ameur-Zaimeche, O.; Kechiched, R.; Heddam, S.; Wood, D.A. Real-time porosity prediction using gas-while-drilling data and machine learning with reservoir associated gas: Case study for Hassi Messaoud field, Algeria. *Mar. Pet. Geol.* **2022**, *140*, 105631. [CrossRef]
22. Feng, S.; Xiong, L.; Radwan, A.E.; Xie, R.; Yin, S.; Zhou, W. Accurate identification of low-resistivity gas layer in tight sandstone gas reservoirs based on optimizable neural networks. *Geoenergy Sci. Eng.* **2024**, *241*, 213094. [CrossRef]
23. Eftekhari, S.H.; Memariani, M.; Maleki, Z.; Aleali, M.; Kianoush, P. Electrical facies of the Asmari Formation in the Mansouri oilfield, an application of multi-resolution graph-based and artificial neural network clustering methods. *Sci. Rep.* **2024**, *14*, 5198. [CrossRef]
24. Liu, J.J.; Liu, J.C. An intelligent approach for reservoir quality evaluation in tight sandstone reservoir using gradient boosting decision tree algorithm-A case study of the Yanchang Formation, mid-eastern Ordos Basin, China. *Mar. Pet. Geol.* **2021**, *126*, 104939. [CrossRef]
25. Hua, Y.; Gao, G.; He, D.; Wang, G.; Liu, W. Reservoir fluid identification based on multi-head attention with UMAP. *Geoenergy Sci. Eng.* **2024**, *238*, 212888. [CrossRef]
26. Tan, M.J.; Bai, Y.; Zhang, H.T.; Li, G.R.; Wei, X.P.; Wang, A.D. Fluid typing in tight sandstone from wireline logs using classification committee machine. *Fuel* **2020**, *271*, 117601. [CrossRef]
27. Hu, X.D.; Song, G.; Wang, C.M.; Xiao, K.; Yuan, H.; Leng, W.F.; Wei, Y.M. An application study of machine learning methods for lithological classification based on logging data in the Permafrost Zones of the Qilian Mountains. *Processes* **2025**, *13*, 1475. [CrossRef]
28. Ren, Q.; Zhang, H.; Zhang, D.; Zhao, X.; Yan, L.; Rui, J.; Zeng, F.; Zhu, X. A framework of active learning and semi-supervised learning for lithology identification based on improved naive Bayes. *Expert Syst. Appl.* **2022**, *202*, 117278. [CrossRef]
29. Wang, J.; Cao, J. A lithology identification approach using well logs data and convolutional long short-term memory networks. *IEEE Geosci. Remote Sens. Lett.* **2023**, *20*, 7506405. [CrossRef]
30. Eftekhari, S.H.; Memariani, M.; Maleki, Z.; Aleali, M.; Kianoush, P.; Shirazy, A.; Shirazi, A.; Pour, A.B. Employing statistical algorithms and clustering techniques to assess lithological facies for identifying optimal reservoir rocks: A case study of the Mansouri oilfields, SW Iran. *Minerals* **2024**, *14*, 233. [CrossRef]
31. Zhou, K.; Zhang, J.; Ren, Y.; Huang, Z.; Zhao, L. A gradient boosting decision tree algorithm combining synthetic minority oversampling technique for lithology identification. *Geophysics* **2020**, *85*, WA147–WA158. [CrossRef]
32. Zhang, J.; He, Y.; Zhang, Y.; Li, W.; Zhang, J. Well-logging-based lithology classification using machine learning methods for high-quality reservoir identification: A case study of Baikouquan formation in Mahu Area of Junggar Basin, NW China. *Energies* **2022**, *15*, 3675. [CrossRef]
33. Lee, J.; Byun, J.; Kim, B.; Yoo, D.G. Delineation of gas hydrate reservoirs in the Ulleung Basin using unsupervised multi-attribute clustering without well log data. *J. Nat. Gas Sci. Eng.* **2017**, *46*, 326–337. [CrossRef]
34. Zhu, L.; Zhou, X.; Sun, J.; Liu, Y.; Wang, J.; Wu, S. Reservoir classification and log prediction of gas hydrate occurrence in the Qiongdongnan Basin, South China Sea. *Front. Mar. Sci.* **2023**, *10*, 1055843. [CrossRef]
35. Tian, D.; Yang, S.; Gong, Y.; Geng, M.; Li, Y.; Hu, G. A comparative study of machine learning methods for gas hydrate identification. *Geoenergy Sci. Eng.* **2023**, *223*, 211564. [CrossRef]
36. Riedel, M.; Collett, T.S.; Malone, M.; Scientists, E. Expedition 311 synthesis: Scientific findings. *Proc. Integr. Ocean. Drill. Program* **2010**, *311*, 2. [CrossRef]
37. Riddihough, R. Recent movements of the Juan de Fuca plate system. *J. Geophys. Res. Solid Earth* **1984**, *89*, 6980–6994. [CrossRef]
38. Riedel, M.; Collett, T.S. Cascadia margin gas hydrates. *IODP Prelim. Rep.* **2005**, *311*, 9. [CrossRef]
39. Riedel, M.; Collett, T.S.; Malone, M.J.; Scientists, E. Expedition 311 Summary. *Proc. Integr. Ocean. Drill. Progr.* **2010**, *311*, 2–26. [CrossRef]
40. Rasmussen, C.E. Gaussian processes in machine learning. In *Summer School on Machine Learning*; Springer: Berlin/Heidelberg, Germany, 2003; Volume 3176, pp. 63–71. [CrossRef]
41. Liu, H.; Ong, Y.S.; Yu, Z.; Cai, J.; Shen, X. Scalable Gaussian process classification with additive noise for non-Gaussian likelihoods. *IEEE Trans. Cybern.* **2021**, *52*, 5842–5854. [CrossRef]
42. Seeger, M. Gaussian processes for machine learning. *Int. J. Neural Syst.* **2004**, *14*, 69–106. [CrossRef]
43. Vapnik, V.N. An overview of statistical learning theory. *IEEE Trans. Neural Netw.* **1999**, *10*, 988–999. [CrossRef]
44. Chang, C.C.; Lin, C.J. LIBSVM: A library for support vector machines. *ACM Trans. Intell. Syst. Technol. (TIST)* **2011**, *2*, 1–27. [CrossRef]
45. Hassoun, M.H. *Fundamentals of Artificial Neural Networks*; MIT Press: Cambridge, MA, USA, 1996; Volume 84, p. 906. [CrossRef]

46. Rao, M.B. Feedforward Neural Network Methodology. *Technometrics* **2000**, *42*, 432–433. [CrossRef]
47. Cox, D.R. The regression analysis of binary sequences. *J. R. Stat. Soc. Ser. B Stat. Methodol.* **1958**, *20*, 215–232. [CrossRef]
48. Blanc-Valleron, M.M.; Pierre, C.; Bartier, D.; Rouchy, J.M. Data report: Bulk carbonate content of sediments and mineralogy of authigenic carbonates along an east-west transect in the northern Cascadia margin, IODP Expedition 311. *Proc. IODP* **2009**, *311*, 2. [CrossRef]
49. Chen, M.A.; Riedel, M.; Spence, G.D.; Hyndman, R.D. Data report: A downhole electrical resistivity study of northern Cascadia marine gas hydrate. In *Integrated Ocean Drilling Program*; Texas A & M University: College Station, TX, USA, 2008; Volume 311. [CrossRef]
50. Riedel, M.; Collett, T.S.; Malone, M.J. Expedition 311 Scientists. Site U1325. *Proc. Integr. Ocean. Drill. Program* **2006**, *311*, 202. [CrossRef]
51. Pierre, C.; Blanc-Valleron, M.M.; Rouchy, J.M.; Bartier, D. Data report: Stable isotope composition of authigenic carbonates from the northern Cascadia margin, IODP Expedition 311, Sites U1325–U1329. *Proc. Integr. Ocean. Drill. Program* **2009**, *311*, 2. [CrossRef]
52. Riedel, M.; Collett, T.S.; Malone, M.J. Expedition 311 Scientists. Site U1326. *Proc. Integr. Ocean. Drill. Program* **2006**, *311*, 6. [CrossRef]
53. Riedel, M.; Collett, T.S.; Malone, M.J. Expedition 311 Scientists. Site U1327. *Proc. Integr. Ocean. Drill. Program* **2006**, *311*, 16–22. [CrossRef]
54. Riedel, M.; Collett, T.S.; Malone, M.J. Expedition 311 Scientists. Site U1328. *Proc. Integr. Ocean. Drill. Program* **2006**, *311*, 25. [CrossRef]
55. Hutter, F.; Kotthoff, L.; Vanschoren, J. *Automated Machine Learning: Methods, Systems, Challenges*; Springer Nature: Cham, Switzerland, 2019; p. 219. [CrossRef]

Journal of
Marine Science
and Engineering

MDPI

Article

Electrical Resistivity Tomography Methods and Technical Research for Hydrate-Based Carbon Sequestration

Zitian Lin [1], Qia Wang [1], Shufan Li [1], Xingru Li [1], Jiajie Ye [1], Yidi Zhang [1], Haoning Ye [1], Yangmin Kuang [1,2,3,*] and Yanpeng Zheng [1,2,3,*]

[1] College of Marine Science and Technology, China University of Geosciences, Wuhan 430074, China; zitianlin@cug.edu.cn (Z.L.); 20221000514@cug.edu.cn (Q.W.)
[2] Hubei Key Laboratory of Marine Geological Resources, China University of Geosciences, Wuhan 430074, China
[3] Laboratory for Marine Geology, Qingdao Marine Science and Technology Center, Qingdao 266237, China
* Correspondence: kuangyangmin@cug.edu.cn (Y.K.); zhengyanpeng@cug.edu.cn (Y.Z.)

Abstract

This study focuses on the application of electrical resistivity tomography (ERT) for monitoring the growth process of CO_2 hydrate in subsea carbon sequestration, aiming to provide technical support for the safety assessment of marine carbon storage. By designing single-target, dual-target, and multi-target hydrate samples, convolutional neural networks (CNNs), recurrent neural networks (RNNs), and residual neural networks (ResNets) were constructed and compared with traditional image reconstruction algorithms (e.g., back-projection) to quantitatively analyze ERT imaging accuracy. The experiments used boundary voltage as the input and internal conductivity distribution as the output, employing the relative image error (RIE) and image correlation coefficient (ICC) to evaluate algorithmic performance. The results demonstrate that neural network algorithms—particularly RNNs—exhibit superior performance compared to traditional image reconstruction methods due to their strong noise resistance and nonlinear mapping capabilities. These algorithms significantly improve the edge clarity in target identification, enabling the precise capture of the hydrate distribution during carbon sequestration. This advancement effectively enhances the monitoring capability of CO_2 hydrate reservoir characteristics and provides reliable data support for the safety assessment of hydrate reservoirs.

Keywords: offshore carbon sequestration; CO_2 hydrates; electrical resistivity tomography; image reconstruction; neural networks

1. Introduction

Natural gas hydrates (NGHs), also known as combustible ice, are non-stoichiometric ice-like crystalline compounds [1]. Various types of naturally occurring hydrates (CH_4, C_2H_6, CO_2, etc.) are primarily distributed at the edges of deep-sea continental shelves and in inland permafrost regions [2]. This distribution is determined by the phase equilibrium conditions for hydrate formation (low temperatures and high pressures) and geological environments (gas sources and water). Under standard conditions, one unit-volume of natural gas hydrate can decompose to produce up to one hundred sixty-four unit-volumes of methane gas. Calculations suggest that the reserves of methane and other gases stored in hydrate form in the natural world are twice the size of those of other fossil fuels [3]. Natural gas hydrates are viewed as an important alternative energy source in the 21st century due to their superior energy storage capabilities, driving the progress of marine combustible ice from small-scale trials to large-scale applications. In addition to being a high-quality replaceable energy source, natural gas hydrates play crucial roles in various energy and environmental industrial applications due to their unique physicochemical properties [4].

The stable existence of natural gas hydrates on the seafloor has been known for thousands of years, so researchers wonder whether CO_2 can also be stored long term beneath the seafloor, like CH_4 gas. The ocean provides a vast area for large-scale CO_2 sequestration. Theoretically, the storage of CO_2 hydrates is feasible, as the low-temperature and high-pressure conditions in the deep sea can create ideal conditions for the formation of CO_2 hydrates.

Currently, CO_2 in the form of hydrates is mainly stored in the seabed in two primary forms. The first form occurs when CO_2 is injected into deeper, high-pressure, and low-temperature sedimentary layers. As it migrates and diffuses, it combines with pore water in the sediments to generate solid CO_2 hydrates. This process further blocks the sediment pores, thereby forming self-sealing, non-permeable barriers [5,6]. The second form involves CO_2 displacing natural gas hydrates for extraction [7,8] or using artificial CO_2 hydrate caps to collaboratively extract natural gas hydrates [9]. The two types of hydrate carbon sequestration mechanisms are shown in Figure 1. This technology not only enhances the energy extraction of natural gas hydrates but also plays a role in maintaining the stability of the reservoir and the seabed's geological structure [10].

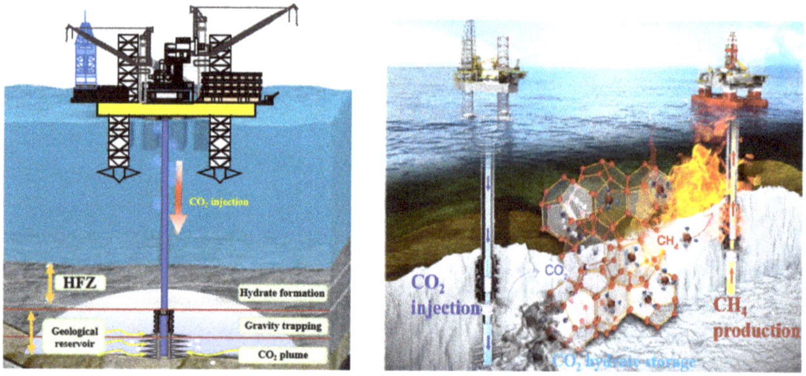

(a) (b)

Figure 1. Two kinds of hydrate-based mechanisms for sub-seabed CO_2 geological storage (**a**) CO_2 sequestration as solid clathrate hydrates in marine sediments; (**b**) CO_2 sequestration via methane (CH_4) displacement in hydrate-bearing reservoirs (modified and reproduced with permission from [11]).

Submarine CO_2 hydrates are typically hosted in porous geological media environments [12]. Their stability and formation mechanisms are influenced by factors such as the consolidation characteristics of reservoir rocks, permeability, and porosity [13]. These factors contribute to the complexity of CO_2 hydrate's stability and formation mechanisms in submarine sediments. Lin et al. [14] conducted experimental and modeling studies, revealing that CO_2 injection induces adsorption-induced swelling in shale. This process alters the pore structure, increases the pore volume, and modifies the permeability while simultaneously weakening the mechanical properties of the shale. Consequently, it compromises the integrity and stability of the reservoir. In the study of the potential CO_2 storage site at Smeaheia, in the offshore Norwegian region, 3D field-scale geomechanical modeling based on seismic data indicates that the reservoir pressure increase induced by CO_2 injection can lead to significant vertical deformation (up to 8 cm) at the reservoir–caprock interface and the seabed surface. Additionally, the spatial variability of the overburden strata mitigates the risks of deformation localization and shear failure through stress redistribution, thereby reducing the likelihood of fault reactivation and subsequent CO_2 leakage [15]. Figure 2a illustrates the gradual increase in the reservoir pressure over a 50-year CO_2 injection period, with time steps delineated every 10 years. Figure 2b depicts the spatial distribution of the pressure, which is concentrated in the injection zone of the Vette–Øygarden fault block and constrained by the fault boundaries.

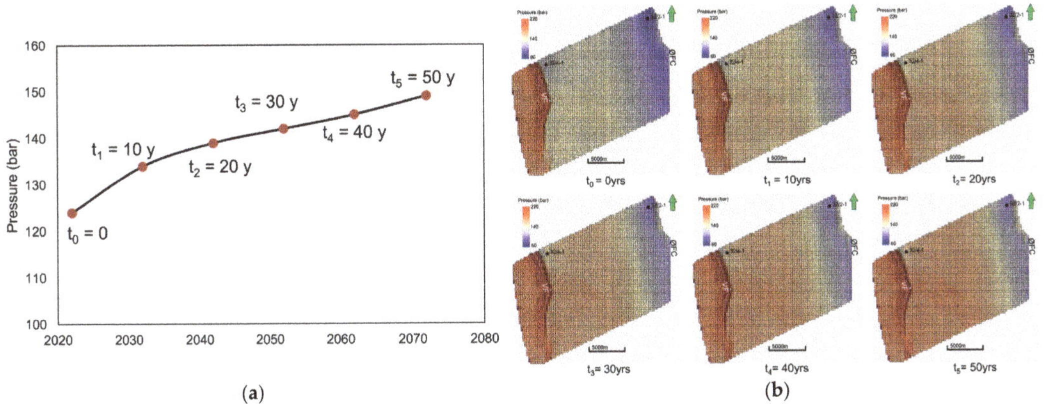

Figure 2. (**a**) The temporal evolution of the reservoir pressure over a 50-year CO_2 injection period, with increments at 10-year intervals (modified and reproduced with permission from [15]); (**b**) the spatial distribution of the pressure concentration within the Vette–Øygarden fault block (modified and reproduced with permission from [15]; black dots indicate the coordinates of two key wells within the study area, while green arrows represent the pressure change trends).

Non-homogeneous porous media are a critical factor in assessing the safety of seabed carbon storage. This understanding directly determines the safe and effective implementation of hydrate-based seabed carbon storage.

The formation of hydrates typically involves two stages: nucleation and crystal growth. In the nucleation stage, disordered or liquid-like molecules develop into a solid crystalline phase. In the crystal growth stage, as the particle size increases, the solid hydrate crystals gradually grow, as illustrated in Figure 3. Monitoring the hydrate formation process at a laboratory scale is an effective and low-cost method that can study the nucleation mechanisms of hydrates, providing theoretical support for in situ hydrate-based carbon sequestration. Currently, the main in situ visualization techniques commonly used in laboratory settings to monitor hydrate growth processes include computed tomography (CT), scanning electron microscopy, magnetic resonance imaging (MRI), and nuclear magnetic resonance (NMR). Zhang et al. utilized CT technology to measure hydrate growth processes and studied the relationship between the hydrate morphology and changes in permeability within the reservoir [16]. Simonetti et al. observed the microcrystalline structures of hydrate samples from different locations in the Gulf of Mexico, using scanning electron microscopy, and concluded that hydrates coexist with liquid oil [17]. Kuang et al. employed NMR to measure the in situ formation of CO_2 hydrates in porous media. They found that CO_2 hydrates predominantly form in larger pores, which further hinders gas–water contact in isolated pores during subsequent stages [18]. While these techniques can all be used to observe hydrate growth, they still have limitations. During the hydrate formation process, various material properties continually change, but the aforementioned technologies cannot satisfy the requirement for dynamic measurements of the hydrate growth process. As a result, they fail to effectively capture the instantaneous states of hydrate formation.

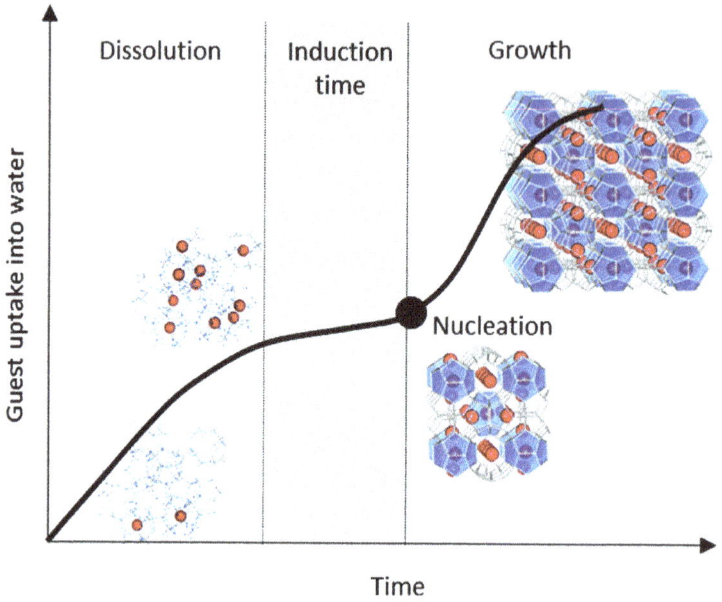

Figure 3. Evolution of the hydrate crystallization process over time (reproduced with permission from [19]).

Electrical resistivity tomography (ERT) [20], an important branch of process tomography (PT) [21], operates on the principles of injecting a weak current into the target medium, measuring boundary voltages via surface electrodes, and combining finite element forward modeling with inverse problem reconstruction to generate cross-sectional images of the medium's internal resistivity or conductivity distribution.

In terms of safety features, ERT technology, with its non-radiative and non-invasive characteristics, enables the long-term and repeated monitoring of sensitive marine geological environments, providing an ideal approach for the non-destructive observation of hydrate formation processes. Regarding resolution and dynamic capture capabilities, neural network models demonstrate sub-second temporal sampling and sub-centimeter spatial precision, allowing for the clear delineation of hydrate edge features. For adaptability to porous media and multi-target resolution, ERT combined with neural network technology exhibits superior performance over traditional algorithms in handling the diverse and complex morphologies of CO_2 hydrates within subsea sediments.

Although ERT has numerous advantages, its limitations, such as the restricted depth of penetration and lower sensitivity to high-resistivity targets, make it less effective in detecting deep permafrost and natural gas hydrates. Therefore, ERT is more suitable for use in combination with other geophysical methods (e.g., the transient electromagnetic method, TEM) to compensate for its shortcomings and achieve more comprehensive subsurface imaging [22].

Since 2014, resistivity tomography technology has gradually gained widespread attention from researchers in the field of hydrate formation dynamics. We have investigated and analyzed recent academic studies using ERT for hydrate imaging, as shown in Table 1.

Unlike the studies mentioned earlier, this paper focuses on the numerical modeling and algorithmic development for hydrate imaging in subsea reservoirs. The core contributions involve generating multi-morphology hydrate data through finite element simulations and enhancing electrical resistivity tomography (ERT) imaging's accuracy by integrating neural network algorithms, thereby providing theoretical support for subsequent in situ monitoring technology development. Specifically, this study achieves innovations in the following aspects: 1. Modeling and Prediction: It simulates the growth state of gas hydrates in the reservoir and employs neural networks to predict ERT resistivity imaging. 2. Methodological Comparison: It utilizes technical approaches, such as the linear back-projection method and the conjugate gradient method, for ERT resistivity-imaging predictions, followed by a comparative analysis with the neural network method.

Table 1. Electrical resistivity tomography (ERT) technology's recent research in the field of hydrate imaging.

Authors	Research Methods
Priegnitz et al., 2014 [23]	Three-dimensional electrical resistivity tomography is utilized to dynamically monitor the formation and dissociation processes of hydrates in a high-pressure low-temperature reservoir simulator (LARS). The measured data are subsequently processed using the inversion software tool Boundless to generate imaging results through inversion.
Li et al., 2020 [24]	Two-dimensional electrical resistivity tomography is used to monitor the formation process of hydrates in sandy sediments, and the ITS2000 industrial fault-scanning system is employed for imaging.
Zhao J. et al., 2022 [25]	Utilizing ERT technology, three-dimensional resistivity images of blocky and layered hydrates were established. It was found that the resistivity of blocky hydrates is significantly higher than that of layered hydrates, and their formation characteristics are influenced by the distribution of the pore water and the microstructure of hydrate pores.
Buddo et al., 2022 [22]	The integration of ERT with other geophysical methods, such as the transient electromagnetic method (TEM), improves the penetration depth and sensitivity to high-resistivity targets.
Liu et al., 2024 [26]	A cross-hole electrical resistivity tomography (CHERT) technique was designed to satisfy a wide range of resistivity measurements ranging from a few ohm-meters to thousands of ohm-meters, consistent with the resistivity responses of actual hydrate reservoirs.

The findings demonstrate that the neural network method, when combined with ERT, offers robust technical support for microscopic hydrate imaging and generation dynamics research.

2. Experimental Theory

ERT is an advanced and widely used geophysical subsurface imaging technique, with applications in civil engineering, environmental surveys, hydrological exploration, mineral exploration, and archeology [27]; ERT has the advantage of being a low-cost, non-invasive geophysical method that allows for the rapid observation of subsurface resistivity variations over the entire length of a line or profile [28]. It can be used to determine the soil lithology, saturation levels, fracture zones, and the presence of groundwater [29].

The ERT system comprises three components: the sensor array, the data acquisition module, and the image reconstruction module, which determine the internal resistivity distribution of the measured object according to the boundary voltage data measured after the injection of a weak current, and visualize it on a computer. The basic principle of the system is shown in Figure 4.

In the ERT system, the computer control unit sends data acquisition commands to the data acquisition system. The data acquisition system applies current excitation to the sensor electrodes according to a specific pattern and transmits the measured voltage data to the data-processing unit for preprocessing. The processed data are then used as input for the image reconstruction algorithm, ultimately yielding a grayscale conductivity image of the object under testing.

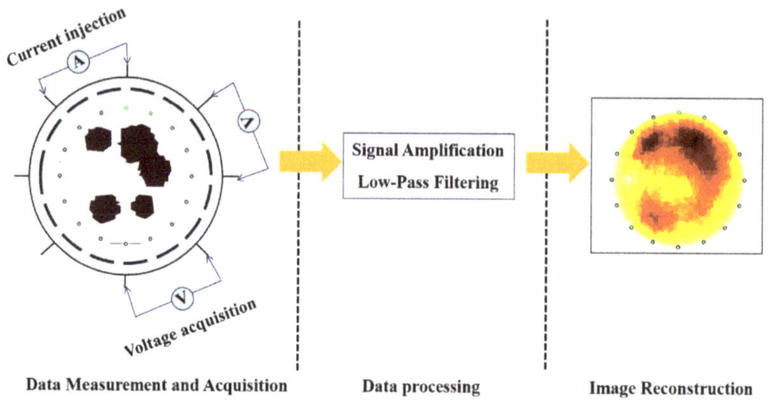

Figure 4. Main components of the ERT technology system.

In the ERT data acquisition system, there are two excitation modes: adjacent current injection and opposite current injection. The adjacent current injection mode ensures uniform information acquisition within the sensitive field and provides a moderate number of measured voltage values. Therefore, this study adopts the adjacent current injection mode. The operational principle of this mode is as follows: First, any pair of adjacent electrodes is selected as the starting point and current is injected into them. Next, the voltage signals at other adjacent electrodes are sequentially measured in a clockwise direction, excluding the electrodes immediately adjacent to the current-injecting pair. After completing a full cycle of measurements for all the electrodes, the next adjacent electrode pair in the clockwise direction is selected and the voltage measurement process is repeated until all the electrode pairs have been excited. Finally, the voltage dataset is saved to the computer. Assuming the number of electrodes is N, a total of $N(N-3)/2$ independent voltage values can be obtained. In this study, the sensor employs 16 electrodes, and a single dataset comprises 108 voltage values.

3. Experimental Methods

3.1. Solution Methods for the ERT Forward Problem

The primary approaches for solving the ERT forward problem include analytical methods and numerical computation methods. The analytical methods require the establishment of an accurate field model to conduct theoretical derivations and solve the analytical expressions of the potential distribution within the field. However, this approach involves complex derivations and is only applicable to scenarios where the geometric configuration and medium distribution are highly uniform. It struggles to address non-uniform fields. Numerical computation methods, including the finite difference method (FDM), finite element method (FEM), boundary element method (BEM), and element-free Galerkin method (EFGM), are widely employed. In this study, the FEM is utilized to solve the forward problem of electrical resistivity tomography (ERT).

The variational principle serves as the theoretical foundation of the finite element method. Its core idea involves discretizing a continuous field into numerous small elements. Calculations are performed on these individual elements to establish local equations, which are then combined to form the global equation system of the original field. Solving this system yields discrete solutions for the continuous field. In this work, the FEM parameters for the sensitivity analysis include 16 electrodes and 1600 imaging pixels, as in Figure 5.

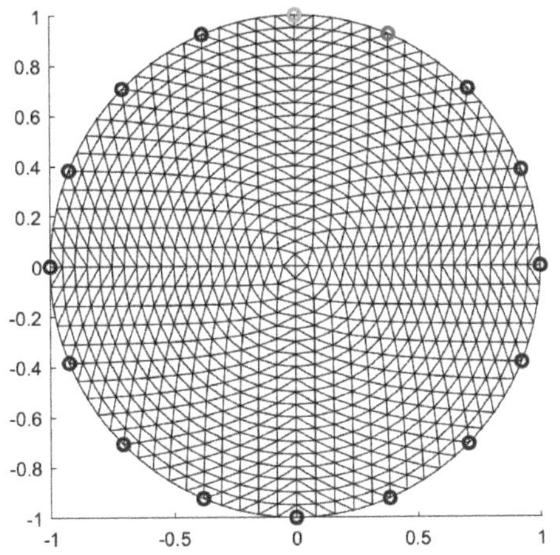

Figure 5. The finite element discretization employed in this study (green circles: locations of applied electrodes).

3.2. Solution Methods for the ERT Inverse Problem

The inverse problem of the ERT refers to the process of reconstructing the resistivity distribution of a region of interest by measuring its boundary voltage. The process is typically "ill-posed" and "nonlinear" [30] and is affected by many factors, such as the geometry of the electrodes and the electrical properties of the measurement target itself. Currently, methods to solve the ERT inverse problem include the linear back-projection (LBP) method [31], conjugate gradient method [32], as well as emerging neural-network-based algorithms [33–35], which have been widely used in imaging.

3.2.1. The Conjugate Gradient Algorithm [36]

The distribution of the boundary voltage and the medium resistivity in the sensitive field obtained by resistive tomography measurements is a nonlinear function. From the principle of the FEM in the positive problem's solution above, it can be seen that the boundary voltage value of the sensitive field and the sensitivity can be approximately transformed to a linear relationship through finite element division, which is expressed by the matrix shown in Formula (1) as follows:

$$SG = Z \tag{1}$$

where S is an n × m-order sensitivity matrix, G is an m × 1-order resistivity vector matrix in the sensitive field, and Z is an n × 1-order boundary voltage vector matrix. The image reconstruction of the ERT is the solution of the G matrix, which is computed as follows using the conjugate gradient algorithm to solve the ERT inverse problem:

1. **Regularization of the sensitivity matrix.** The sensitivity matrix calculated above is not a symmetric positive definite matrix, so its transpose (S^T) at both ends of the formula should be multiplied simultaneously in order to convert the coefficient matrix to a symmetric positive definite matrix; otherwise, the conjugate gradient algorithm cannot converge. The regularization of the sensitivity matrix yields Formula (2) as follows:

$$S^T SG = S^T Z \tag{2}$$

By making A = S^TS, x = G, and b = S^TZ, Equation (2) can be transformed to the calculation of the linear equation Ax = b.

2. **Iterative calculation.** In accordance with the principle of calculating the conjugate gradient above, for any initial value of x_0, the first iteration of the direction of $r_0 = b - A_{X_0}, p_0 = r_0$ and then the directions of the subsequent conjugate gradients are, in turn, iteratively calculated in accordance with Formulas (3)–(7) as follows:

$$\alpha_k = \frac{(r_k, r_k)}{(A_{P_k}, p_k)} \tag{3}$$

$$x_{k+1} = x_k + \alpha_k p_k \tag{4}$$

$$r_{k+1} = r_k - \alpha_k A p_k \tag{5}$$

$$\beta_k = \frac{(r_{k+1}, r_{k+1})}{(r_k, r_k)} \tag{6}$$

$$p_{k+1} = r_{k+1} - \beta_k p_k \tag{7}$$

Among them, P_k represents the direction of the kth iteration, and each iteration direction directly satisfies the conjugate relationship, as shown in Formula (8):

$$(p_i, Ap_j) = 0, \quad i \neq j \tag{8}$$

After several iterations, the residuals are less than the set value, at which point the iteration is terminated and the optimal solution is found.

3.2.2. The Back-Projection Method

The inverse problem in resistive tomography is typically an ill-posed problem, where small perturbations in the measured boundary voltages can lead to significant changes in the reconstructed grayscale estimates of the image. Therefore, regularization methods are usually used to stabilize the solution of the inverse problem. Common regularization methods include least squares and Tikhonov regularization.

The reconstruction process of resistive tomography often needs to be solved using numerical methods, such as iterative methods and filtered back-projection (FBP). Filtered back-projection combines filtering techniques to effectively reduce the blurring effect in simple back-projection.

The goal of the ERT is to recover the resistivity, $\sigma(x, y)$, from known voltage data, $P(\theta, t)$, the process of which can be represented by the following back-projection equation, as shown in Formula (9):

$$\sigma_{rc}(x, y) = \int_0^{\pi} P(\theta, xcos\theta + ysin\theta) d\theta \tag{9}$$

In resistive tomography, $P(\theta, t)$ represents the projection data for the voltage, while $\sigma(x, y)$ represents the resistivity distribution. θ is the angle of the electrode pair, and t is the projection path associated with the electrode array.

1. **The difference between the actual measured voltage and the background voltage.** The actual measured voltage is $V_{ij}^{measured}$, and the background voltage ($V_{ij}^{measured}$) is calculated from the forward model. The voltage difference is shown in Formula (10) as follows:

$$\Delta V_{ij} = V_{ij}^{measured} - V_{ij}^{background} \tag{10}$$

2. **Constructing the sensitivity matrix and voltage difference vector.** The sensitivity matrix (S) is arranged in two dimensions, with rows corresponding to the measure-

ment pairs (i,j) and columns corresponding to the pixels (k). The voltage difference (ΔV) is organized as a vector.

3. **Using inverse projection to calculate the conductivity change.** The resistivity change $(\Delta\sigma)$ is obtained by multiplying the sensitivity matrix's transpose with the voltage difference, as shown in Formula (11):

$$\Delta\sigma = S^T \Delta V \tag{11}$$

That is, the amount of the change per pixel k is shown in Formula (12) as follows:

$$\Delta\sigma_k = \sum_{i,j} S_{ijk} \Delta V_{ij} \tag{12}$$

4. **Normalization.** Normalization to $\Delta\sigma_k$ is carried out as shown in Formula (13):

$$\Delta\sigma_k^{norm} = \frac{\Delta\sigma_k}{\Sigma_{ij}\left|S_{ijk}\right| + \varepsilon} \tag{13}$$

where ε is a small constant to avoid dividing by zero.

5. **The reconstruction of the result output.** The $\Delta\sigma$ value is output as a resistivity change image.

3.2.3. Neural Network Methods

The Preparation of the Training Dataset

In this work, a voltage–conductivity dataset was designed using the EIDORS software platform (EIDORS 3.8) [37] to simulate the gas hydrate generation situation in an autoclave.

In the preparation of the training dataset, we have comprehensively considered the natural occurrence states of gas hydrates within reservoirs. Based on their morphological characteristics, gas hydrates can be classified into five occurrence modes: blocky, nodular, layered, vein-like, and dispersed. According to their filling mechanisms, they are further categorized into five types: contact type, cementation type, pore-filling type, supporting type (or bridging type), and plaque type, as illustrated in Figure 6. Building on this framework, this study systematically accounts for diverse gas hydrate occurrence patterns. We established two primary categories of samples based on their dominant occurrence modes: (1) blocky/nodular gas hydrate models and (2) layered/vein-like gas hydrate models, as depicted in Figures 7 and 8.

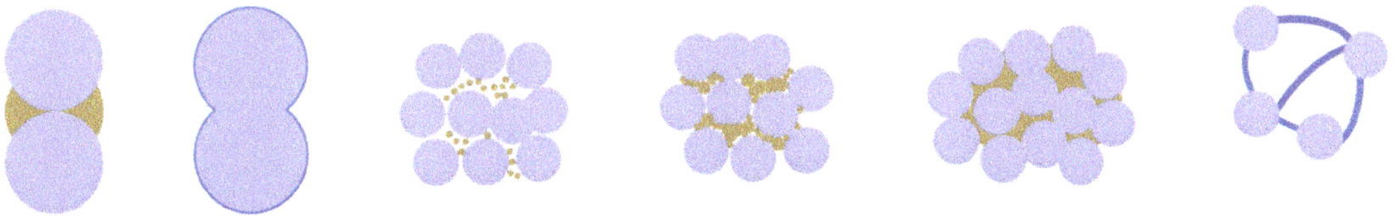

(a) **(b)** **(c)** **(d)** **(e)** **(f)**

Figure 6. The positions and occurrence forms of hydrates in sediments: the pore-filling type and fracture-filling type. (**a**) Contact mode, (**b**) cementation mode, (**c**) pore-filling mode, (**d**) load-bearing/supporting mode, (**e**) patchy mode, and (**f**) bridging mode (modified from [38,39]; reproduced with permission from [38]).

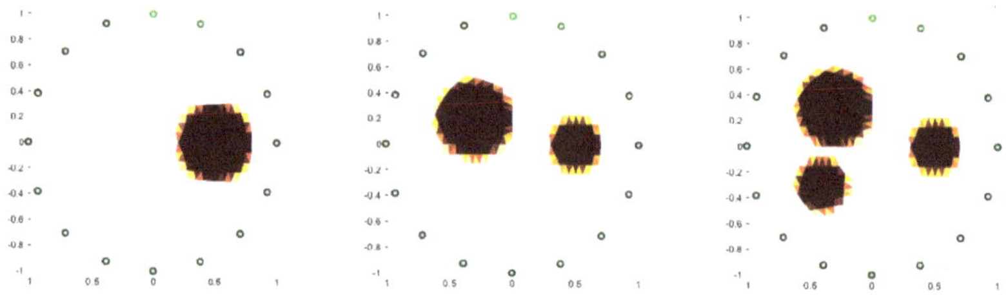

Figure 7. Blocky or nodular hydrate model (green circles: locations of applied electrodes).

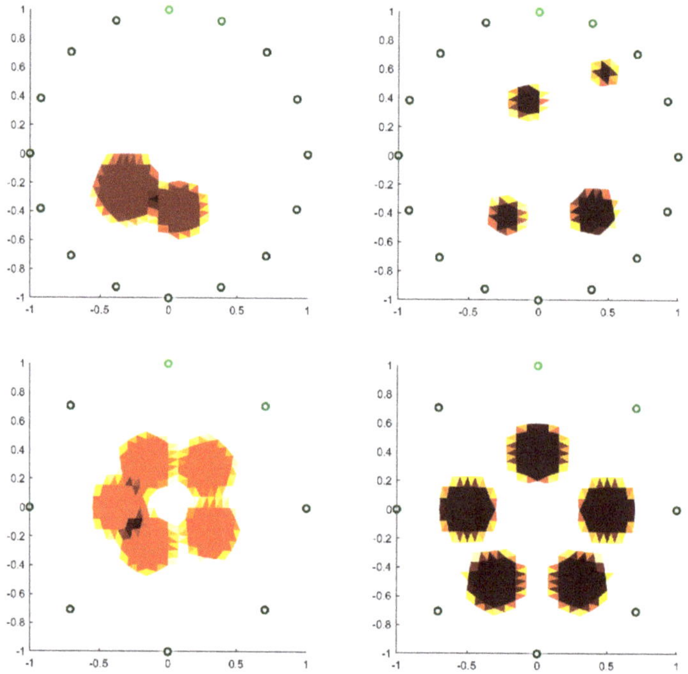

Figure 8. Layered or vein-type hydrate model.

(1) **Blocky or Nodular Hydrate Model:**

Blocky and nodular hydrates are large-scale gas hydrate structures characterized by blocky and nodular morphologies, often with minor sand grain inclusions. These features are exemplified by the hydrate deposits identified in China's Shenhu Area [40], as illustrated in Figure 7.

(2) **Layered or Vein-type Hydrate Model:**

Layered and vein-type hydrates refer to hydrate formations that grow within fractures, primarily found in inland seas and permafrost regions, such as the permanent permafrost layer of the Qilian Mountains in China [41]. This type of hydrate generally has a smaller shape and often exhibits irregular shapes during formation. In response to this, we designed simulated samples by adjusting the positions and numbers of high-resistance objects (hydrates) in the model, as shown in Figure 8.

Because neural network training requires large-scale datasets, we generated 14,508 sets of data by varying the size and location of the generated heterogeneous bodies. The number of each type of data is shown in Table 2.

Table 2. Two-dimensional ERT voltage–resistivity simulation dataset distribution.

Sample Type	One Heterostructure	Two Heterostructures	Three Heterostructures	Four Heterostructures	Five Heterostructures
Training Sample	3000	3000	2000	3000	3000
Validation Sample	100	100	100	100	100
Testing Sample	1	1	1	1	1

Because there are certain differences between the simulated and measured datasets, which, to some extent, affect the performance capability of the neural network model in the measured dataset, we carried out a normalization operation on the voltage data in the simulated dataset and used the normalized voltage data for model training. The data normalization formula is expressed as shown in Formula (14):

$$x^* = \frac{x - x_{min}}{x_{max} - x_{min}} \tag{14}$$

3.2.4. Network Algorithms and Models

1. **CNN**[1]

Because CNNs are suitable for processing image data, and the collected data in this work are measured voltage data, which do not match the shape of the picture data, before the data were input into the model, the data's shape was transformed. The original sequence of 1 × 208 (as a 13 × 16 matrix) was zero-padded (Zero Padding), and the output was turned into a matrix of (18, 18, 1). Then, three convolutional operations were carried out using 8, 16, and 32 convolutional filters of a size of 3 × 3 and a step size of 1, and after each convolution, after the BN (batch normalization) layer on the feature map, to enhance the stability of the model. After the convolution, the shape of the output data was 18 × 18 × 256. Because the final number of prediction results for the internal conductivity was 1600, in this paper, the flattened layer and fully connected layer were used to pass through the data after the convolution; after flattening, the data shape was converted to (1, 82, 944), and further, after one fully connected layer with 512 neurons, the data shape was converted to (1512). After going through the last fully connected layer with 1600 neurons, the data shape was converted to (1, 1600). Because in this study, the prediction is for the presence or absence of hydrate as a 0−1 binary classification problem, after the last fully connected layer, a sigmoid function is used as the activation function for the output layer to map the 1600 input values between 0 and 1. The changes in the dimensions of the data matrix throughout this process are detailed in Table 3.

Table 3. The changes in the dimensions of the data matrix throughout this process.

Step	Layer(s)	Output Size
Input	Reshaping + Zero Padding	(1, 18, 18, 1)
1	Conv2d/BN	(1, 18, 18, 64)
2	Conv2d/BN	(1, 18, 18, 256)
3	Conv2d/BN	(1, 2000)
4	Flattening	(1, 82, 944)
5	Fully connected	(1, 512)
6	Fully connected	(1, 1600)

The CNN architecture used in this paper is shown in Figure 9.

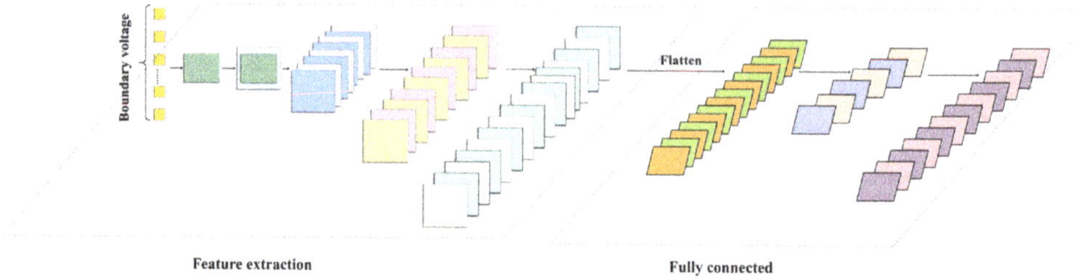

Figure 9. CNN architecture used in this paper.

2. **ResNet**[2]

Compared with the convergence speeds of CNNs and other model architectures, those of the ResNet network are faster, and at the same time, it is beneficial to alleviate the problem of the gradient's instability. Based on the deep residual network, a single-input, dual-output "h-Net" model architecture was designed [42], which can simultaneously achieve manifold discrimination and Electrical Resistivity Tomography (ERT) imaging. In this paper, targeting the imaging problem of saturated layer reservoirs, the "h-Net" model was simplified, and the simplified model was trained on sample data. He et al. proposed two base residual modules in ResNet50, Conv_block and Identity_block (as shown in Figure 10), which have the following model structure [43]:

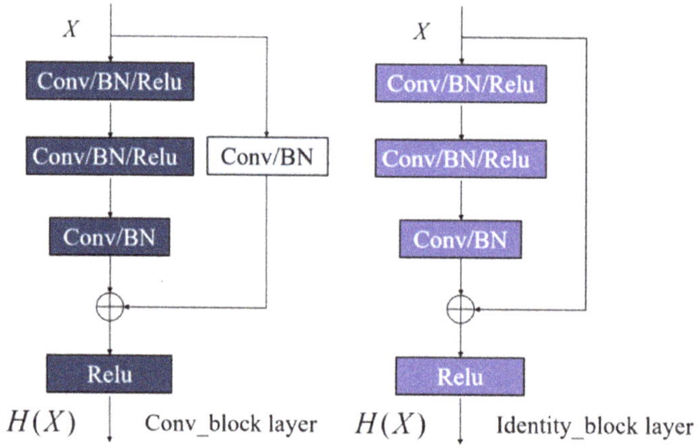

Figure 10. Two types of residual model architectures.

As in the case of predictions using CNN models, in this paper, the original sequence was first subjected to a shape transformation and a complementary zero operation so that the shape of the data was changed to (1, 18, 18, 1); Following this, the data passed through a convolutional layer with 16 filters and astride of 2, followed by a Batch Normalization (BN) layer, resulting in an output shape of (1, 17, 17, 16). This output is then fed into a Conv_block layer, which main-tains the data shape without any changes. Next, the data undergoes flattening and is processed by a fully connected layer with 1800 neurons. A Dropout layer is applied afterward, followed by another fully connected layer with 1600 neurons. Finally, a Sigmoid activation function is used to map the output values to the range [0, 1], producing the final prediction. Based on the ResNet network, the model structure in this paper is as shown in Figure 11.

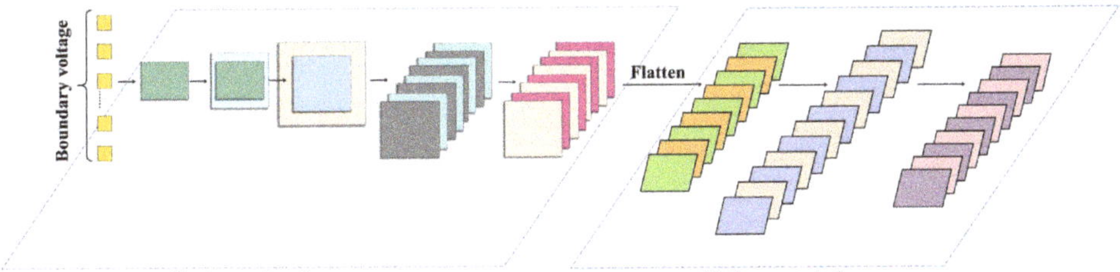

Figure 11. The ResNet network structure constructed in this work.

The changes in the dimensions of the data matrix throughout this process are detailed in Table 4.

Table 4. The changes in the dimensions of the data matrix throughout this process.

Step	Layer(s)	Output Size
Input	Reshaping + Zero Padding + Upsampling/BN	(1, 36, 36, 1)
1	Conv2d/BN	(1, 17, 17, 16)
2	Conv_block layer/BN	(1, 17, 17, 16)
3	Flattening	(1, 4624)
5	Fully connected	(1, 1800)
6	Fully connected	(1, 1600)

3. **RNN**[3]

Unlike CNNs, RNNs are more suitable for processing sequence-based data and have been used in bioinformatics to analyze bioinformatic sequences, such as DNA, RNA, and proteins [44]. For example, Li et al. utilized RNNs for gene prediction and protein structure prediction, demonstrating the ability of RNNs to capture dependency relationships in biological sequences and providing insights into genetic information and biological processes [45]. In this study, the input data comprise boundary voltage sequences, and the output data correspond to resistivity sequences. To compare the predictive performances of the RNN, CNN, and ResNet while enhancing the prediction accuracy, we employ the RNN for resistivity image reconstruction.

The input to the network consists of voltage measurement sequences, requiring no additional shape transformation. That is, the sequence of the input network is (208, 1), which first passes through an RNN layer with 16 neurons, at which point the sequence shape is (208, 16). Thereafter, the output sequence is paved and passes through the first fully connected layer with 1024 neuron nodes, at which point the output is (1024, 1). After this, 30% of the neurons are discarded to prevent possible overfitting, and, finally, a fully connected neural network with 1600 neuron nodes is used and activated with a sigmoid function to make the output data shape (1, 1600). The changes in the dimensions of the data matrix throughout this process are detailed in Table 5.

Table 5. Changes in the dimensions of the data matrix throughout this process.

Step	Layer	Output Size
Input	——	(208, 1)
1	RNN × 16	(208, 16)
2	Flattening	(1, 3328)
3	Fully connected	(1, 2000)
4	Fully connected	(1, 1600)

The RNN architecture used in this paper is shown in Figure 12 below.

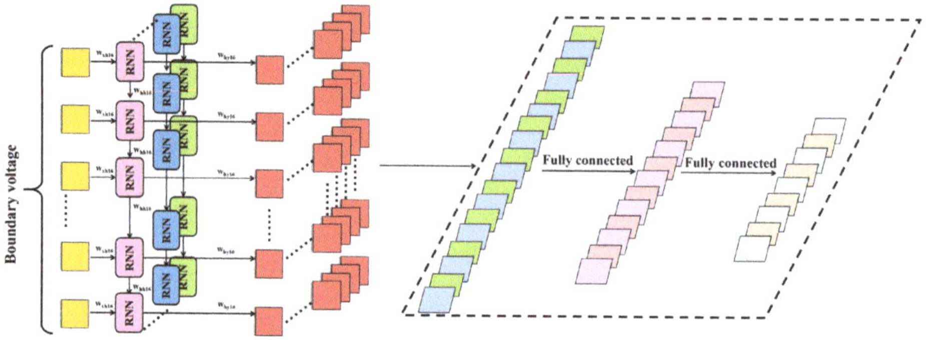

Figure 12. The RNN model architecture constructed in this work.

3.2.5. Network-Training-Related Parameter Settings

The computer operating system used for the experiment is Windows 11, the memory is 16 GB, the training framework used is TensorFlow 2.15, and the loss function used in the model compilation is the binary cross-entropy function, which portrays the distance between two probability distributions. The lower the cross-entropy is, the closer the two function distributions are, which is applicable to the binary classification problem. The function's expression is shown in Formula (15) as follows:

$$binary\_crossentro\,py(Y^{pred}, Y^{true}) = \frac{1}{N}\sum_{i=1}^{N}(Y_i^{true}log(Y_i^{pred}) + (1 - Y_i^{true})log(1 - Y_i^{pred})) \quad (15)$$

Adaptive moment estimation (Adam) is one of the most commonly used optimization algorithms, which combines the advantages of the AdaGrad and RMSProp methods and is able to generate adaptive learning rates for different parameters in the model. At the same time, updating of the parameters is not affected by the scalar transformation of the gradient, and because the updating step is independent of the gradient size, this ensures robust convergence during optimization. Adam is able to better handle noisy samples and is suitable for tasks with large-scale data and parameters. Therefore, in this paper, we use Adam as the optimizer of the loss function and set the initial learning rate to 0.001.

The detailed metrics for each of the neural network trainings performed in this paper are shown in Table 6.

Table 6. Neural network training configurations in this paper.

Computer Operating System	Windows 11
Training framework	Keras
Number of training samples	14,000
Number of calibration samples	500
Batch_size	64
Learning rate	0.01
Total training rounds	50
Loss function	Binary cross-entropy

The loss variations in the above three models during the training process are shown in Figures 13–15.

From the loss changes in the three models (as shown in Figures 13–15), it can be seen that for the loss values of the three models in the training process, the training set and the calibration set data show a converging trend, and there is no noticeable increase. This indicates that the three models perform better on the training set data and that the models do not appear to exhibit the overfitting phenomenon.

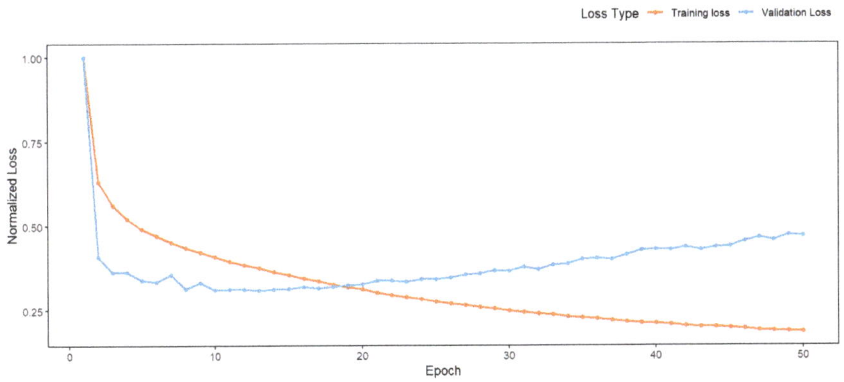

Figure 13. Variation in loss values for the training set and validation set during CNN-based algorithmic training (normalized).

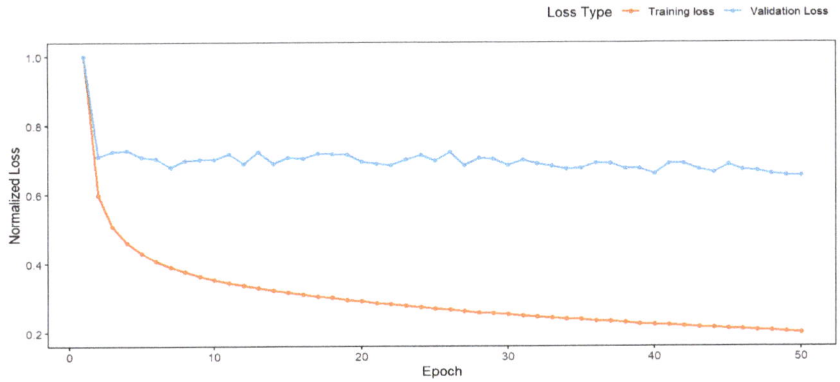

Figure 14. Variation in loss values for the training set and validation set during ResNet-based algorithmic training (normalized).

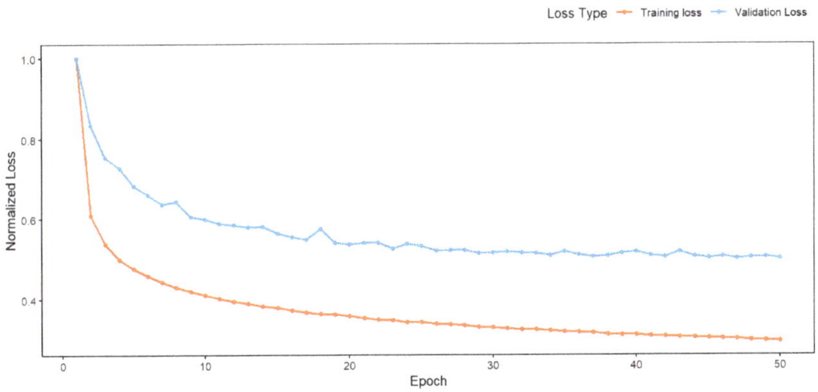

Figure 15. Variation in loss values for the training set and validation set during RNN-based algorithmic training (normalized).

3.3. Model Evaluation Indicators

The image reconstruction quality's evaluation criteria in this paper are the relative image error (RIE) and the image correlation coefficient (ICC), as shown in Equations (16) and (17):

$$RIE = \frac{\|y^* - y\|_2}{\|y\|_2} \tag{16}$$

$$ICC = \frac{cov(y^*, y)}{\sqrt{var(y^*)var(y)}} \tag{17}$$

where y^* and y are the estimated conductivity sequence and target conductivity sequence, respectively, and $cov(\ )$ and $var(\ )$ denote the covariance and mean-square deviation, respectively. The higher the ICC value is, the higher the correlation between the reconstructed image and the target image [46].

The lower the RIE value is, the more similar the reconstructed image is to the target image.

4. Results and Discussion

This study models the growth dynamics of CO_2 hydrates in reservoir environments and applies neural-network-based image reconstruction to characterize their spatial distributions. Simulated test samples, featuring circular heterogeneous bodies with variable sizes, quantities, and positions, were generated using EIDORS software(EIDORS 3.8). Both a neural network algorithm and a conventional reconstruction method were employed to predict electrical conductivity distributions, after which their performances were systematically compared across different target configurations, including varying body sizes and numbers.

4.1. Comparison of Neural Networks and Traditional Image Reconstruction Algorithms for Recognizing a Single-Target Body

In comparing the accuracies of different algorithms for single-target recognition, the original single-target model is a good conductor located in the lower right corner. The imaging results and various metrics for the different image reconstruction algorithms are shown in Figure 16.

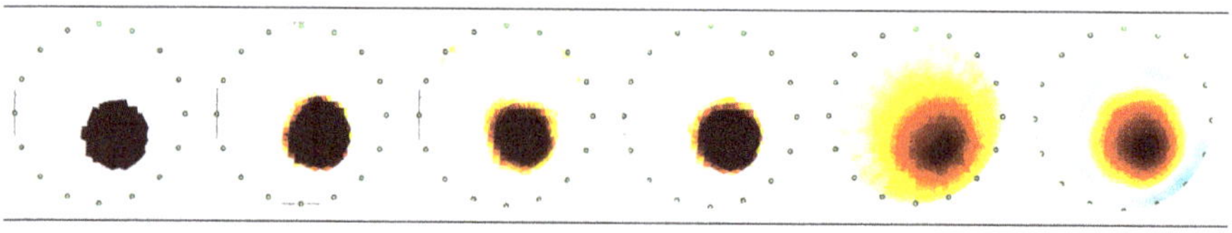

Original model	CNN	ResNet	RNN	Conjugate Gradient	Back projection
RIE	0.102	0.132	0.092	1.148	0.910
ICC	0.964	0.941	0.972	-0.881	-0.782

Figure 16. Prediction effects of different image reconstruction algorithms on single-target simulation samples (green circles: locations of applied electrodes).

The relative image errors (RIEs) of traditional image reconstruction algorithms are significantly higher than those of the neural network method, and RNN-FCNN shows the maximum and minimum values of the RIE, indicating that RNN-FCNN has a significant advantage in image reconstruction. For the image correlation coefficient (ICC), RNN-FCNN and the conjugate gradient method show the maximum and minimum ICC values, respectively, which indicates that RNN-FCNN is more capable of maintaining the image quality, whereas image reconstruction algorithms, such as the conjugate gradient method, with ICC values below 0, are still deficient in image recognition tasks.

4.2. Comparison of Recognition Performances Between Neural Networks and Traditional Image Reconstruction Algorithms for Multiple Objects

By varying the number of and distance between heterogeneities, we generated dual-heterogeneity samples at different distances, as well as triple- and quintuple-heterogeneity samples. Among these, the recognition performances of the different imaging algorithms for the dual-heterogeneity samples at varying distances are shown in Figure 17.

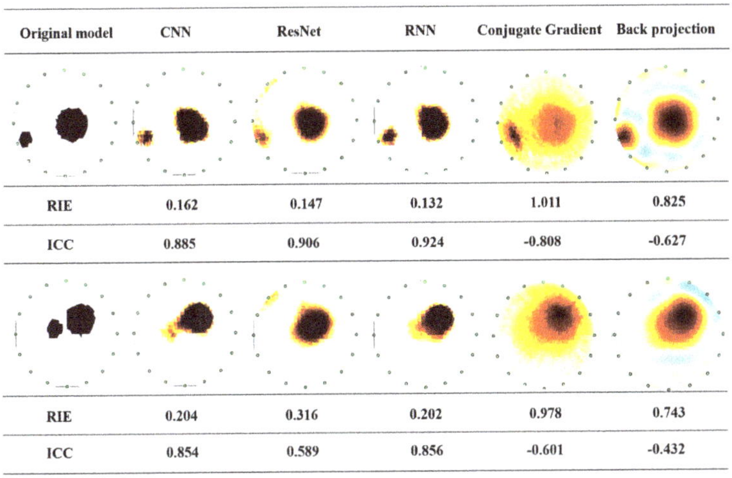

Figure 17. Effectiveness of different imaging algorithms in recognizing heterogeneous-body targets at different distances (green circles: locations of applied electrodes).

The analysis of the experimental results shows that various types of image reconstruction algorithms exhibit significant distance correlations when dealing with heterogeneous bodies. Specifically, the algorithms as a whole show higher ICC values and lower RIE values when the heterogeneous bodies are farther apart, while they show lower ICC values and higher RIE values when the heterogeneous bodies are closer together. This indicates that the spatial resolution of the heterogeneous bodies directly affects the accuracy of the target recognition, and the variation in the spacing plays an important role in the quality of the image reconstruction.

When comparing the neural-network-based image reconstruction algorithms with the traditional image reconstruction algorithms, a notable conclusion is that even under the challenging condition of the close spacing of the heterogeneous bodies, the neural network method still exhibits lower RIE values and higher ICC values, demonstrating greater robustness and adaptability. Among them, the overall performance of the RNN-FCNN model is the most prominent, and this model outperforms the other algorithms.

Next, we analyzed the recognition performances of the different image reconstruction algorithms for the triple- and quintuple-heterogeneity targets; the imaging and metric calculation results are shown in Figure 18.

Analyzing the above results, the neural-network-based algorithms perform well in terms of the RIE in the multi-target recognition task and maintain a low overall level. Among them, the RNN-FCNN model performs the best, with the lowest RIE value. In contrast, the RIE value of the traditional algorithm is about ten times that of the neural network method, which is significantly higher than the latter. In terms of the ICC, the neural network method also generally outperforms the traditional image reconstruction algorithms, among which the RNN-FNCC model has the highest average ICC value, demonstrating excellent image reconstruction quality.

Original model	CNN	ResNet	RNN	Conjugate Gradient	Back projection

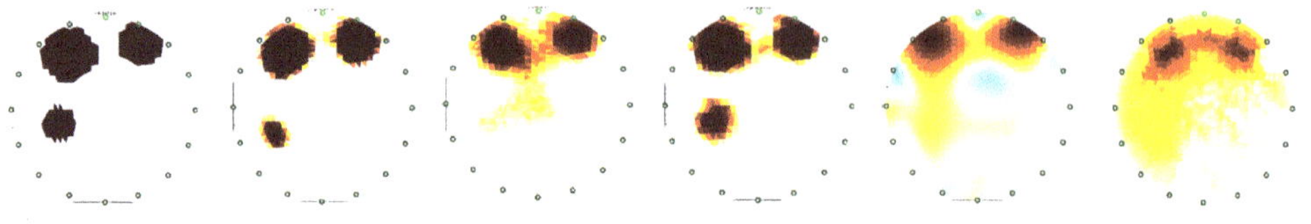

RIE	0.208	0.252	0.142	1.278	1.405
ICC	0.869	0.800	0.941	-0.779	-0.681

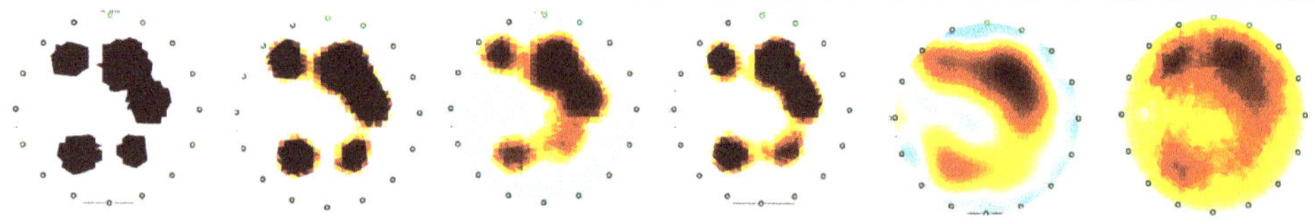

RIE	0.195	0.302	0.208	1.182	0.899
ICC	0.923	0.801	0.915	-0.732	-0.693

Figure 18. Effectiveness of different imaging algorithms in recognizing multi-heterogeneous targets (green circles: locations of applied electrodes).

In summary, the following conclusions can be drawn: (1) All the image reconstruction algorithms show the optimal reconstruction performance, i.e., the highest ICC value and the lowest RIE value, in single-target detection scenarios. (2) Compared with the traditional image reconstruction algorithms, the neural network image reconstruction algorithms show better overall performances, especially in the case of the close proximity of heterogeneous bodies and irregular shapes, with a generally higher ICC value and a lower RIE value. (3) Among the neural network algorithms, the RNN model has the most outstanding performance; whether in single-target, dual-target, or multi-target reconstruction tasks, its ICC value can reach more than 0.85, and its RIE value is the lowest, which fully proves that the RNN model has the highest reconstruction accuracy and the most stable performance in image reconstruction tasks.

4.3. Phantom Experiment

To further test the performance of the neural-network-based image reconstruction, this study uses data acquired from the Kuopio electrical impedance tomography device (KIT4) [41] to evaluate the trained network.

The KIT4 EIT system was developed by the University of Eastern Finland (UEF). The voltage-controlled current source operates at 15 kHz, with an output impedance of up to 17 MΩ. During long-term experiments, the KIT4 system achieved a signal-to-noise ratio (SNR) of up to 97.5 dB, which is relatively high.

The boundary voltages measured using the KIT4 system were used as input to the network model to predict the conductivity distribution and reconstruct the image. The imaging performances of the conjugate gradient method and the back-projection method were compared with that of the neural network method, as shown in Figure 19.

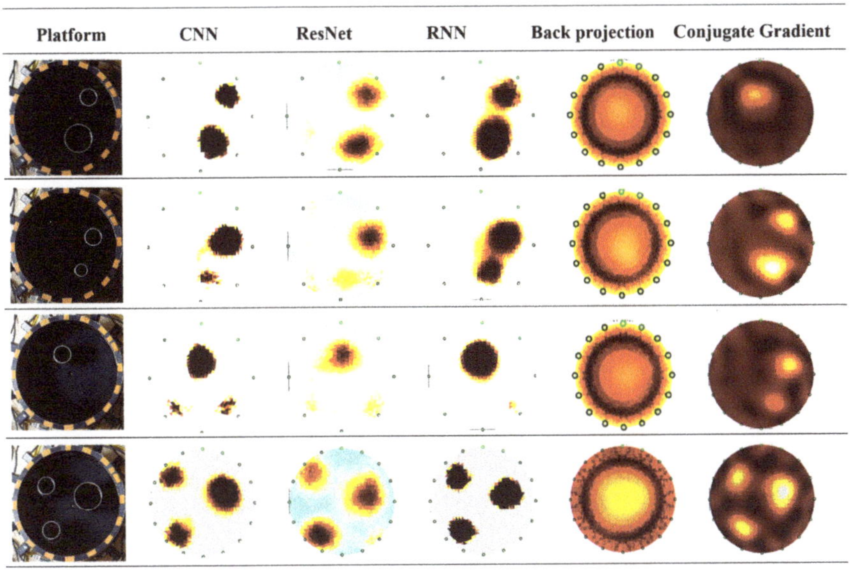

Figure 19. Performances of the different image reconstruction algorithms with KIT4 (green circles: locations of applied electrodes).

The results show that the neural network model trained using the simulated dataset is able to correctly identify the shape and position of the target metal object in the KIT4 dataset, and the model has better generalization. Meanwhile, compared with the imaging effect of the traditional algorithm, the neural network imaging results have clearer edge contours, fewer artifacts, and higher imaging quality. Therefore, the neural network image reconstruction algorithm proposed in this paper is capable of the real-time monitoring and imaging of detection targets in different industrial tasks.

5. Conclusions

This study aimed to enhance the precision of electrical resistivity tomography (ERT)-based image reconstruction, providing a novel and effective solution for imaging the growth process of CO_2 hydrates in laboratory experiments. By systematically comparing various ERT image reconstruction algorithms, including traditional iterative methods, back-projection methods, and emerging neural network algorithms, the research evaluated their imaging performances based on key metrics. The goal was to identify algorithms with superior performances and propose optimization recommendations to address their limitations. This will enable ERT technology to provide more reliable and efficient solutions for the dynamic imaging of gas hydrate formation at the microscale. Using both simulated datasets and the KIT4 experimental dataset for sample-imaging analysis, the following conclusions were drawn: **1. Neural network models outperform traditional image reconstruction algorithms in terms of imaging accuracy and reduced artifacts.** The neural-network-based image reconstruction algorithm significantly improves the issues of multiple artifacts and blurred edges present in traditional ERT-imaging algorithms, thereby enhancing the quality of the reconstructed images. Compared to traditional algorithms, neural network algorithms show significant advantages in key metrics, such as RIE and ICC, particularly in multi-target recognition tasks, where neural networks are more precise in identifying target positions and sizes. **2. Compared to CNN and ResNet, RNN performs the best, with the highest image reconstruction quality.** In the tasks involving different numbers of targets, RNN consistently achieves ICC values of 0.85 or higher and the lowest RIE values, while ResNet exhibits relatively poor imaging performance, with lighter artifacts in the backgrounds of sample images from the KIT dataset.

Based on the comprehensive analysis, future research will focus on the following directions:

1. **Designing a multi-electrode seabed simulation system based on ERT to measure dynamic data of hydrate formation.**
   Design and assemble a multi-layer reactor suitable for ERT research, with embedded electrode plates for voltage measurements during hydrate formation, providing a data foundation for ERT imaging.
2. **Advancing ERT technology from 2D to 3D.**
   By adjusting electrode placements, optimizing algorithmic structures, and developing more precise 3D ERT technology, the system can process large amounts of data more efficiently and generate high-precision resistivity distribution images.
3. **Exploring fusion strategies between neural networks and traditional image reconstruction algorithms.**
   Using the results of traditional algorithms as input to neural networks, this study aims to leverage the prior information of traditional algorithms to enhance the learning capabilities of neural networks, thereby improving the model's generalization and robustness across different experimental datasets.

Author Contributions: Writing—original draft, investigation, and formal analysis, Z.L.; writing—review and editing, Q.W.; writing—review and editing, S.L.; writing—review and editing, X.L.; writing—review and editing, J.Y.; writing—review and editing, Y.Z. (Yidi Zhang); writing—review and editing, H.Y.; conceptualization, methodology, and supervision, Y.K.; supervision and project administration, Y.Z. (Yanpeng Zheng). All authors have read and agreed to the published version of the manuscript.

Funding: This study was supported by the National Natural Science Foundation of China (Grant No. 42306238), the National Key Research and Development Program (2023YFC2811201-02), the Natural Science Foundation of Hubei Province of China (Grant No. 2023AFB005), and the Guangdong Basic Applied Basic Research Foundation (Grant No. 2022A1515110999), which are gratefully acknowledged.

Data Availability Statement: The original contributions presented in this study are included in the article. Further inquiries can be directed to the corresponding author.

Conflicts of Interest: The authors declare that they have no known competing financial interests or personal relationships that could appear to have influenced the work reported in this paper.

Notes

1 CNN: Convolutional Neural Network.
2 Resnet: Residual Neural Network.
3 RNN: Recurrent Neural Network.

References

1. Wei, N.; Pei, J.; Li, H.; Zhou, S.; Zhao, J.; Kvamme, B.; Coffin, R.B.; Zhang, L.; Zhang, Y.; Xue, J. Classification of natural gas hydrate resources: Review, application and prospect. *Gas Sci. Eng.* **2024**, *124*, 205269. [CrossRef]
2. Sloan, E.D. Fundamental principles and applications of natural gas hydrates. *Nature* **2003**, *426*, 353–359. [CrossRef] [PubMed]
3. Suess, E.; Torres, M.E.; Bohrmann, G.; Collier, R.W.; Rickert, D.; Goldfinger, C.; Linke, P.; Heuser, A.; Sahling, H.; Heeschen, K.; et al. Sea Floor Methane Hydrates at Hydrate Ridge, Cascadia Margin. In *Natural Gas Hydrates: Occurrence, Distribution, and Detection*; AGU Publications: Washington, DC, USA, 2001; pp. 87–98.

4. Song, Y.; Yang, L.; Yang, M. Natural gas hydrate, far beyond a vast resource of natural gas. *Innov. Energy* **2025**, 2, 100074. [CrossRef]
5. Teng, Y.; Zhang, D. Long-term viability of carbon sequestration in deep-sea sediments. *Sci. Adv.* **2018**, *4*, eaao6588. [CrossRef]
6. White, C.M.; Strazisar, B.R.; Granite, E.; Hoffman, J.S.; Pennline, H.W. Separation and Capture of CO_2 from Large Stationary Sources and Sequestration in Geological Formations—Coalbeds and Deep Saline Aquifers. *J. Air Waste Manag. Assoc.* **2003**, *53*, 645–715. [CrossRef]
7. Li, M.; Rojas Zuniga, A.; Stanwix, P.L.; Aman, Z.M.; May, E.F.; Johns, M.L. Insights into CO_2-CH_4 hydrate exchange in porous media using magnetic resonance. *Fuel* **2022**, *312*, 122830. [CrossRef]
8. Zhang, L.; Yang, L.; Wang, J.; Zhao, J.; Dong, H.; Yang, M.; Liu, Y.; Song, Y. Enhanced CH_4 recovery and CO_2 storage via thermal stimulation in the CH_4/CO_2 replacement of methane hydrate. *Chem. Eng. J.* **2017**, *308*, 40–49. [CrossRef]
9. Sun, L.; Meng, X.; Wang, J.; Li, G.; Kan, G.; Liu, B.; Lu, D. Research on the sediment acoustic properties based on a water coupled laboratory measurement system. *Mar. Georesour. Geotechnol.* **2019**, *38*, 595–603. [CrossRef]
10. Zhao, J.; Xu, K.; Song, Y.; Liu, W.; Lam, W.; Liu, Y.; Xue, K.; Zhu, Y.; Yu, X.; Li, Q. A Review on Research on Replacement of CH4 in Natural Gas Hydrates by Use of CO_2. *Energies* **2012**, *5*, 399–419. [CrossRef]
11. Koh, D.-Y.; Kang, H.; Lee, J.-W.; Park, Y.; Kim, S.-J.; Lee, J.; Lee, J.Y.; Lee, H. Energy-efficient natural gas hydrate production using gas exchange. *Appl. Energy* **2016**, *162*, 114–130. [CrossRef]
12. Boswell, R.; Hancock, S.H.; Yamamoto, K.; Collett, T.S.; Pratap, M.; Lee, S.-R.J. *Natural Gas Hydrates*; Elsevier: Amsterdam, The Netherlands, 2020.
13. Kumar, A.; Maini, B.; Bishnoi, P.R.; Clarke, M.; Zatsepina, O.; Srinivasan, S. Experimental determination of permeability in the presence of hydrates and its effect on the dissociation characteristics of gas hydrates in porous media. *J. Pet. Sci. Eng.* **2010**, *70*, 114–122. [CrossRef]
14. Lin, S.; Chang, X.; Wang, K.; Yang, C.; Guo, Y. Swelling damage characteristics induced by CO_2 adsorption in shale: Experimental and modeling approaches. *J. Rock Mech. Geotech. Eng.* 2024; *in press*. [CrossRef]
15. Rahman, M.J.; Fawad, M.; Chan Choi, J.; Mondol, N.H. Effect of overburden spatial variability on field-scale geomechanical modeling of potential CO_2 storage site Smeaheia, offshore Norway. *J. Nat. Gas Sci. Eng.* **2022**, *99*, 104453. [CrossRef]
16. Zhang, L.; Ge, K.; Wang, J.; Zhao, J.; Song, Y. Pore-scale investigation of permeability evolution during hydrate formation using a pore network model based on X-ray CT. *Mar. Pet. Geol.* **2020**, *113*, 104157. [CrossRef]
17. Simonetti, A.; Knapp, J.H.; Sleeper, K.; Lutken, C.B.; Macelloni, L.; Knapp, C.C. Spatial distribution of gas hydrates from high-resolution seismic and core data, Woolsey Mound, Northern Gulf of Mexico. *Mar. Pet. Geol.* **2013**, *44*, 21–33. [CrossRef]
18. Kuang, Y.; Zhang, L.; Zheng, Y. Enhanced CO_2 sequestration based on hydrate technology with pressure oscillation in porous medium using NMR. *Energy* **2022**, *252*, 124082. [CrossRef]
19. Hassanpouryouzband, A.; Joonaki, E.; Vasheghani Farahani, M.; Takeya, S.; Ruppel, C.; Yang, J.; English, N.; Schicks, J.; Edlmann, K.; Mehrabian, H.; et al. Gas hydrates in sustainable chemistry. *Chem. Soc. Rev.* **2020**, *49*, 5225–5309. [CrossRef]
20. Geselowitz, D.B. An Application of Electrocardiographic Lead Theory to Impedance Plethysmography. *IEEE Trans. Biomed. Eng.* **1971**, *BME-18*, 38–41. [CrossRef]
21. Isaksen, O.; Nordtvedt, J.E. A new reconstruction algorithm for process tomography. *Meas. Sci. Technol.* **1993**, *4*, 1464. [CrossRef]
22. Buddo, I.; Misyurkeeva, N.; Shelokhov, I.; Chuvilin, E.; Chernikh, A.; Smirnov, A. Imaging Arctic Permafrost: Modeling for Choice of Geophysical Methods. *Geosciences* **2022**, *12*, 389. [CrossRef]
23. Priegnitz, M.; Thaler, J.; Spangenberg, E.; Schicks, J.M.; Abendroth, S. Spatial resolution of gas hydrate and permeability changes from ERT data in LARS simulating the Mallik gas hydrate production test. In *EGU General Assembly Conference Abstracts*; EGU General Assembly: Vienna, Austria, 2014.
24. Li, Y.; Sun, H.; Meng, Q.; Liu, C.; Chen, Q.; Xing, L. 2-D electrical resistivity tomography assessment of hydrate formation in sandy sediments. *Nat. Gas Ind. B* **2020**, *7*, 278–284. [CrossRef]
25. Zhao, J.; Liu, C.; Chen, Q.; Zou, C.; Liu, Y.; Bu, Q.; Kang, J.; Meng, Q. Experimental Investigation into Three-Dimensional Spatial Distribution of the Fracture-Filling Hydrate by Electrical Property of Hydrate-Bearing Sediments. *Energies* **2022**, *15*, 3537. [CrossRef]
26. Liu, Y.; Chen, Q.; Li, S.; Wang, X.; Zhao, J.; Zou, C. Characterizing spatial distribution of ice and methane hydrates in sediments using cross-hole electrical resistivity tomography. *Gas Sci. Eng.* **2024**, *128*, 205378. [CrossRef]
27. Zhou, B. Electrical Resistivity Tomography: A Subsurface-Imaging Technique. In *Applied Geophysics with Case Studies on Environmental, Exploration and Engineering Geophysics*; IntechOpen: London, UK, 2018.
28. Mojica, A.; Pérez, T.; Toral, J.; Miranda, R.; Franceschi, P.; Calderón, C.; Vergara, F. Shallow electrical resistivity imaging of the Limón fault, Chagres River Watershed, Panama Canal. *J. Appl. Geophys.* **2017**, *138*, 135–142. [CrossRef]
29. Ducut, J.D.; Alipio, M.; Go, P.J.; Concepcion, R., II; Vicerra, R.R.; Bandala, A.; Dadios, E. A Review of Electrical Resistivity Tomography Applications in Underground Imaging and Object Detection. *Displays* **2022**, *73*, 102208. [CrossRef]

30. Aleardi, M.; Vinciguerra, A.; Hojat, A. A convolutional neural network approach to electrical resistivity tomography. *J. Appl. Geophys.* **2021**, *193*, 104434. [CrossRef]
31. Huang, S.M.; Xie, C.G.; Beck, M.S.; Thorn, R.; Snowden, D. Design of sensor electronics for electrical capacitance tomography. *IEE Proc. G (Circuits Devices Syst.)* **1992**, *139*, 83–88. [CrossRef]
32. Wang, Q.; Wang, H.; Cui, Z.; Xu, Y.; Yang, C. Fast reconstruction of electrical resistance tomography (ERT) images based on the projected CG method. *Flow Meas. Instrum.* **2012**, *27*, 37–46. [CrossRef]
33. Tan, C.; Lv, S.; Dong, F.; Takei, M. Image Reconstruction Based on Convolutional Neural Network for Electrical Resistance Tomography. *IEEE Sens. J.* **2019**, *19*, 196–204. [CrossRef]
34. Yang, W.; Tan, C.; Dong, F. Electrical Resistance Tomography Image Reconstruction Based on Res2net4 Network. In Proceedings of the 2021 40th Chinese Control Conference (CCC), Shanghai, China, 26–28 July 2021; pp. 6459–6465.
35. Ammaiappan, S.; Liang, G.; Tan, C.; Dong, F. A Priority-Based Adaptive Firefly Optimized Conv-BiLSTM Algorithm for Electrical Resistance Image Reconstruction. *IEEE Sens. J.* **2024**, *24*, 624–634. [CrossRef]
36. Xiao, L.Q. Finite Element Model and Image Reconstruction Algorithm for Electrical Resistance Tomography; Finite Element Model and Image Reconstruction Algorithm for Electrical Resistance Tomography. Ph.D. Thesis, Tianjin University, Tianjin, China, 2014. (In Chinese).
37. Rao, S. Electrical Impedance Tomography Image Reconstruction using MATLAB based freeware EIDORS. MATLAB Central File Exchange. 2025. Available online: https://www.mathworks.com/matlabcentral/fileexchange/59263-electrical-impedance-tomography-image-reconstruction-using-matlab-based-freeware-eidors (accessed on 15 June 2025).
38. Wu, Y.; Tahmasebi, P.; Liu, K.; Lin, C.; Kamrava, S.; Liu, S.; Fagbemi, S.; Liu, C.; Chai, R.; An, S. Modeling the physical properties of hydrate-bearing sediments: Considering the effects of occurrence patterns. *Energy* **2023**, *278*, 127674. [CrossRef]
39. Lyu, C.; Sun, Q.; Li, G.; Lyu, W.; Xie, P.; Hu, J.; Sun, W.; Cong, C. P-wave velocity evolution and interpretation of seepage mechanisms during CO_2 hydrate formation in quartz sands: Perspective of hydrate pore habits. *Energy* **2024**, *306*, 132536. [CrossRef]
40. Ye, J.; Wei, J.; Liang, J.; Lu, J.; Lu, H.; Zhang, W. Complex gas hydrate system in a gas chimney, South China Sea. *Mar. Pet. Geol.* **2019**, *104*, 29–39. [CrossRef]
41. Kourunen, J.; Savolainen, T.; Lehikoinen, A.; Vauhkonen, M.; Heikkinen, L. Suitability of a PXI platform for an electrical impedance tomography system. *Meas. Sci. Technol.* **2008**, *20*, 015503. [CrossRef]
42. Tong, W.G.; Zeng, S.C.; Zhang, L.F. Electrical Resistance Tomography and Flow Pattern Identification Method Based on Deep Residual Neural Network. *J. Syst. Simul.* **2022**, *34*, 2028–2036.
43. He, K.; Zhang, X.; Ren, S.; Sun, J. Deep Residual Learning for Image Recognition. In Proceedings of the 2016 IEEE Conference on Computer Vision and Pattern Recognition (CVPR), Las Vegas, NV, USA, 27–30 June 2016; pp. 770–778.
44. Mienye, I.D.; Swart, T.G.; Obaido, G.J.I. Recurrent Neural Networks: A Comprehensive Review of Architectures, Variants, and Applications. *Information* **2024**, *15*, 517. [CrossRef]
45. Cho, K.; Merrienboer, B.v.; Gülçehre, Ç.; Bahdanau, D.; Bougares, F.; Schwenk, H.; Bengio, Y. Learning Phrase Representations using RNN Encoder–Decoder for Statistical Machine Translation. In Proceedings of the Conference on Empirical Methods in Natural Language Processing, Doha, Qatar, 25–29 October 2014.
46. Zuo, C.L.; Li, J. Study on Electrical Impedance Imaging Based on Hybrid Total Variation Regularization Algorithm. *J. Sens. Microsyst.* **2021**, *40*, 40–43+46. (In Chinese)

Journal of
Marine Science and Engineering

MDPI

Article

Seismic Prediction of Shallow Unconsolidated Sand in Deepwater Areas

Jiale Chen, Yingfeng Xie *, Tong Wang *, Haoyi Zhou, Zhen Zhang, Yonghang Li, Shi Zhang and Wei Deng

Guangzhou Marine Geological Survey, China Geological Survey, Guangzhou 511458, China;
chenjialegmgs@163.com (J.C.); haoyi17@mails.jlu.edu.cn (H.Z.); zhangzhenhnpy@163.com (Z.Z.);
liyonghang@mail.cgs.gov.cn (Y.L.); zhang_s0823@foxmail.com (S.Z.); hahens@163.com (W.D.)
* Correspondence: xieyingfeng_gmgs@163.com (Y.X.); wangtong321@163.com (T.W.)

Abstract: Recently, shallow gas fields and hydrate-bearing sand in the deepwater area of the northern South China Sea have been successively discovered, and the accurate prediction of shallow sands is an important foundation. However, most of the current prediction methods are mainly for deep oil and gas reservoirs. Compared with those reservoirs with high degree of consolidation, shallow sandy reservoirs are loose and unconsolidated, whose geophysical characteristics are not well understood. This paper analyzes the logging data of shallow sandy reservoirs recovered in the South China Sea recently, which show that the sand content has a significant influence on Young's modulus and Poisson's ratio of the sediments. Therefore, this paper firstly constructs a new petrophysical model of unconsolidated strata targeting sandy content and qualitatively links the mineral composition and the elastic parameters of the shallow marine sediments and defines a new indicator for sandy content: the modified brittleness index (MBI). The effectiveness of MBI in predicting sandy content is then verified by measured well data. Based on pre-stack seismic inversion, the MBI is then inverted, which will identify the sandy deposits. The method proposed provides technical support for the subsequent shallow gas and hydrate exploration in the South China Sea.

Keywords: shallow unconsolidated sand; gas hydrate; seismic inversion; South China Sea

1. Introduction

Marine shallow sandy sediments are favorable reservoirs for shallow gas and gas hydrates [1–4], because when free gas enters a reservoir, free gas preferentially flows into the coarse-grained layers because the coarse-grained layers have low capillary entry pressures [5] (e.g., <0.1 MPa). However, free gas cannot enter the interbedded muds because the muds usually have much higher capillary entry pressures [5,6] (e.g., >1 MPa). In 2019–2023, the China Geological Survey discovered sandy hydrates at a shallow sand near at about 130 m below the seafloor (mbsf) in the Qiongdongnan basin, South China Sea [7], and in 2024 CNOOC discovered the LS36-1 shallow gas field in western Qiongdongnan Basin, where the gas layer is mainly located at 170–300 mbsf [8], accurate prediction of shallow sand is an important foundation for realizing the above exploration breakthroughs. Additionally, hydrate-bearing sand and shallow gas systems store significant amounts of methane as a huge marine carbon sink, which also supply gas for cold seep worldwide. They are a part of the global carbon cycle interacting with the global climate [9–12]. Moreover, large-scale gases release from hydrate dissociation can trigger submarine landslides [13]. Thus, predicting shallow sand layers holds great scientific importance.

Currently, the identification of sandy reservoir mainly relies on the interpretation of seismic sedimentary facies [14], but the results are subject to some uncertainties due to the complexity of the seismic response. Generally, sand layers exhibit higher P-wave velocity and density than surrounding clays, resulting in continuous positive-polarity reflections on seismic profiles [15]. However, shallow unconsolidated sand with higher porosity may also have lower velocities than surrounding clays depending on the sand content [16,17], leading to negative-polarity reflections. Additionally, gas or gas hydrates may be present in these sands. The velocities of gas- or hydrate-bearing sands with different saturation cause varying degrees of velocity changes [18], making interpretation more challenging. Seismic inversion may be an effective means to accurately predict sandy reservoir [19–22]. Previous studies have shown that the mechanical properties of sediments are sensitive to sand content, and the main component of sand is quartz, which is one of the brittle minerals. Rickman et al. [23] performed statistics on several wells in the Barnett Shale area and obtained a map of the relationship between formation brittleness and rock mineral fractions. Sondergeld et al. [24,25] proposed a relationship between formation brittleness and rock mineral fractions that a high brittleness is associated with a larger content of brittle minerals, and quartz, feldspar, calcite, and hydrate-associated pyrite and carbonatite are all related to brittleness. However, the shallow marine sediments have not yet been consolidated, and the geophysical characteristics are of special nature, especially for the ultra-shallow sand bodies in the deep-water area.

In addition, hydrates may occur in shallow marine sands, and the velocity of hydrate-bearing sands increase [26,27], and fee gas causes velocity reduction and frequency amplitude attenuation [28], which will affect the evaluation of sand content using seismic inversion [29]. These characteristics can be modeled by rock physics. Hydrates cementing sand grains will significantly increase the P-wave velocity, and the classic cementation model [30] can be used to link the hydrate saturation and cementing radius to elastic velocities. We can also use effective medium theory [31] or the three-phase Biot equation (TPBE) [32] to model the velocity of the hydrate-bearing sand, assuming that hydrate is part of the matrix. However, these models ignore the intergrain friction across the Hertzian contact area [33], and it often overpredicts the S-wave velocity. There was also a lack of actual logging and drilling data before, and these models are not well constrained [34,35]. Therefore, unconsolidated characteristics and the occurrences of hydrate and free gas need to be taken into account for the prediction of shallow marine sands.

In this paper, we first consider the geophysical characteristics of unconsolidated sediments by integrating the Hertz–Mindlin (H-M) contact model [36] and the anisotropic petrophysical model, then we analyze the impacts of sandy content, grain size, and gas and hydrate saturation on the elastic parameter, and we propose a new indicator for sandy reservoir: the modified brittleness index (MBI). The scientific validity of this indicator is analyzed by well log data acquired recently in the South China Sea. Finally, we use pre-stack seismic inversion to obtain the MBI, and the results are consistent with logging and drilling results.

2. Materials and Methods

2.1. Geological Settings

The Qiongdongnan Basin (QNDB), situated in the western sector of the northern South China Sea (Figure 1a). This basin hosts proven hydrocarbon accumulations including the LS17-2 gas field and LS36-1 shallow gas field [8,37]. The Guangzhou Marine Geological Survey (GMGS) has also carried out successive hydrate drilling expeditions (GMGS5-GMGS8, 2018–2022) for the shallow marine gas hydrate occurrences targeting Neogene-Quaternary sedimentary sequences [38,39].

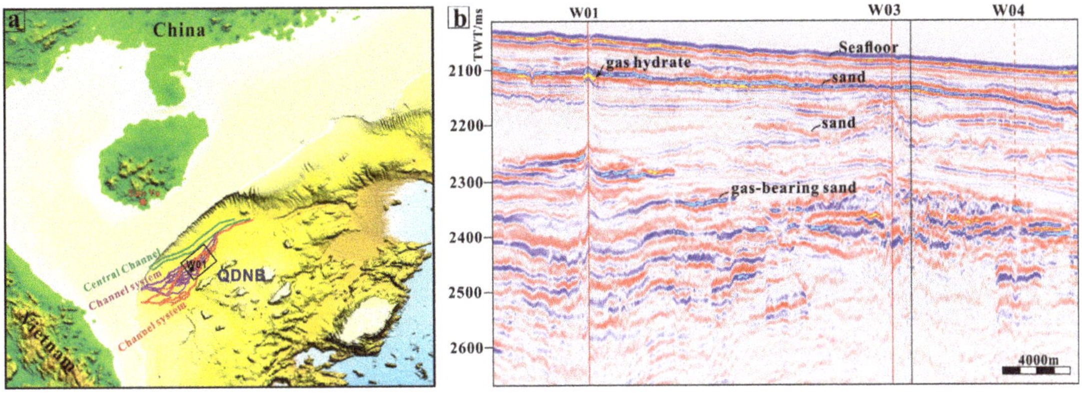

Figure 1. The location and channel system in the QDNB. (**a**) The location of QDNB. The distribution of channel system is from Meng et al. (2021) [15]. (**b**) The seismic section crossing W01, W03, and W04.

Hydrate-bearing core samples were obtained from Pleistocene strata in wells W01 and W03 (Figure 1b) [40]. The QDNB hosts syntectonic channel complexes and associated depositional architectures within Neogene sequences, as evidenced by extensive seismic stratigraphic analyses [41]. Modern sedimentological investigations reveal stacked six Late Quaternary channel-levee systems in shallow sedimentary packages, comprising four principal lithofacies [42,43]. These stratigraphic conduits serve as hydrocarbon migration pathways, with buoyancy-driven fluid migration preferentially utilizing high-permeability channel sands for vertical hydrocarbon transport.

2.2. Well Logs and Seismic Data

A comprehensive set of high-quality logging-while-drilling (LWD) measurements was successfully obtained during the operation. The acquired dataset comprised key measurements including caliper, penetration rate, natural gamma radiation, resistivity, bulk density, neutron porosity, acoustic waveforms, and nuclear magnetic resonance (NMR). Following data acquisition, an integrated analytical workflow was executed to characterize hydrate-bearing formations through sequential operations: elastic wave velocity analysis and elemental analysis (ELAN)-based reservoir assessment, combined with specialized processing and interpretation of both nuclear magnetic and resistivity image data. Additional petrophysical evaluations incorporated elemental spectral analysis and neutron capture cross-section measurements, all performed according to Schlumberger's established protocols to quantify hydrate saturation, pore space distribution, and lithological constituents.

The 3D seismic acquisition employed a triple-source, 12-streamer configuration with 5100 m streamers deployed at 75 m lateral intervals, maintaining a 5 m tow depth. Airgun arrays with a total volume of 2030 in^3 were positioned at 4 m subsurface depth, generating seismic energy sampled at 1 ms intervals with 68-fold coverage.

The processing sequence incorporated zero-phase wavelet correction, ghost wavefield attenuation, surface-related multiple elimination, followed by both pre-stack time and depth migration techniques. The dataset exhibits an operational bandwidth of 5–85 Hz with a dominant frequency peak at 58 Hz, preserving positive polarity seabed reflections critical for identifying key geological features including bottom-simulating reflectors (BSR), hydrate phase boundaries, sand deposits, and stratigraphic discontinuities.

2.3. Rock Physics Modeling

Unconsolidated clastic sandstone has a granular structure. It consists of particles like quartz and feldspar, which form a rock framework through certain contact modes. In essence, it is a soft condensed material of many discrete particles. Adjacent particles interact via contact deformation to transmit energy. A key feature of unconsolidated sediments is that particles can slide relative to each other under stress. Under specific boundary conditions, tangential forces do not just cause tangential movement for local equilibrium. Instead, they generate torques on particle centers. Under these torques, contacting particles rotate and rearrange locally without tangential displacement increments. These particle motions can also make the resultant force on particles zero, achieving local equilibrium and causing stress relaxation in the medium. The H-M particle contact model is used to characterize the equivalent modulus of unconsolidated sediments in this case [44,45]. Based on the physical property of hydrate reservoirs, we built an unconsolidated matrix with multiple minerals including quartz, hydrate, clay. The modulus and densities of different components are shown in Table 1 [31].

Table 1. Geological parameters used to construct the model.

	Mineral Components	Bulk Modulus (Gpa)	Shear Modulus (Gpa)	Density (g/cm^3)
matrix	quartz	36.6	45	2.65
	shale	20.9	6.85	2.58
	dolomite	61.5	41.1	2.79
	calcite	76.8	32	2.71
fluid	water	2.55	-	1.05
	gas	0.01	-	0.1
hydrate	-	7.7	3.2	0.92

① Modulus of the matrix

The bulk modulus and shear modulus of unconsolidated sediments can be calculated using the H-M contact model as Equations (1) and (2).

$$K_{HM} = \frac{n(1-\phi)}{12\pi R} S_n \tag{1}$$

$$G_{HM} = \frac{n(1-\phi)}{20\pi R}(S_n + 1.5S_t) \tag{2}$$

where n is the number of contacts, ϕ is porosity, R is the radius of the grain, and S_n and S_t are the normal and tangential stiffness of the contact, respectively. The unconsolidated tensor $C_{unconsolid}$ can be obtained by Equation (3).

$$C_{unconsolid} = \begin{bmatrix} C'_{11} & C'_{11}-2C'_{44} & C'_{11}-2C'_{44} & 0 & 0 & 0 \\ C'_{11}-2C'_{44} & C'_{11} & C'_{11}-2C'_{44} & 0 & 0 & 0 \\ C'_{11}-2C'_{44} & C'_{11}-2C'_{44} & C'_{11} & 0 & 0 & 0 \\ 0 & 0 & 0 & C'_{44} & 0 & 0 \\ 0 & 0 & 0 & 0 & C'_{44} & 0 \\ 0 & 0 & 0 & 0 & 0 & C'_{44} \end{bmatrix} \tag{3}$$

where C'_{ij} is unconsolidated (isotropic) elastic stiffness tensor of each component, and

$$C'_{11} = K_{HM} + \frac{4}{3}G_{HM}, C'_{44} = G_{HM}.$$

For sandy reservoir, grain size is very important. Figure 1 shows a schematic diagram of the grain contacts, and we simply divide the grain contact relationship into two categories: (1) uniform grain size (Figure 2a); (2) non-uniform grain size (Figure 2b).

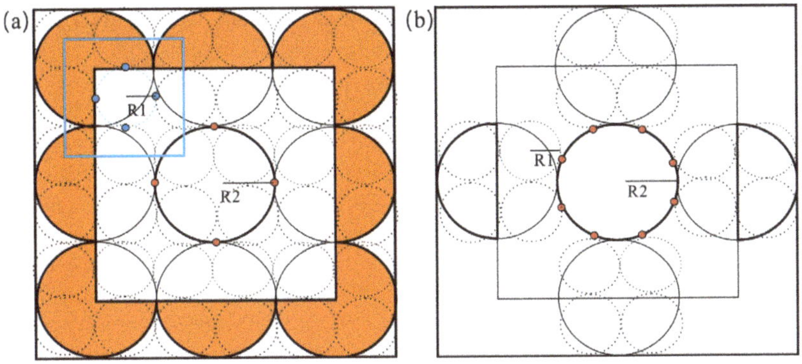

Figure 2. Schematic diagram of grain contact relations; (**a**) contact relations of grains with the same size; (**b**) contact relations of grains with different size. The red dots indicate contact points.

(1) The effect of grain size on modulus

Assuming that all grains are of the same radius, R_0, and the number of contacts of individual grain is n_0, as shown in Figure 2a, the bulk modulus and shear modulus can be achieved from Equations (4)–(7).

$$K_{HM} = \frac{n_0(1-\phi)}{12\pi R_0}\frac{4G}{1-v}[\frac{3\pi(1-v)P}{2n_0(1-\phi)G}]^{\frac{1}{3}}\frac{R_0R_0}{R_0+R_0} = \frac{1}{2}[n_0(1-\phi)]^{\frac{2}{3}}K_0 \quad (4)$$

$$K_0 = \frac{1}{12\pi}\frac{4G}{1-v}[\frac{3\pi(1-v)P}{2G}]^{\frac{1}{3}} \quad (5)$$

$$G_{HM} = \frac{n_0(1-\phi)}{20\pi R_0}[\frac{3\pi(1-v)P}{2n_0(1-\phi)G}]^{\frac{1}{3}}\frac{R_0R_0}{R_0+R_0}(\frac{4G}{1-v}+1.5\frac{8G}{2-v}) = \frac{1}{2}[n_0(1-\phi)]^{\frac{2}{3}}G_0 \quad (6)$$

$$G_0 = \frac{1}{20\pi}[\frac{3\pi(1-v)P}{2G}]^{\frac{1}{3}}(\frac{4G}{1-v}+1.5\frac{8G}{2-v}) \quad (7)$$

It can be seen that if the grain size is uniform, the modulus is independent from the grain size for a fixed porosity. However, in general, the larger the grain size, the larger the porosity, so the modulus of coarse sand reservoir is smaller than the modulus of fine sand reservoir.

(2) Effect of sorting on modulus

The grain size distribution of actual sandy reservoirs is within a certain range, the better the degree of sorting, the smaller the range of grain size distribution, and the average grain size of coarse sandstone is larger than that of fine sand. For the H-M model, different degrees of sorting affect the number of contact surfaces of individual grains, porosity, and grain contact radius. As shown in Figure 2b, assuming that the radius of two grains are R_1, R_2, respectively, we can achieve the bulk modulus and shear modulus in Equations (8) and (9).

$$K_{HM}' = 2\frac{\overline{R}}{R}[\frac{n(1-\phi)}{n_0(1-\phi_0)}]^{\frac{2}{3}}K_{HM}{}^0 = \sqrt{\frac{1}{R}}\sqrt{2\overline{R}[\frac{n(1-\phi)}{n_0(1-\phi_0)}]^{\frac{2}{3}}}K_{HM}{}^0 = F_RF_SK_{HM}{}^0 \quad (8)$$

$$G_{HM}' = 2\frac{\overline{R}}{R}[\frac{n(1-\phi)}{n_0(1-\phi_0)}]^{\frac{2}{3}}G_{HM}{}^0 = \sqrt{\frac{1}{R}}\sqrt{2\overline{R}[\frac{n(1-\phi)}{n_0(1-\phi_0)}]^{\frac{2}{3}}}G_{HM}{}^0 = F_RF_SG_{HM}{}^0 \quad (9)$$

where ${K_{HM}}^0$, ${G_{HM}}^0$, n_0, ϕ_0 are the bulk modulus, shear modulus, number of contacts, and porosity calculated for the same grain size (best case for sorting). F_R is only related to grain size, and F_s is related to sorting. $\overline{R} = \frac{R_1 R_2}{R_1 + R_2}$. If grain size is large and the sorting is good, F_R and F_s are both small, so the modulus is small. If hydrates form in such reservoirs and cause velocity increase, the velocity increase is less pronounced than in reservoirs with small grain sizes and poor sorting. The load distribution within the hydrate-bearing sediment depends on the relative size of the hydrate granules and sediment grains [28]. For reservoirs with good sorting, hydrates are uniformly distributed in pores. As hydrate saturation increases, the velocity of hydrate-bearing sediment increases gradually. When saturation exceeds 40%, hydrate particles contact sediment particles, forming a load-bearing morphology [28], and velocity increases more significantly. In poorly sorted reservoirs, hydrates form a non-uniform in sediments with varying pore sizes, forming a patchy distribution. The velocity increases more rapidly at high hydrate saturation when hydrate begins bridging between multiple sediment grains across a pore of fully hydrate-saturated patches [46]. The number of hydrate patches will increase the bulk stiffness of the sediments, which reflect the interaction among patches as the number of patches increase and their relative distance becomes less than two times than patch size.

② Vertical transversely isotropic (VTI) matrix and modulus calculation

The most common anisotropic model is the transversely isotropic (VTI) model. This model is frequently employed to describe the characteristics of layered sedimentation. White et al. [47] studied the stiffness tensor of the model, as shown in Equation (10).

$$C_{VTI} = \begin{bmatrix} C_{11} & C_{12} & C_{13} & 0 & 0 & 0 \\ C_{12} & C_{11} & C_{13} & 0 & 0 & 0 \\ C_{13} & C_{13} & C_{33} & 0 & 0 & 0 \\ 0 & 0 & 0 & C_{44} & 0 & 0 \\ 0 & 0 & 0 & 0 & C_{44} & 0 \\ 0 & 0 & 0 & 0 & 0 & 0.5(C_{11} - C_{12}) \end{bmatrix} \quad (10)$$

According to the Backus average [48], each elastic coefficient in the tensor is calculated by Equation (11).

$$\begin{gathered} C_{11} = \left\langle \frac{4\mu(\lambda+\mu)}{\lambda+2\mu} \right\rangle + \left\langle \frac{1}{\lambda+2\mu} \right\rangle^{-1} \left\langle \frac{\lambda}{\lambda+2\mu} \right\rangle^2,\ C_{12} = \left\langle \frac{2\mu\lambda}{\lambda+2\mu} \right\rangle + \left\langle \frac{1}{\lambda+2\mu} \right\rangle^{-1} \left\langle \frac{\lambda}{\lambda+2\mu} \right\rangle^2 \\ C_{13} = \left\langle \frac{1}{\lambda+2\mu} \right\rangle^{-1} \left\langle \frac{\lambda}{\lambda+2\mu} \right\rangle,\ C_{33} = \left\langle \frac{1}{\lambda+2\mu} \right\rangle^{-1},\ C_{44} = \left\langle \frac{1}{\mu} \right\rangle^{-1} \end{gathered} \quad (11)$$

where λ and μ refer to the Lame parameters of each isotropic thin layer (or each component) obtained by equations $K_{HM} = \lambda + \frac{2}{3}\mu$ and $G_{HM} = \lambda + 2\mu$. Operator $< >$ represents the weighted average of the variables in Equation (11).

③ Fluid substitution

The Brown–Korringa equation [49] was used to add fluid to the connected dry micropores. Fluid substitution for anisotropic rocks can be performed with Equation (12).

$$s_{ijkl}^{sat} = s_{ijkl}^{dry} - \frac{(s_{ijaa}^{dry} - s_{ijaa}^{0})(s_{bbkl}^{dry} - s_{bbkl}^{0})}{\phi'(\beta_{f1} - \beta_0) + (s_{ccdd}^{dry} - s_{ccdd}^{0})} \quad (12)$$

where s_{ijkl}^{sat}, s_{ijkl}^{dry} are the compliance tensors for saturated rock and dry rock, respectively. s_{ijkl}^{0} is compliance tensor for a mineral. β_{f1}, β_0 are compressibility for the fluid and mineral, respectively. ϕ' is nuclear magnetic porosity that will reduce with an increase in hydrate saturation. The total porosity, ϕ, nuclear magnetic porosity, ϕ', and hydrate saturation (S_{gh}) should satisfy Equation (13),

$$\phi' = \phi - S_{gh} \cdot \phi \tag{13}$$

2.4. Seismic Inversion

AVO (Amplitude versus offset) analysis and inversion are effective methods to quantify stratigraphic physical properties [50–53]. Subsequent analyses in this study will employ Young's modulus and Poisson's ratio as key petrophysical parameters for quantitative estimation of sand fraction in clastic formations, but there are some challenges to the inversion of Young's modulus and Poisson's ratio. On the one hand, Downton [54] pointed out that compared with other elastic parameters, the stability of density inversion is poor, and there is a large uncertainty even in the case of large angles, so it is difficult to invert the density in the indirect calculation of Young's modulus. On the other hand, indirect calculations themselves have large cumulative errors [55,56]. To address this issue, Zong et al. [57,58] derived the YPD (Young's modulus–Poisson's ratio–Density) elastic impedance equation of based on the Aki approximation [59,60], and the direct seismic extraction of Young's modulus and Poisson's ratio can be realized based on the equation, as shown in Equations (14)–(16).

$$\begin{aligned} R(\theta) = \left(\tfrac{1}{4}sec^2\theta - 2ksin^2\theta\right)\tfrac{\Delta E}{E} + \left(\tfrac{1}{4}sec^2\theta\tfrac{(2k-3)(2k-1)^2}{k(4k-1)} + 2ksin^2\theta\tfrac{1-2k}{3-4k}\right)\tfrac{\Delta\sigma}{\sigma} \\ +\left(\tfrac{1}{2} - \tfrac{1}{4}sec^2\theta\right)\tfrac{\Delta\rho}{\rho} \end{aligned} \tag{14}$$

Based on this equation, the objective functional is established under the framework of Bayesian theory. Assuming that the parameters to be inverted obey the Cauchy distribution, the posterior probability distribution of the parameters is as follows:

$$p(R|D) \propto \prod_{i=1}^{M}\left[\frac{1}{1+\frac{R^2}{\delta_m{}^2}}\right] \times exp\left[-\frac{(D-GR)^T(D-GR)}{2\delta_n{}^2}\right] \tag{15}$$

where $\delta_m{}^2$, $\delta_n{}^2$ are the variances of the inversion parameters and noise, respectively. R is the matrix of reflection coefficients related to Young's modulus, Poisson's ratio, and density. D is the observed seismic data. Under the maximum posterior probability, after adding the initial model constraints, the target parameters become

$$F(R) = (D-GR)^T(D-GR) + 2\delta_n{}^2\sum_{i=1}^{M} ln\left(1+\frac{R_i{}^2}{\delta_m{}^2}\right) + S \tag{16}$$

S is the constraint coefficient related to Young's modulus, Poisson's ratio, and density, and the solution is optimized by the damped least squares algorithm.

3. Results

3.1. The Elastic Characteristic of Sandy Sediments

Based on the above model, we change the grain size (0.02 mm~0.1 mm) and quartz content (0~70%) and analyze the effect of grain size and quartz content on Young's modulus and Poisson's ratio (Figure 3). The larger the grain size, the smaller Young's modulus and the larger the Poisson's ratio. The higher the quartz content, the larger the velocity and the smaller the Poisson's ratio. This understanding is in line with the conventional understanding, but the shallow marine sands in the South China Sea show some variability.

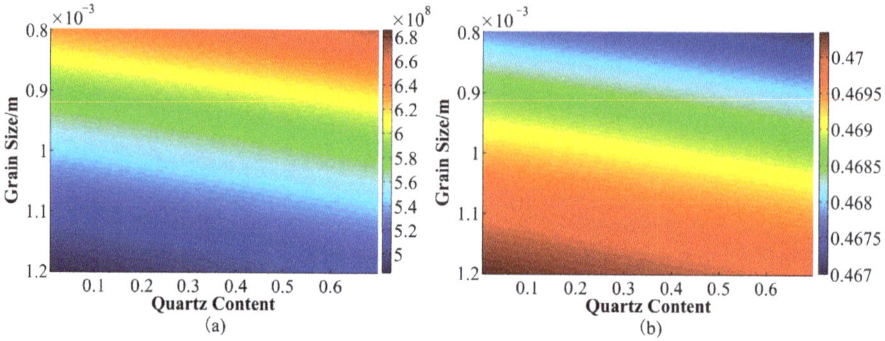

Figure 3. Effect of grain size and quartz content on Young's modulus (**a**) and Poisson's ratio (**b**). The color axis is Young's modulus (**a**) and Poisson's ratio (**b**), respectively.

Well W04 is a well drilled in the Qiongdongnan basin, where several sets of sands were discovered (Figure 4). The sandy layers exhibit lower gamma, lower Young's modulus, and higher Poisson's ratio, which are quite different from the elastic characteristics of conventional sands [57].

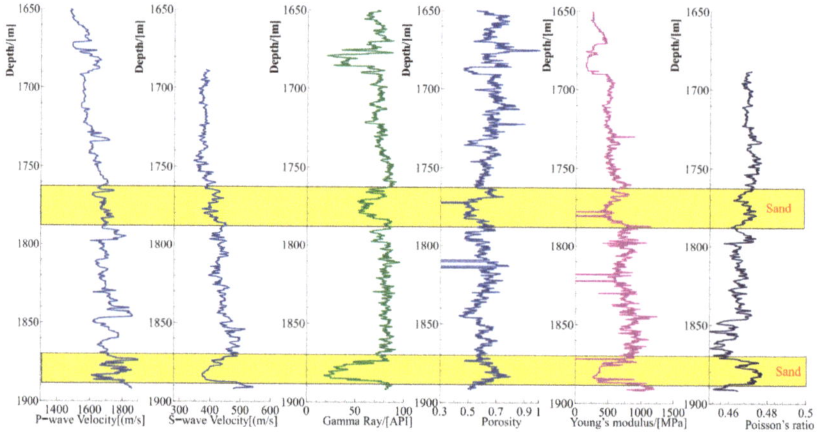

Figure 4. Well logs in well W04. The well logs are P-wave velocity, S-wave velocity, Gamma Ray, Porosity, Young's modulus, and Poisson's ratio, respectively.

The Quartz content, derived from logging interpretation, serves as an indicator of sand content. A Quartz content exceeding 60% suggests a sand layer. The P-wave velocity maintains a positive correlation with the S-wave velocity, while the quartz content of the high-velocity layer is low (Figure 5a). The negative correlation between Young's modulus and quartz content and Poisson's ratio is obvious (the arrow in Figure 5b). And overall, Young's modulus is negatively correlated with the porosity (the arrow in Figure 5c). Therefore, the low Young's modulus and high Poisson's ratio of the highly porous sand layer contradicts the understanding that the modulus of sand is larger and Poisson's ratio is smaller than the surrounding clays, since quartz grains are not prone to deformation compared with clay grains, so this feature is not caused by lithology, and there should be other reasons.

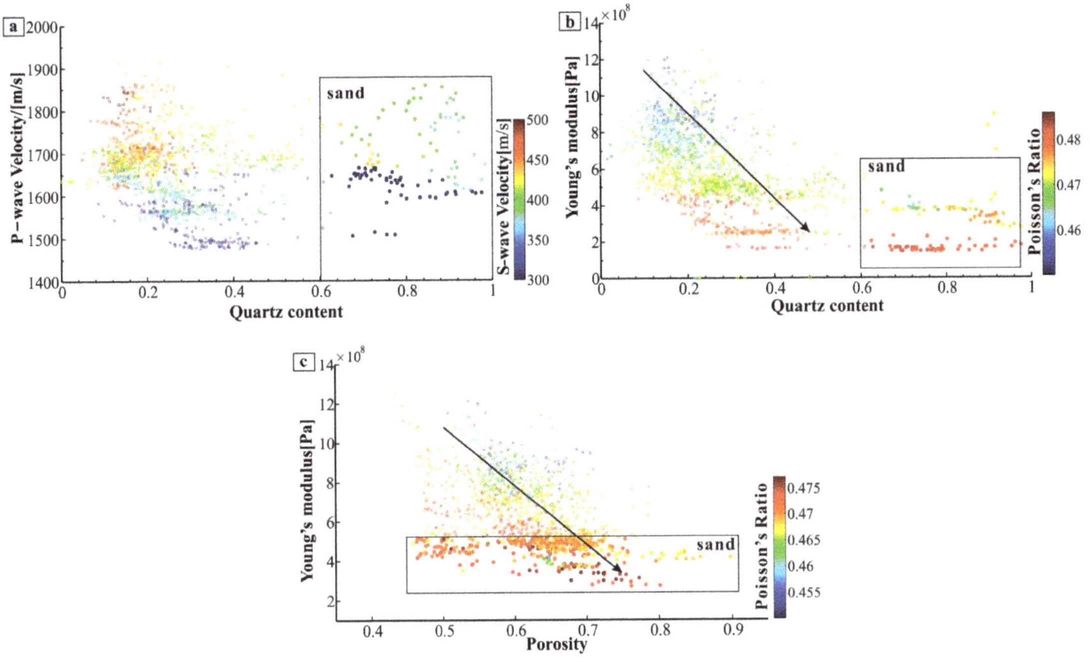

Figure 5. The crossplot of different parameters. (**a**) The crossplot of quartz content and P-wave velocity. The color bar is S-wave velocity. (**b**) the crossplot of quartz content and Young's modulus; The color bar is Poisson's ratio. (**c**) the crossplot of Young's modulus and porosity. The color bar is Poisson's ratio.

In Figure 3, we simply changed the grain size versus quartz content without noticing the relationship between grain size and quartz content. In fact, a suitable relationship between quartz content and grain size should be considered in the model. In this paper, this relationship was obtained by fitting the core data (Figure 6a). Under this relationship, the relationship between quartz content and Young's modulus and Poisson's ratio is obtained again based on the above modeling, and as shown in Figure 6b,c, the higher the quartz content, the lower the Young's modulus and the higher the Poisson's ratio, which is consistent with the actual logging data. This also verifies the rationality of the above model at the same time.

3.2. Elastic Indicator of Sand Content

3.2.1. Water Saturated Sandy Sediments

The brittle index (BI) of the rock can be calculated from Young's modulus and Poisson's ratio [23], as shown in Equation (17).

$$BI = 0.5\frac{E - E_{\min}}{E_{\max} - E_{\min}} + 0.5\frac{\sigma_{\max} - \sigma}{\sigma_{\max} - \sigma_{\min}} \tag{17}$$

However, the above equation is no longer suitable for the shallow marine sandy sediment in the SCS. Based on the model as well as the analysis of well logging, we newly define a modified brittle index (MBI) and rewritten Equation (17) as Equation (18) in this paper, and MBI is sensitive to the sand content of unconsolidated reservoir.

$$MBI = a\frac{E_{\max} - E}{E_{\max} - E_{\min}} + b\frac{\sigma - \sigma_{\min}}{\sigma_{\max} - \sigma_{\min}} \tag{18}$$

where a and b equal to 0.5, generally, but also need to be adjusted according to the real situation. BI and MBI in well W2 were calculated using Equations (17) and (18), and the results are shown in Figure 7. The BI could not identify the sand layers, and the relationship between BI and porosity was not obvious (Figure 7a). However, the MBI well distinguished the high quartz content layer from the low quartz content layer, and in the high quartz content layer section, the MBI was higher, and the porosity was also larger (the circle in Figure 7b). After obtaining Young's modulus and Poisson's ratio by seismic inversion, the MBI is calculated according to Equation (18).

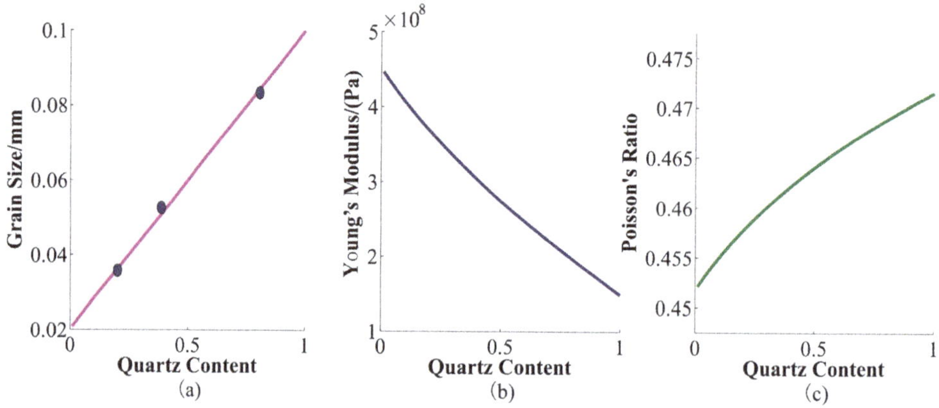

Figure 6. Quartz content versus Young's modulus and Poisson's ratio after considering the relationship between grain size and quartz content; (**a**) Grain size versus quartz content. The black dots refer to the cores obtain in the study area; (**b**) Quartz content versus Young's modulus; (**c**) Quartz content versus Poisson's ratio.

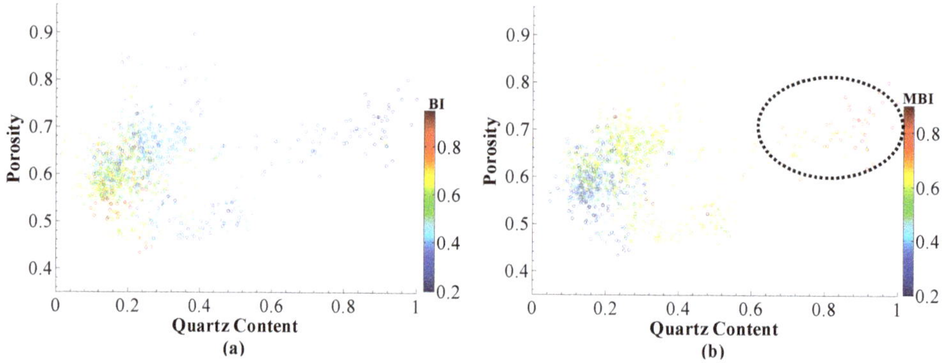

Figure 7. The crossplots of quartz content and porosity. The color axis is BI (**a**) and MBI (**b**).

3.2.2. Gas-Bearing Sediments

Free gas will have a large effect on Poisson's ratio; does it affect MBI? A gas-bearing layer was recovered in well W09-2019 [39] in the Qiongdongnan basin (Figure 8). The gas layer is at about 1870~1907 m. The Poisson's ratio of the gas-bearing layer is obviously reduced, and Young's modulus is a medium-low value (Figure 9). It can be seen that the section of high MBI still corresponds to the high sand content and high porosity (the red circle in Figure 10). This is because gas occurrence reduces Young's modulus and Poisson's ratio, and MBI can offset the effect of gas influence to a certain extent. By adjusting a and b in Equation (18), we can better eliminate the influence of the gas on MBI.

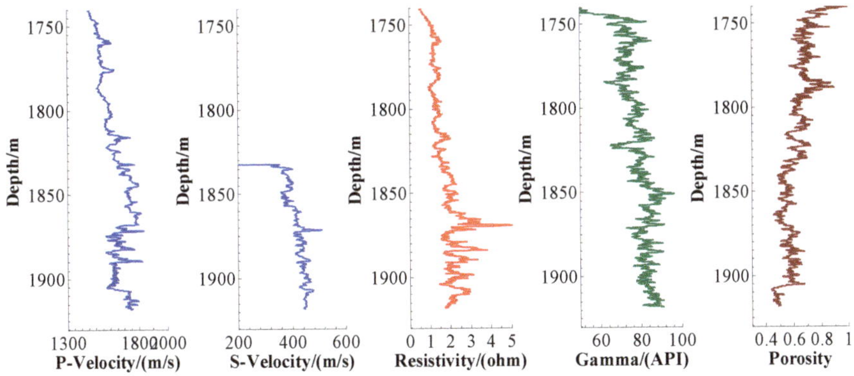

Figure 8. Well logs in well W09-2019. The well logs are P-wave velocity, S-wave velocity, Resistivity, Gamma Ray, Porosity, respectively.

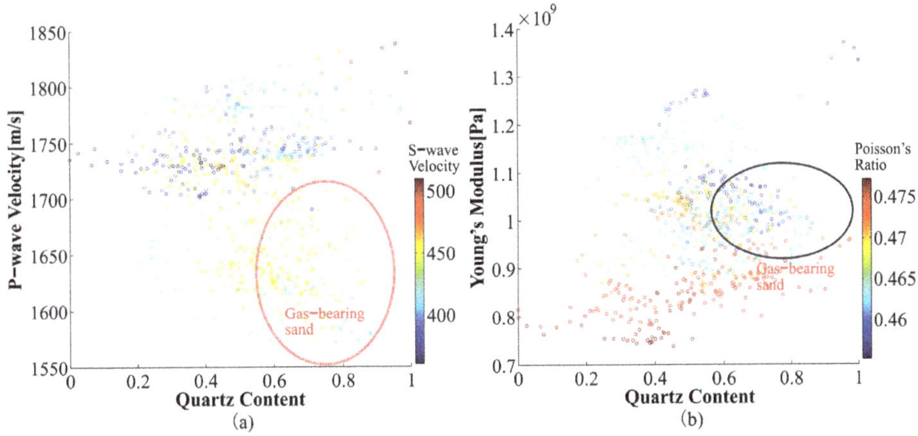

Figure 9. The crossplot of different parameters in well W09-2019. (**a**) The crossplot of quartz content and P-wave velocity; (**b**) The crossplot of quartz content and Young's modulus.

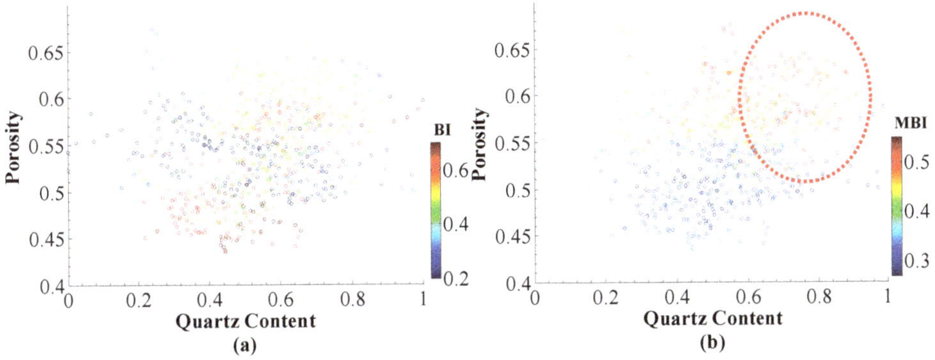

Figure 10. The crossplots of quartz content and porosity in well W09-2019. The color axis is BI (**a**) and MBI (**b**).

3.2.3. Hydrate-Bearing Sand

Based on the above model, we analyze how hydrate saturation and quartz content influence Young's modulus and Poisson's ratio. As shown in Figure 11, Young's modulus increases, and Poisson's ratio decreases with increasing hydrate saturation. The change rate of Young's modulus and Poisson's ratio caused by hydrate saturation is almost the same, while the change rate caused by quartz content is smaller. Therefore, the MBI of the hydrate layer will be smaller. Assuming that the MBI of water saturated sand is MBI_1, MBI_1 is proportional to the quartz content. Due to the hydrate occurrence, the MBI of the hydrate-bearing sand is reduced to MBI_2, which is almost inversely proportional to the hydrate saturation. Therefore, the hydrate saturation has a greater impact on MBI. If near the bottom of the hydrate layer, the MBI is obviously smaller than the surrounding sediments, which can be used as an indicator of hydrate presence. The smaller the MBI, the higher the hydrate saturation. If the MBI is larger, it indicates that the quartz content has a greater effect on the increase in the MBI than the reduction effect of hydrate saturation, and in this case, the quartz content is high.

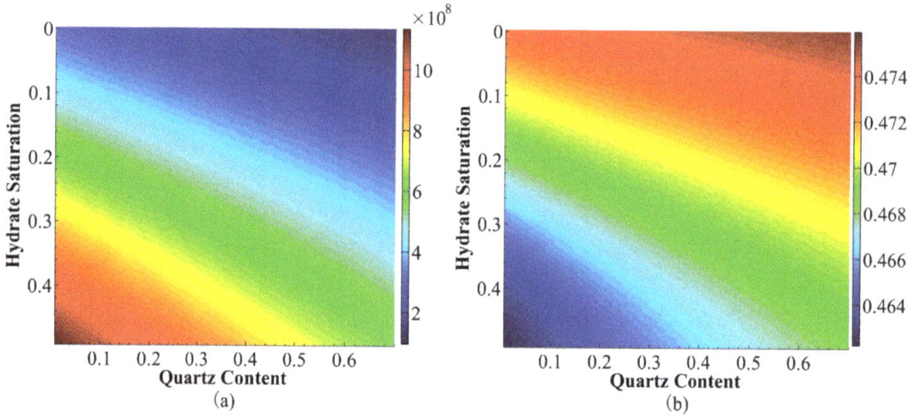

Figure 11. Effect of hydrate saturation and quartz content on Young's modulus (**a**) and Poisson's (**b**).

A hydrate-bearing sand layer at about 1568~1582 m is discovered in well W01-2019 [31] (Figure 12). The hydrate-bearing sand has a large Young's modulus and a low Poisson's ratio. Young's modulus is proportional to the quartz content (arrow 1 in Figure 13a), and Poisson's ratio is inversely proportional to the quartz content (arrow 1 in Figure 13b). However, Young's modulus at the non-hydrate layer and the low-saturation hydrate layer (resistivity 1.5~5 ohm) is low, and Young's modulus is almost unchanged with the quartz content (arrow 2 in Figure 13a), the Poisson's ratio increases slightly (arrow 2 in Figure 13b), and the quartz content of this layer is also low.

Therefore, in the hydrate interval, MBI cannot predict well the quartz content. However, in practical applications, we target those formations with high saturation and high sand content, and the MBI in those sections are relatively obvious. According to the results of petrophysical model analysis, if there is a high saturation hydrate (greater than 30%), the MBI decreases significantly. If the hydrate saturation is low, but MBI is large at the same time, it indicates that the quartz content is high.

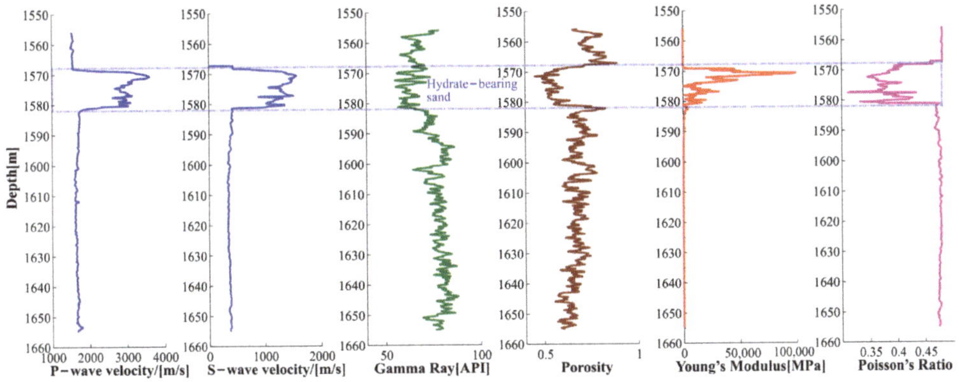

Figure 12. Well logs in well W01-2019. The well logs are P-wave velocity, S-wave velocity, Gamma Ray, Porosity, Young's modulus, and Poisson's Ratio, respectively.

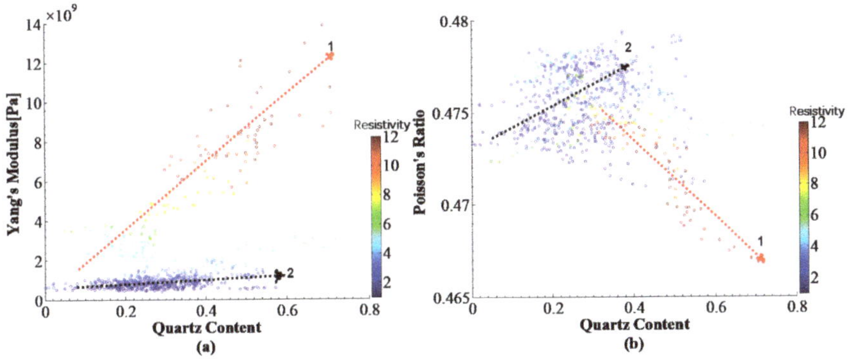

Figure 13. The crossplots of quartz content versus Young's modulus (**a**) and Poisson's ratio (**b**) in well W3. The color axis is resistivity.

3.3. Seismic Inversion Results

The study area is located in the Lingnan Low uplift in the deep-water area of QDNB [7,47]. The water depth of the study area ranges from 1500 to 1665 m, and the submarine topography is relatively flat. The previous geological survey shows that the many layers of shallow sand is widely distributed. In this paper, one of the seismic lines was selected for actual data testing. Figure 14 shows the overlay display of the post-stack seismic profile and the inverted P-wave impedance. There are obvious amplitude anomalies in the range of 2600 ms~2800 ms at CDP 1~1500, and there are obvious high and low impedance anomalies, which indicate the occurrences of free gas and hydrate in sands.

Recent studies in sedimentology indicate that multi-stage channel-natural levee sedimentary systems are developed in the shallow sediments. Channel systems are identified in the Quaternary strata, and the channel-related sedimentary facies include channel-filling facies, levee facies, crevasse splay facies, and lobes facies [32]. It is suspected that multiple small channels are developed at 2700 ms between CDP 150~1650 (Figure 15), so the sand content is large.

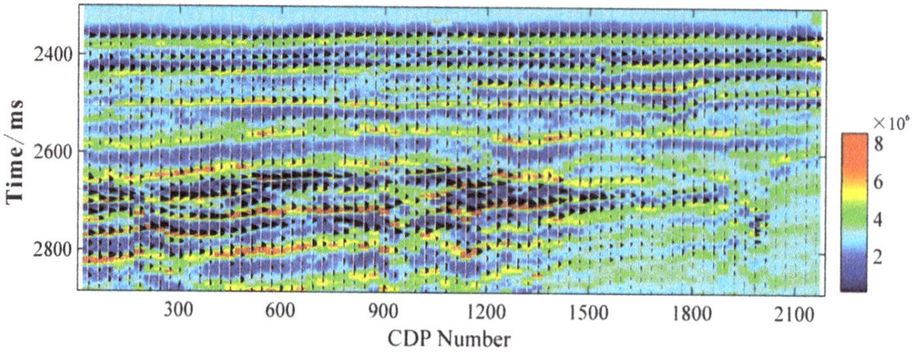

Figure 14. The overlay display of the post-stack seismic profile and the inverted P-wave impedance.

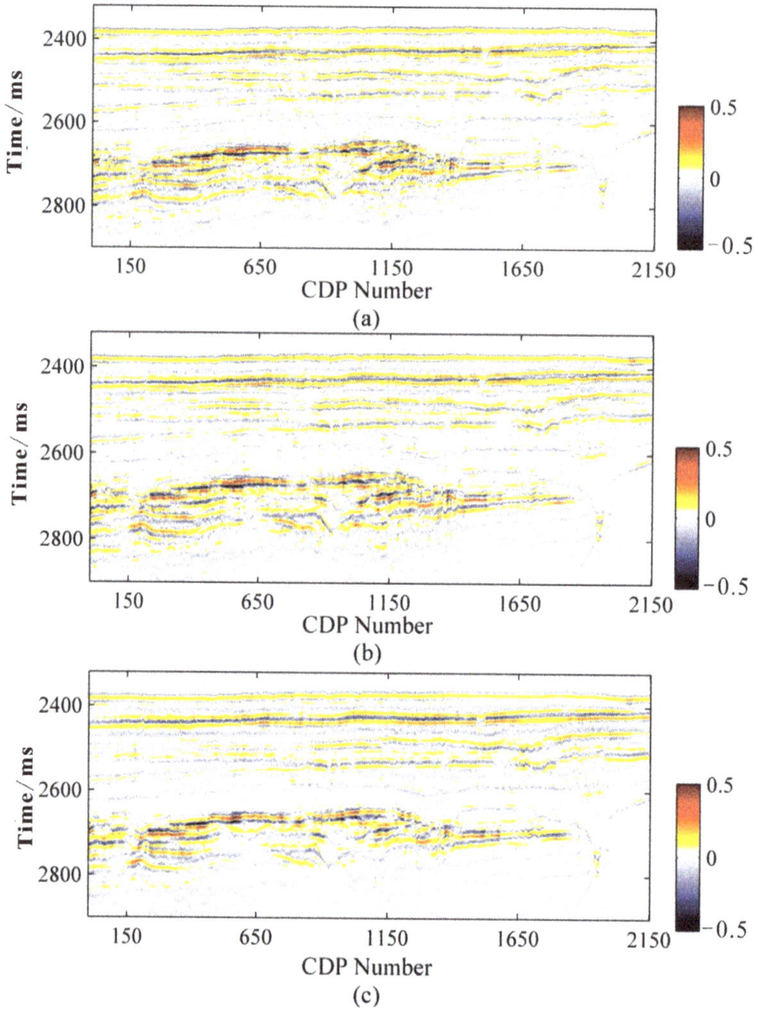

Figure 15. Angle stack seismic profiles with different angle ranges. (**a**) 1~11°; (**b**) 11~21° (**c**) 22~35°.

Seismic waves experience varying travel paths, attenuation, and reflection characteristics at different incidence angles, altering the frequency and phase of seismic wavelet. Wide-angle waves encounter stronger formation absorption and dispersion, causing wavelet broadening and frequency attenuation. Thus, amplitude varies in stacked data from different angles. When extracting seismic wavelet, the Aki–Richards equation is used to calculate reflection coefficients at diverse angles using the well logs (P-velocity, S-velocity and density) and then using the well-seismic tie to extract wavelets for each angle. Then, seismic inversion was carried out. Figure 16 shows the inverted MBI, and the MBI are generally high, with anomalies in both the shallow and deep layers. MBI has good lateral continuity and good stratigraphy. MBI can depict more information about mineral composition, which is less affected by other factors. There are certain high sand anomalies in the shallow layer and even on the surface of the seabed, and the anomalies are more obvious in the impedance anomaly area, which may be related to the development of small channels. However, in seismic data, we only notice some anomalies at 2600 ms~2800 ms, but in fact, there are many layers of sand even in the shallower section. Compared with seismic data, MBI inversion has significant advantages in the prediction of sand layers.

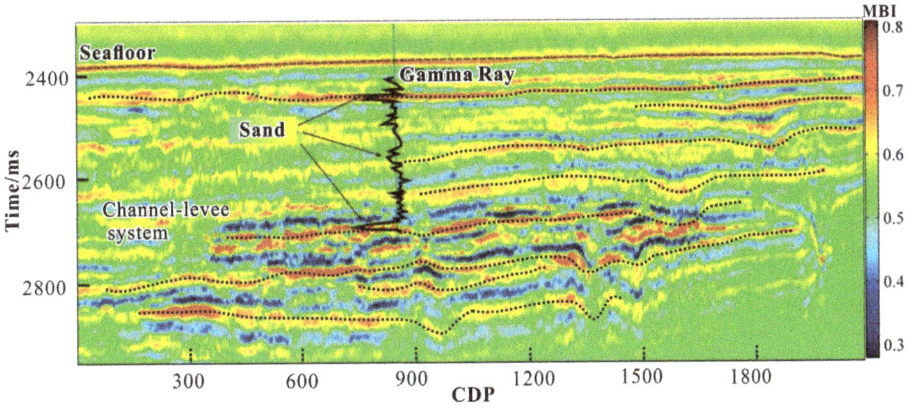

Figure 16. The inverted MBI profile.

4. Discussion

4.1. *Effectiveness of MBI*

In general, due to the higher density of sandy particles compared to muddy sediment particles, sand layers typically exhibit higher acoustic velocities and produce positive polarity reflections on seismic profiles when their porosity is similar to that of muddy sediments. However, the porosity of shallow sandy sediments may be larger or lower than that of muddy sediments. As shown in Figure 4, the elastic properties of the two sand layers differ significantly, indicating that using only velocity or density parameters is ineffective for identifying sand layers. Figure 17 shows the inversion results for P-wave velocity and density. These results reveal no obvious anomalies in the shallow sand layer, suggesting that P-wave velocity and density parameters alone cannot effectively distinguish shallow sandy sedimentary layers. Previous studies have shown that the study area contains multiple sand bodies from deep to shallow depths, with sand almost throughout the entire area. This is well reflected in the MBI inversion results, which also shows better stratification, further validating the effectiveness of MBI in identifying sand layers.

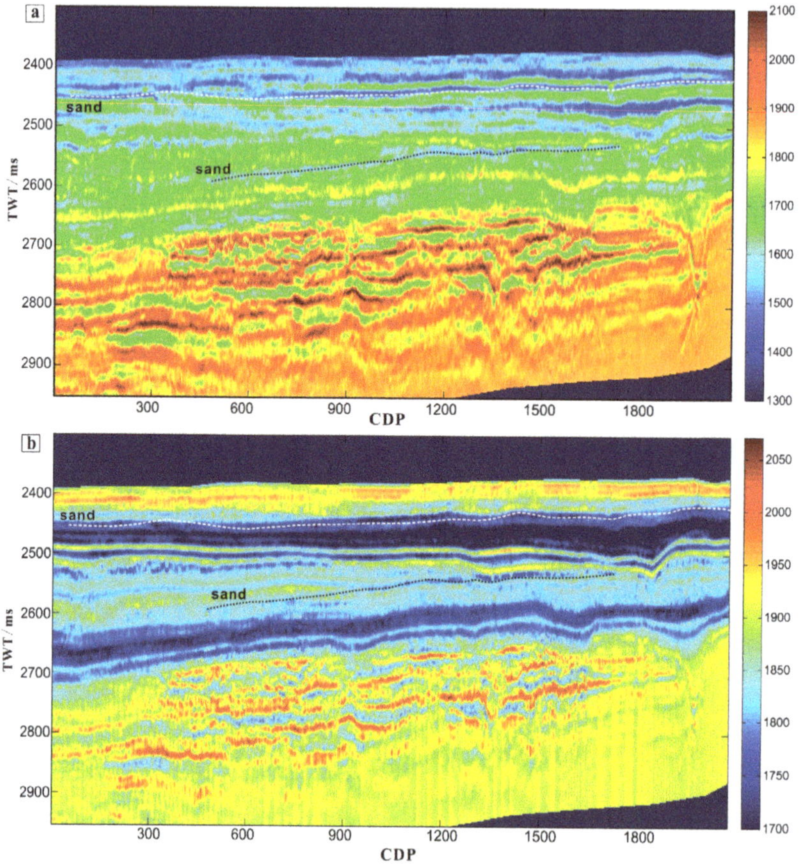

Figure 17. The inverted P-wave velocity (**a**) and density (**b**).

4.2. Geological Implications

Shallow sands are excellent reservoirs both for shallow gas and gas hydrates, and shallow sandy sediments and their associated hydrocarbon–hydrate systems may be important targets for future offshore exploration and development. The discovery of LS36-1 will significantly accelerate the exploration and development of deep-sea shallow gas and gas hydrate [8]. As the basis for exploring shallow gases and hydrates, identifying shallow sand bodies has been challenging due to insufficient drilling data. Current predictions mostly depend on sedimentary facies interpretation. This study uses actual drilling data from multiple sites in the South China Sea to develop a seismic identification technique for shallow sand layers. It supplements well the existing seismic interpretation methods, effectively predicts the distribution of shallow sand bodies, and offers a qualitative evaluation of their sand content.

The seismic prediction technique proposed in this paper is based on seismic data and leverages pre-stack seismic inversion to derive Young's modulus and Poisson's ratio. This method can achieve reliable prediction results even in the absence of well logging data, making it particularly suitable for exploration in new areas [37,61].

Additionally, the elastic properties of shallow sandy sediments differ significantly from those of deeper conventional oil and gas sand reservoirs. However, the study on the elastic characteristics of shallow sands is still weak, which may be attributed to the limited availability of well logging and lithological data for shallow sandy sediments in previous studies. This paper highlights that shallow sands may exhibit varying characteristics at different depths, which are closely related to factors such as effective stress levels and mineral composition. These aspects warrant further in-depth investigation.

5. Conclusions

The properties and elastic characteristics of conventional deep-buried sand layers are well-known. However, the features of shallow unconsolidated sand layers remain poorly understood. Typically, sand layers have high velocity and porosity. Yet, our logging data analysis shows that unconsolidated sand may have high or low velocities, making it hard to identify shallow sand layers using these parameters. In this paper, we focus on the physical properties of the shallow unconsolidated sandy sediments and construct a new petrophysical model to link sand content with elastic parameters, and we derive a new indicator of shallow sand: modified brittleness index (MBI). The theoretical model and the well logs show that MBI has a good indication of quartz content compared with other inversion parameters. Finally, MBI extraction is realized based on pre-stack seismic inversion, and the inversion results are consistent with geological understandings and drilling results. This method can be used in the future shallow gas and hydrate explorations.

This paper offers a solution by developing MBI that incorporates the mechanical properties of shallow sand layers, providing an effective way to identify them. However, this method is only suitable for shallow layers. Also, MBI alone cannot identify hydrate-bearing sand layers. When sediments contain hydrates, their elastic parameters like Young's modulus and Poisson's ratio change significantly. In such cases, identification requires both the velocity and MBI. Sandy hydrate-bearing sediments have high velocity and low MBI, while high sand content is indicated by low velocity and low MBI.

Author Contributions: Conceptualization, J.C. and Y.X.; methodology, T.W.; software, H.Z.; validation, H.Z. and Z.Z.; formal analysis, J.C. and Y.X.; investigation, Y.L.; data curation, S.Z.; writing—original draft preparation, J.C. and Y.X.; writing—review and editing, T.W. and W.D. All authors have read and agreed to the published version of the manuscript.

Funding: The research works are financially supported by the National Natural Science Foundation of China (Grant numbers U2244224 and 42206067), the National Key Research and Development Program of China (Grant number 2021YFC2800901), and the geological survey projects of China Geological Survey (Grant number DD20221700 and DD20230063).

Data Availability Statement: Data are available on request by contacting the corresponding author.

Acknowledgments: This research used samples and data provided by the Guangzhou Marine Geological Survey (GMGS). We also really appreciate the associate editor and anonymous reviewers for providing important insights.

Conflicts of Interest: The authors declare no conflicts of interest.

References

1. Dickens, G.R.; Quinby, H.M.S. Methane hydrate stability in seawater. *Geophys. Res. Lett.* **1994**, *21*, 2115–2118. [CrossRef]
2. Kvenvolden, K.A. Methane hydrate a major reservoir of carbon in the shallow geosphere? *Chem. Geol.* **1988**, *71*, 41–51. [CrossRef]
3. Kvenvolden, K.A. Gas hydrates: Geological perspective and global change. *Rev. Geophys.* **1993**, *31*, 173–187. [CrossRef]
4. Collett, T.S.; Lee, M.W.; Zyrianova, M.V.; Mrozewski, S.A.; Guerin, G.; Cook, A.E.; Goldberg, D.S. Gulf of Mexico gas hydrate Joint industry Project leg II logging-while-drilling data acquisition and analysis. *Mar. Petrol. Geol.* **2012**, *34*, 41–61. [CrossRef]
5. Fang, Y.; Flemings, P.B.; Daigle, H.; Phillips, S.; Meazell, K.; You, K. Petrophysical properties of the Green Canyon Block 955 hydrate reservoir inferred from reconstituted sediments: Implications for hydrate formation and production. *AAPG Bull.* **2020**, *104*, 1997–2028. [CrossRef]

6. Colwell, F.; Matsumoto, R.; Reed, D. A review of the gas hydrates, geology, and biology of the Nankai Trough. *Chem. Geol.* **2004**, *250*, 391–404. [CrossRef]
7. Kuang, Z.; Cook, A.; Ren, J.; Deng, W.; Cao, Y.; Cai, H. A flat-lying transitional free gas to gas hydrate system in a sand layer in the Qiongdongnan Basin of the South China Sea. *Geophys. Res. Lett.* **2023**, *50*, e2023GL105744. [CrossRef]
8. Xu, C.; Wu, K.; Pei, J.; Hu, L. Enrichment mechanisms and accumulation model of ultra-deep water and ultra-shallow gas: A case study of Lingshui 36-1 gas field in Qiongdongnan Basin, South China Sea. *Pet. Explor. Dev.* **2025**, *51*, 50–63. [CrossRef]
9. Ruppel, C. Permafrost-associated gas hydrate: Is it really approximately 1% of the global system? *J. Chem. Eng. Data* **2015**, *60*, 429–436. [CrossRef]
10. Kessler, J.D.; Ruppel, C.D. *Characterizing Ocean Acidification and Atmospheric Emission caused by Methane Released from Gas Hydrate Systems along the US Atlantic Margin (No. DOE-ROCHESTER-0028980)*; Univeristy of Rochester, Ed.; National Energy Technology Laboratory (NETL): Pittsburgh, PA, USA; Morgantown, WV, USA; Albany, OR, USA, 2020.
11. Farahani, M.V.; Guo, X.; Zhang, L.; Yang, M.; Hassanpouryouzband, A.; Zhao, J.; Yang, J.; Song, Y.; Tohidi, B. Effect of thermal formation/dissociation cycles on the kinetics of formation and pore-scale distribution of methane hydrates in porous media: A magnetic resonance imaging study. *Sustain. Energy Fuels* **2021**, *5*, 1567–1583. [CrossRef]
12. Farahani, M.V.; Hassanpouryouzband, A.; Yang, J.; Tohidi, B. Insights into the climate-driven evolution of gas hydrate-bearing permafrost sediments: Implications for prediction of environmental impacts and security of energy in cold regions. *RSC Adv.* **2021**, *11*, 14334–14346. [CrossRef] [PubMed]
13. Maslin, M.; Owen, M.; Day, S.; Long, D. Linking continental-slope failures and climate change: Testing the clathrate gun hypothesis. *Geology* **2004**, *32*, 53–56. [CrossRef]
14. Meng, M.; Li, L.; Liang, J.; Xu, J.; Feng, J.; Kuang, Z.; Zhang, W.; Huang, W.; Ren, J.; Deng, W.; et al. Quantifying the relative provenance contributions to submarine channel systems in the Qiongdongnan Basin since the Miocene: Implications for tectonic responses and channel migration. *Basin Res.* **2024**, *36*, 1365–2117.
15. Meng, M.; Liang, J.; Lu, J.A.; Zhang, W.; Kuang, Z.; Fang, Y.; He, Y.; Deng, W.; Huang, W. Quaternary Deep-Water Sedimentary Characteristics and Their Relationship with the Gas Hydrate Accumulations in the Qiongdongnan Basin, Northwest South China Sea. *Deep Sea Res. Part I Oceanogr. Res. Pap.* **2021**, *177*, 103628. [CrossRef]
16. Yang, Y.; Aplin, A.C. Influence of lithology and compaction on the pore size distribution and modelled permeability of some mudrocks from the Norwegian margin. *Mar. Petrol. Geol.* **1998**, *15*, 163–175. [CrossRef]
17. Yang, Y.; Aplin, A.C. Definition and practical application of mudstone porosity-effective stress relationships. *Petrol. Geosci.* **2004**, *10*, 153–162. [CrossRef]
18. Waite, W.F.; Santamarina, J.C.; Cortes, D.D.; Dugan, B.; Espinoza, D.N.; Germaine, J.; Jang, J.; Jung, J.W.; Kneafsey, T.J.; Shin, H.; et al. Physical properties of hydrate-bearing sediments. *Rev. Geophys.* **2009**, *47*, RG4003. [CrossRef]
19. Harris, N.B.; Miskimins, J.L.; Mnich, C.A. Mechanical anisotropy in the Woodford Shale, Permian Basin: Origin, magnitude, and scale. *Lead. Edge* **2011**, *30*, 284–291. [CrossRef]
20. Sena, A.; Castillo, G.; Chesser, K.; Voisey, S.; Estrada, J.; Carcuz, J.; Carmona, E.; Schneider, R.V. Seismic reservoir characterization in resource shale plays: "Sweet spot" discrimination and optimization of horizontal well placement. In *SEG Technical Program Expanded Abstracts*; Society of Exploration Geophysicists: Houston, TX, USA, 2011; p. 4424.
21. Mehrabi, A.; Bagheri, M.; Bidhendi, M.N.; Delijani, E.B.; Behnoud, M. Improved porosity estimation in complex carbonate reservoirs using hybrid CRNN deep learning model. *Earth Sci. Inf.* **2024**, *17*, 4773–4790. [CrossRef]
22. Riahi, Z.T.; Sarkarinejad, K.; Faghih, A.; Soleimany, B.; Payrovian, G.R. Frature detection using multi seismic attributes ant-tracking in the Rag-e-Sefid oilfield, SW Iran. *EGU* **2024**, *21*, 12039.
23. Rickman, R.; Mullen, M.; Petre, E.; Grieser, B.; Kundert, D. A practical use of shale petrophysics for stimulation design optimization: All shale plays are not clones of the Barnett Shale. In Proceedings of the SPE Annual Technical Conference and Exhibition, Denver, CO, USA, 21–24 September 2008; p. 115258.
24. Sondergeld, C.H.; Newsham, K.E.; Comisky, J.T.; Rice, M.C.; Rai, C.S. Petrophysical consideration in evaluation and producing shale gas resources. In Proceedings of the SPE Unconventional Gas Conference, Calgary, AB, Canada, 19–21 October 2010.
25. Shelander, D.; Dai, J.; Bunge, G. Predicting saturation of gas hydrates using pre-stack seismic data, Gulf of Mexico. *Mar Geophys. Res.* **2010**, *31*, 39–57. [CrossRef]
26. Lee, M.W.; Hutchinson, D.R.; Collett, T.S.; Dillon, W.P. Seismic velocities for hydrate-bearing sediments using weighted equation. *J. Geophys. Res.-Solid Earth* **1996**, *101*, 20347–20358. [CrossRef]
27. Lee, M.W.; Collett, T.S.; Inks, T.L. Seismic-attribute analysis for gas-hydrate and free-gas prospects on the North Slope of Alaska. In *Natural Gas Hydrates-Energy Resource Potential and Associated Geologic Hazards*; Collett, T., Johnson, A., Knapp, C., Boswell, R., Eds.; AAPG Memoir: Tulsa, OK, USA, 2009; Volume 89, pp. 541–554.
28. Waite, W.F.; Helgerud, M.B.; Nur, A.; Pinkston, J.C.; Stern, L.A.; Kirby, S.H.; Durham, W.B. Laboratory measurements of compressional and shear wave speeds through methane hydrate. *Ann. N. Y. Acad. Sci.* **2000**, *912*, 1003–1010. [CrossRef]

29. Yun, T.S.; Francisca, F.M.; Santamarina, J.C.; Rupple, C. Compressional and shear wave velocities in uncemented sediment containing gas hydrate. *Geophys. Res. Lett.* **2005**, *32*, L10669. [CrossRef]
30. Dvorkin, J.; Nur, A. Elasticity of high-porosity sandstones: Theory for two North Sea data sets. *Geophysics* **1996**, *61*, 1363–1370. [CrossRef]
31. Helgerud, M.B.; Dvorkin, J.; Nur, A.; Sakai, A.; Collett, T. Elastic-wave velocity in marine sediments with gas hydrates: Effective medium modeling. *Geophys. Res. Lett.* **1999**, *26*, 2021–2024. [CrossRef]
32. Lee, M.W. Modifed Biot-Gassmann theory for calculating elastic velocities for unconsolidated and consolidated sediments. *Mar. Geophys. Res.* **2002**, *23*, 403–412. [CrossRef]
33. Duffaut, K.; Landrø, M.; Sollie, R. Using Mindlin theory to model friction-dependent shear modulus in granular media. *Geophysics* **2010**, *75*, E143–E152. [CrossRef]
34. Fujii, T.; Saeki, T.; Kobayashi, T.; Inamori, T.; Hayashi, M.; Takano, O.; Takayama, T.; Kawasaki, T.; Nagakubo, S.; Nakamizu, M.; et al. Resource assessment of methane hydrate in the eastern Nankai Trough, Japan. In Proceedings of the Offshore Technology Conference, Houston, TX, USA, 5–8 May 2008; OnePetro: Richardson, TX, USA, 2008.
35. Pan, H.; Li, H.; Chen, J.; Zhang, Y.; Cai, S.; Huang, Y.; Zheng, Y.; Zhao, Y.; Deng, J. A unified contact cementation theory for gas hydrate morphology detection and saturation estimation from elastic-wave velocities. *Mar. Petrol. Geol.* **2020**, *113*, 104146. [CrossRef]
36. Mindlin, R.D. Compliance of elastic bodies in contact. *J. Appl. Mech.* **1949**, *16*, 259–268. [CrossRef]
37. Zhu, W.; Huang, B.; Mi, L.; Wilkins, R.W.; Fu, N.; Xiao, X. Geochemistry, origin, and deep-water exploration potential of natural gases in the Pearl River Mouth and Qiongdongnan basin, South China Sea. *AAPG Bull.* **2009**, *93*, 741–761. [CrossRef]
38. Liang, J.; Zhang, W.; Lu, J.A.; Wei, J.; Kuang, Z.; He, Y. Geological occurrence and accumulation mechanism of natural gas hydrates in the eastern Qiongdongnan Basin of the South China Sea: Insights from site GMGS5-W9-2018. *Mar. Geol.* **2019**, *418*, 106042. [CrossRef]
39. Deng, W.; Kuang, Z.; Liang, J.; Yan, P.; Lu, J.; Meng, M. The seismic and rock-physics evidences of the different migration efficiency between different types of gas chimneys. *Deep Sea Res. Part I: Oceanogr. Res. Pap.* **2023**, *191*, 103942. [CrossRef]
40. Deng, W.; Liang, J.; Zhang, W.; Kuang, Z.; Zhong, T.; He, Y. Typical characteristics of fracture-filling hydrate-charged reservoirs caused by heterogeneous fluid flow in the Qiongdongnan Basin, northern South China Sea. *Mar. Pet. Geol.* **2021**, *124*, 104810.
41. Wei, J.; Liang, J.; Lu, J.; Zhang, W.; He, Y. Characteristics and dynamics of gas hydrate systems in the northwestern South China Sea-results of the fifth gas hydrate drilling expedition. *Mar. Pet. Geol.* **2019**, *110*, 287–298. [CrossRef]
42. Xie, X.; Müller, R.D.; Ren, J.; Jiang, T.; Zhang, C. Stratigraphic architecture and evolution of the continental slope system in offshore Hainan, northern South China Sea. *Mar. Geol.* **2008**, *247*, 129–144. [CrossRef]
43. Meng, M.; Liang, J.; Kuang, Z.; Ren, J.; He, Y.; Deng, W.; Gong, Y. Distribution characteristics of Quaternary channel systems and their controlling factors in the Qiongdongnan basin, South China Sea. *Front. Earth Sci.* **2022**, *10*, 902517. [CrossRef]
44. Winkler, K.W. Contact stiffness in granular and porous meterials: Comparison between theory and experiment. *Geophys. Res. Lett.* **1983**, *10*, 1073–1076. [CrossRef]
45. Walton, K. The effective elastic moduli of a random packing of spheres. *J. Mech. Phys. Solids.* **1987**, *35*, 213–226. [CrossRef]
46. Dai, S.; Santamarina, J.C.; Waite, W.F.; Kneafsey, T.J. Hydrate morphology: Physical properties of sands with patchy hydrate saturation. *J. Geophys. Res. Solid Earth* **2012**, *117*. [CrossRef]
47. White, J.E.; Martineau-Nicholetis, L.; Monash, C. Measured anisotropy in Pierre shale. *Geophys. Prospect.* **1983**, *31*, 709–729. [CrossRef]
48. Backus, G.E. Long-wave elastic anisotropy produced by horizontal layering. *J. Geophys. Res.* **1962**, *67*, 4427–4440. [CrossRef]
49. Brown, R.; Korringa, J. On the dependence of the elastic properties of a porous rock on the compressibility of the pore fluid. *Geophysics* **1975**, *40*, 608–616. [CrossRef]
50. Lu, S.; McMechan, G.A. Elastic impedance inversion of multichannel seismic data from unconsolidated sediments containing gas hydrate and free gas. *Geophysics* **2004**, *69*, 164–179. [CrossRef]
51. Lu, S.; McMechan, G.A.; Liaw, A. Identification of shallow-water-flow sands by Vp/Vs inversion of conventional 3D seismic data. *Geophysics* **2005**, *70*, O29–O37. [CrossRef]
52. Downton, J.E. Seismic Parameter Estimation from AVO Inversion. Ph.D. Thesis, University of Calgary, Calgary, AB, Canada, 2005.
53. Russell, B.H.; Hedlin, K.; Hilterman, F.J.; Lines, L.R. Fluid-property discrimination with AVO: A Biot-Gassmann perspective. *Geophysics* **2003**, *68*, 29–39. [CrossRef]
54. Russell, B.H.; Gray, D.; Hampson, D.P. Linearized AVO inversion and poroelasticity. *Geophysics* **2011**, *76*, C19–C29. [CrossRef]
55. Wang, B.; Yin, X.; Zhang, F. Lamé parameters inversion based on elastic impedance and its application. *Appl. Geophys.* **2006**, *3*, 120–123. [CrossRef]
56. Yin, X.; Zhang, S. Bayesian inversion for effective pore-fluid bulk modulus based on fluid-matrix decoupled amplitude variation with offset approximation. *Geophysics* **2014**, *79*, R221–R232. [CrossRef]

57. Zong, Z.; Yin, X.; Wu, G. Elastic impedance parameterization and inversion with Young's modulus and Poisson's ratio. *Geophysics* **2012**, *77*, N17–N24. [CrossRef]
58. Zong, Z.; Yin, X.; Wu, G. Geofluid discrimination incorporating poroelasticity and seismic reflection inversion. *Surv. Geophys.* **2015**, *36*, 659–681. [CrossRef]
59. Aki, K.; Richards, P.G. *Quantitative Seismology: Theory and Methods*; University Science Books: Herndon, VA, USA, 2002.
60. Connolly, P. Elastic impedance. *Lead. Edge* **1999**, *18*, 438–452. [CrossRef]
61. Shelander, D.; Dai, J.; Bunge, G.; Singh, S.; Eissa, M.; Fisher, K. Estimating saturation of gas hydrates using conventional 3D seismic data, Gulf of Mexico Joint Industry Project Leg II. *Mar. Pet. Geol.* **2012**, *34*, 96–110. [CrossRef]

Journal of
Marine Science and Engineering

MDPI

Article

Dynamic Analysis of Subsea Sediment Engineering Properties Based on Long-Term In Situ Observations in the Offshore Area of Qingdao

Zhiwen Sun [1,2,3,4], Yanlong Li [1,3], Nengyou Wu [1,3,*], Zhihan Fan [2], Kai Li [2], Zhongqiang Sun [2], Xiaoshuai Song [2], Liang Xue [2] and Yonggang Jia [2,*]

1 Key Laboratory of Gas Hydrate, Ministry of Natural Resources, Qingdao Institute of Marine Geology, Qingdao 266237, China; zhiwensun91@163.com (Z.S.); liyanlongupc@163.com (Y.L.)
2 Shandong Key Laboratory of Marine Environmental Geology Engineering, Ocean University of China, Qingdao 266100, China; fanzhihan@ouc.edu.cn (Z.F.); lk@ouc.edu.cn (K.L.); sunzhongqiang@ouc.edu.cn (Z.S.); songxiaoshuai@stu.ouc.edu.cn (X.S.); xueliang@stu.ouc.edu.cn (L.X.)
3 Laoshan Laboratory, Qingdao 266237, China
4 Key Laboratory of Polar Geology and Marine Mineral Resources, China University of Geosciences, Ministry of Education, Beijing 100083, China
* Correspondence: wuny@ms.giec.ac.cn (N.W.); yonggang@ouc.edu.cn (Y.J.)

Abstract: The drastic changes in the marine environment can induce the instability of seabed sediments, threatening the safety of marine engineering facilities such as offshore oil platforms, oil pipelines, and submarine optical cables. Due to the lack of long-term in situ observation equipment for the engineering properties of seabed sediments, most existing studies have focused on phenomena such as the erosion suspension of the seabed boundary layer and wave-induced liquefaction, leading to insufficient understanding of the dynamic processes affecting the seabed environment. In this study, a long-term in situ observation system for subsea engineering geological environments was developed and deployed for 36 days of continuous monitoring in the offshore area of Qingdao. It was found that wave action significantly altered sediment mechanical properties, with a 5% sound velocity increase correlating to 39% lower compression, 7% higher cohesion, 11% greater internal friction angle, and 50% reduced excess pore water pressure at 1.0–1.8 m depth. suggesting sustained 2.2 m wave loads of expelled pore water, driving dynamic mechanical property variations in seabed sediments. This long-term in situ observation lays the foundation for the monitoring and early warning of marine engineering geological disasters.

Keywords: long-term in situ observation; offshore seabed sediments of Qingdao; acoustic/resistivity/excess pore pressure of sediment; sediment physical and mechanical properties; natural marine disasters

1. Introduction

The shallow marine hydrodynamic environment is governed by diverse forces—including waves, tides, and storm surges—that drive sediment transport through bedload migration, suspended advection, and mass movements, posing global risks to subsea infrastructure [1]. Among these, wave action is particularly significant due to its short periodicity and high frequency, making it the dominant hydrodynamic force in shallow seas [2]. Under such intense hydrodynamic conditions, the seabed experiences processes such as erosion, sediment resuspension and deposition [3], liquefaction, debris flow, shear deformation, and submarine landslides [4], all of which pose serious threats to the safety of offshore drilling platforms, subsea pipelines, and optical cables [5]. These processes represent critical risk factors for marine engineering and are urgent issues that need to be addressed in national marine development [6]. The occurrence and development of seabed geological disasters lead to significant changes in the engineering geological properties of seabed sediments [7]. Therefore, conducting long-term in situ observations of dynamic changes in seabed engineering geological parameters is essential for understanding the dynamic behavior of sediment properties. This research will help reveal the mechanisms behind the formation and evolution of seabed geological disasters, providing valuable insights for preventing such disasters and safeguarding the construction of marine engineering projects.

Recent studies on dynamic changes in shallow marine engineering geological environments primarily focus on long-term in situ observations, with an emphasis on wave-induced seabed erosion, resuspension, and liquefaction. In 1967, the NGI-UI pore water pressure sensor, jointly developed by the Norwegian Geotechnical Institute (NGI) and the University of Illinois (UI), was successfully applied in a geological survey of the Wilkinson Basin at a water depth of 278 m [8,9]. Subsequently, Bennett et al. [10] conducted an 8-month long-term observation at the Mississippi River Delta using composite pressure sensors, measuring dynamic changes in pore water pressure under wave action. In the 1980s, a joint Sino-American investigation deployed pore water pressure sensors in the underwater delta region of the Yellow River, capturing pore water pressure variations during three storm surge events [11], lacking multi-parameter and long-period observational verification ([12]). Liu [13] developed an in situ pore water pressure monitoring device for seabed soils, capable of measuring excess pore water pressure at different soil depths up to 15 m below the mud line. A research team led by Professor Jia developed a series of in situ long-term monitoring technologies and equipment for seabed engineering environments, covering areas from shallow coastal zones to deep offshore regions, revealing the dynamic changes in the seabed boundary layer [14]. The Piezometer series equipment, developed by the French Institute for Research and Exploitation of the Sea (Ifremer), is capable of operating at water depths up to 6000 m [15]. In 2006, Yang et al. [16] conducted a 7-day in situ observation in the tidal flat area of Jiaozhou Bay, using ADP-XR and OBS-3A to measure seabed horizontal flow characteristics, backscatter intensity, and suspended sediment concentration, revealing the variation patterns and controlling factors of suspended sediment concentration, which showed a U-shaped trend during each tidal cycle. In the same year, Ma [17] conducted a 5-month in situ observation in the Beibu Gulf region, utilizing a monitoring platform consisting of ADCP, PC-ADP, OBS, an inclinometer, and fan-shaped sonar to acquire continuous wave data, turbulence data, flow profiles, seabed morphology images, and the suspended sediment concentration.

In 2015, Zhang [18] conducted a 4-month in situ observation experiment in the underwater delta of the Yellow River, quantitatively revealing the contribution of wave-induced transient liquefaction and residual liquefaction to sediment erosion and resuspension. In 2021, Feng [19], based on long-term in situ observation data and laboratory experiments, used a deep learning model combining Long Short-Term Memory (LSTM) networks to predict and analyze 3.5-day transient pumping resuspension. Wen [20] conducted a 24-day in situ observation in the Chengdao Sea area of the Yellow River underwater delta, successfully capturing seven storm wave events. This study investigated the changes in the structure of near-bottom suspended sediments and the interaction between bottom seawater, the seabed interface, and shallow sediments under varying dynamic marine conditions.

Shallow marine sediments undergo dramatic physical and mechanical property changes due to external geological forces such as ocean currents, waves, and tides. However, existing in situ long-term observations primarily focus on wave-induced seabed erosion, resuspension, and liquefaction, with limited research on the dynamic changes in the physical and mechanical properties of seabed sediments. Resistivity measurements, as an important means in modern marine geophysical research, can analyze the changes in the engineering geological properties of seafloor sediments with high sensitivity [21]. In recent years, the seafloor resistivity measurement method has become an effective indirect method for in situ soil physical property testing [22] and long-term observation [23]. Through the analysis of the observed resistivity data, the physical state and changes in seafloor sediments can be obtained [24]. Acoustic measurement is an effective indirect method for the non-destructive long-term observation of the changes in the mechanical properties of seafloor sediments. By establishing the internal connection between the acoustic parameters of seafloor sediments and the engineering mechanical characteristics of sediments, the in situ observation of the structure and engineering properties of seafloor sediments can be achieved [25]. The mechanical properties of seafloor sediments cover a lot of aspects. At present, the basic parameters that have attracted widespread attention include compression modulus, shear strength, cohesion, and internal friction angle [26].

Since a single geophysical measurement data point has multiple interpretations, it is impossible to accurately judge the physical and mechanical state of sediments using only one of the acoustic or electrical methods. To address this gap, a newly developed in situ monitoring system for seabed engineering geological environments was deployed for a 36-day long-term observation in the shallow seas off Qingdao. The 36-day observation period can cover two complete spring-neap tide cycles (14.75 days) and record the wave event triggering (such as wave loading) and the lagged response of sediments (such as the accumulation release cycle of pore water pressure). This study revealed the temporal variation patterns of seabed water, sediment resistivity, acoustic parameters, and excess pore water pressure, contributing valuable insights into the dynamic changes in seabed sediments.

2. Study Area

The study area was located at Caodao Island (Figure 1), managed by the Qingdao National Deep Sea Center. This island is situated in a semi-enclosed bay within a temperate maritime zone, influenced by the East Asian monsoon and the South Yellow Sea water mass system. The region experiences a temperate monsoon climate, with an average annual temperature of 12.3 °C. During the autumn and winter, northwesterly winds prevail, while southeasterly winds dominate in spring and summer.

Figure 1. In situ observation sites and columnar sediment samples off the coast of Qingdao.

The study area is characterized by a semi-diurnal tidal regime, with water currents reaching speeds of up to 180 cm/s. The bay exhibits complex circulation patterns, with multiple circulation systems contributing to the intricate hydrodynamics of the region. The water depth in the area is approximately 15 m, and the overlying sediments consist of low-permeability silts and fine sands, ranging from 3 to 5 m in thickness. The density is 1.13–1.23 g/cm^3, water content is 102.7–109.5%, plasticity index is 32.60–76.30, and liquid index is 1.50–3.22 [27]. Therefore, this sediment is vulnerable to wave liquefaction. Meanwhile, the topography of the Yellow Sea continental shelf leads to enhanced wave refraction (with an annual average effective wave height of 1.5–2.5 m), providing a natural experimental site for studying the dynamic response of wave-induced sediments (such as pore water migration and shear strength changes)

The ecological environment of the study area is shaped by both natural processes and human activities, making it a representative site for studying the effects of wave action on the dynamic changes in the engineering geological properties of seabed sediments.

3. Method

The observations in this study were conducted using an in situ long-term monitoring system for subsea engineering geological environments, which was independently developed by our team [14]. This system employs various methods, including acoustic, resistivity, and excess pore water pressure measurements, to monitor the dynamic changes in the physical and mechanical properties of seabed sediments.

The system primarily consists of the following components: an acoustic profile monitoring system, a high-precision seabed resistivity monitoring system, an excess pore water pressure monitoring system, an oceanic environmental monitoring system, a data acquisition and real-time transmission system, and a long-term power supply system. The core of the monitoring system focuses on the long-term observation of changes in the resistivity, acoustic properties, and excess pore water pressure of the seabed sediments (Figure 2), thereby capturing dynamic variations in the physical and mechanical properties of the seabed. The high-precision observation module for the resistivity of seabed sediments adopts the pseudo-random coding emission of electrical signals and characteristic signal recognition technology, realizing the nano-volt level observation of the deep-sea electric field and achieving a resistivity measurement accuracy of 5‰. The multi-channel acoustic section observation technology on the seabed utilizes the long-term transmission technology of high-frequency acoustic signals under the high-pressure environment in the deep sea and the multi-channel acoustic section scanning technology to achieve the construction of multi-channel sound fields on the seabed and high-resolution acoustic section observations. The high-resolution observation module for pore water in seafloor sediments based on the fiber-optic differential pressure structure adopts the deep-sea differential pressure fiber of Bragg grating excess pore water pressure sensing and ultra-high hydrostatic pressure self-balancing technology, realizing the accurate measurement of the excess pore water pressure of sediments in the deep-sea environment (with a measuring range of 500 kPa and an accuracy of 0.1 kPa).

The communication and control system enables bidirectional communication and data transmission between the seabed, the sea surface, and the shore. The power supply system, which utilizes a uniquely designed seawater battery, can meet the system's energy requirements for continuous operation over a year on the seabed.

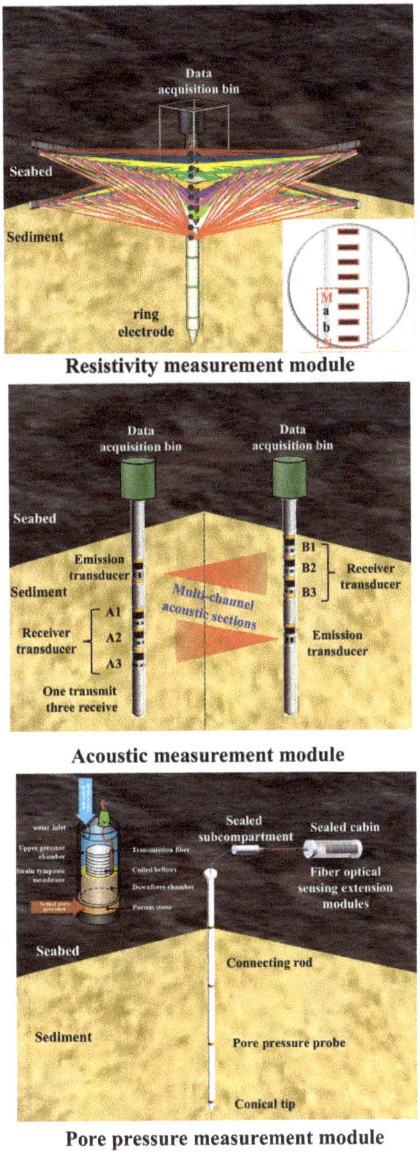

Figure 2. Improved resistivity acoustic pore pressure measurement sensor and measurement scheme (Symmetrical Four-Electrode: Four electrodes are arranged in a straight line in sequence as A-M-N-B. A and B are current electrodes used for injecting current; M and N are potential electrodes used for measuring potential difference; A1/A2/A3, B1/B2/B3 are the receiving electrodes of the A and B probes respectively; The same is below).

This monitoring equipment (Figure 3) has undergone multiple nearshore sea trials and has been successfully deployed aboard research vessels, including Haiyang Geological No. 6, Dongfanghong III, and Zhang Jian, for deep-sea tests in the South China Sea. The results of these trials have been stable, accurate, and reliable [28].

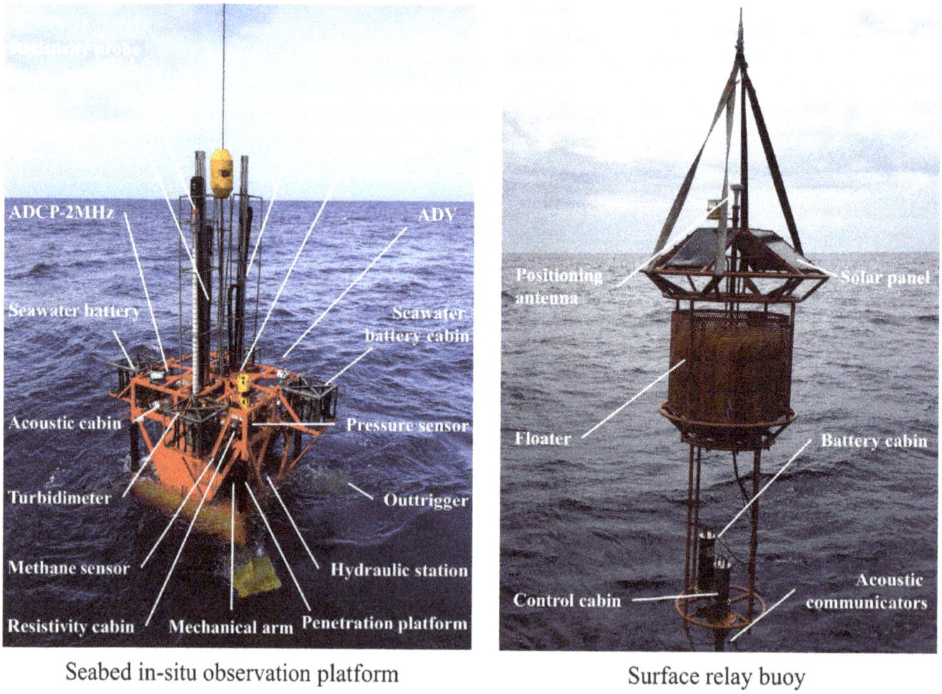

Figure 3. In situ long-term monitoring system for subsea engineering geological environments. The seabed in situ observation platform (**left**) is used to acquire seafloor sediment data and store the data; the surface relay buoy (**right**) receives seabed survey data via underwater acoustic communication and transmits them to the lab via satellite, modified from [28].

In November 2019, a 36-day in situ long-term observation of the seabed engineering geological environment was conducted in the offshore area of Qingdao using the independently developed subsea engineering geological environment monitoring system (Figure 4). The observation platform was equipped with a vertical resistivity probe, two acoustic probes, and an excess pore water pressure probe.

The resistivity probe, which is 4.5 m long, consists of 60 electrode rings spaced 6.5 cm apart, with measurements taken using the Wenner method in a rolling configuration. The acoustic measurement system is divided into two groups, each consisting of one transmitter transducer and three receiver transducers. During operation, the transmitter emits sound waves, while the three receivers capture the waves. The two groups alternate between transmitting and receiving acoustic signals. The excess pore water pressure probe is 5 m long and is equipped with four fiber-optic Bragg grating sensors, each separated by 1 m, to measure differential pressure.

During the observation, the probes were sequentially driven into the seabed sediments to a depth of 1.6 m. It can cover the engineering geological risk layers within the burial depth range of submarine pipelines. The system measured the resistivity, acoustic properties, and excess pore water pressure of the seabed sediments every 2 h over 36 days. These measurements allowed for the acquisition of vertical distribution profiles and dynamic variation patterns of the resistivity, sound speed, and excess pore water pressure in the seabed water and sediments.

Figure 4. In situ observation of the engineering geological environment of Qingdao shallow seabed (the photo on the (**left**) was taken when the in situ seafloor observation platform was deployed, and the photo on the (**right**) was taken when the in situ seafloor observation platform was retrieved).

4. Results

4.1. Characteristics of the Offshore Water Environment in Qingdao

Under varying sea conditions, the extent and structure of the seabed boundary layer undergo dynamic changes. These changes in the boundary layer contribute to geological hazards such as the erosion, deposition, resuspension, transport, disturbance, and disruption of shallow sediments, and even landslides or instability in the seabed. This study conducted a 36-day observation of the dynamic changes in seabed sediments under different sea conditions. During the observation period, there were two instances when the wave height exceeded 1 m and five instances when the wave height was greater than 0.5 m. During the observation period, no waves or tide gauges were installed, so there were no wave height data. Therefore, the data from the adjacent observation stations were used instead. It can be seen from the figure that there are a total of three observation stations near the seafloor observation location. Among them, the Wheat Island Observatory Station is the closest to the observation location, and the absolute distance between them is very short. Hence, the wave height data of the Wheat Island Observatory Station can be approximately regarded as the wave height data of the seafloor observation point (Figure 5). The first event occurred from 11 November to 13 November, lasting for 2 days with a maximum wave height of 1 m. The second event lasted from 16 November to 20 November for 4 days, with a maximum wave height of 2.2 m. The third event took place from 26 November to 27 November, with a maximum wave height of 0.6 m. The fourth event occurred from December 1 to December 2, lasting 2 days with a maximum wave height of 0.5 m. The fifth event took place from 12 December to 14 December, lasting 2 days, with a maximum wave height of 0.6 m.

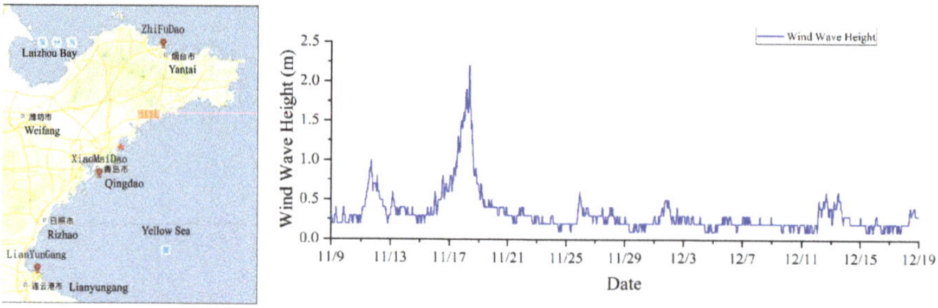

Figure 5. Wind wave height data during shallow sea observation. The (**left**) picture shows the wave observation locations during the shallow sea observation period. Among them, the red five-pointed star indicates the location of seafloor observation, and the balloons represent the adjacent observation stations. The (**right**) picture shows the wave height data during the observation period. Source: National Marine Science Data Center—Wheat Island Observatory.

4.2. In Situ Subsea Resistivity Observation Results

The resistivity monitoring of the subsea environment includes measurements of seawater, the seabed boundary layer, and shallow sediments. The ranges of resistivity changes observed in these areas were as follows: seawater resistivity varied from 0.08 to 0.14 Ω·m; the seabed boundary layer resistivity ranged from 0.07 to 0.24 Ω·m; and the resistivity of the sediments fluctuated between 0.01 and 0.24 Ω·m (Figure 6).

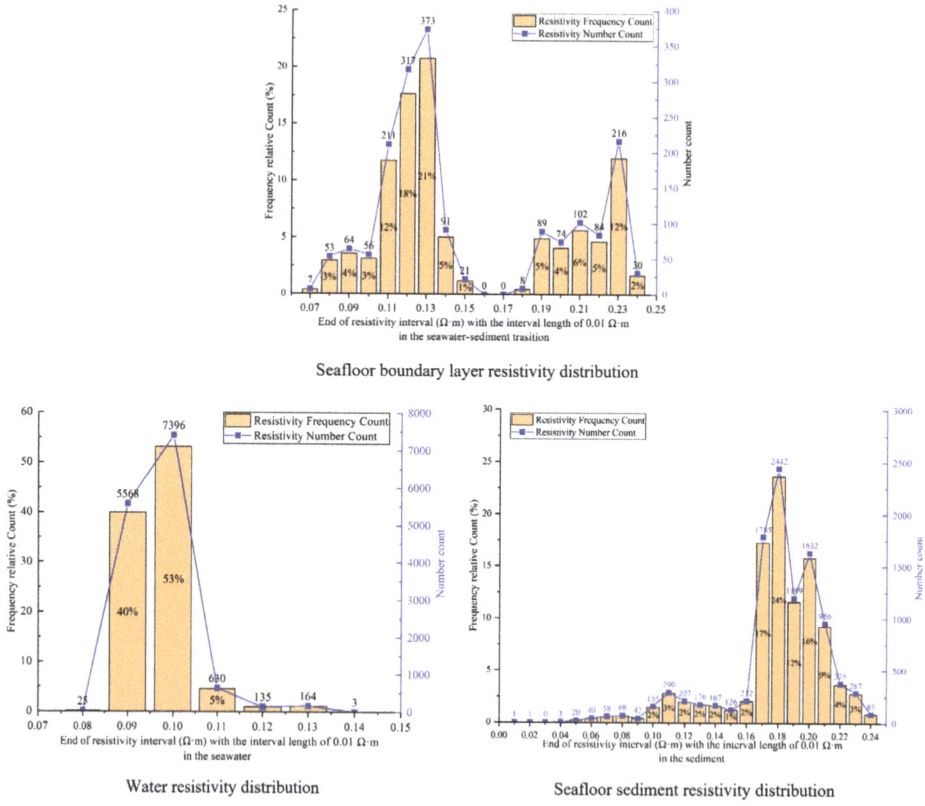

Figure 6. Resistivity distribution in seawater, seabed boundary layer, and sediments.

4.2.1. Vertical Spatial Structure of Seabed Resistivity

The vertical spatial structure of seabed resistivity can be divided into three parts: seawater resistivity, seabed boundary layer resistivity, and sediment resistivity (Figure 7). To illustrate, the vertical resistivity profile on November 9 shows a resistivity range from 0.05 to 0.20 Ω·m.

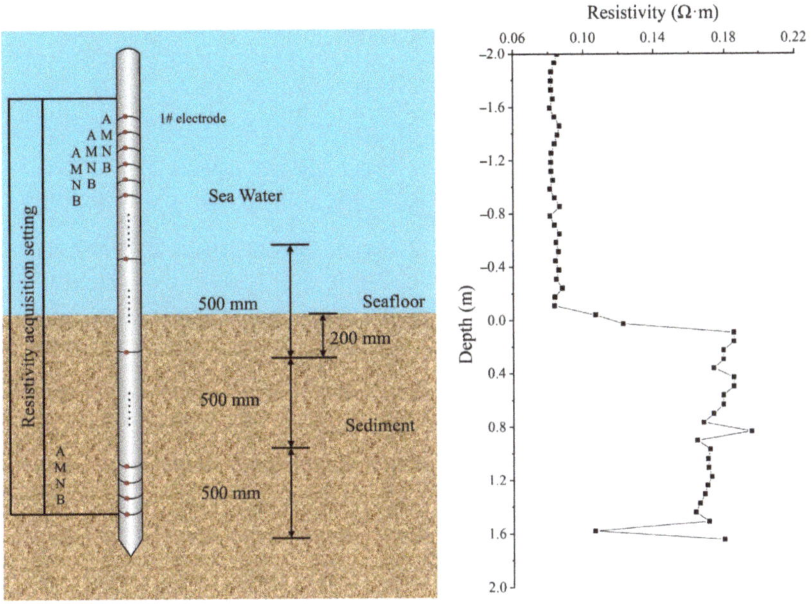

Figure 7. Schematic diagram of seafloor resistivity measurement and measurement results (1# means the 1st electrode).

Seawater Resistivity: Measured up to 2.09 m above the seabed, seawater resistivity ranged from 0.07 to 0.08 Ω·m, with a change of 0.01 Ω·m. The small variation in vertical resistivity indicates good consistency in electrode measurements and suggests that the seawater is relatively homogeneous in the vertical direction.

Seabed Boundary Layer Resistivity: Measured in the 0.2 m range near the seabed, resistivity ranged from 0.07 to 0.18 Ω·m, with a change of 0.11 Ω·m. This measurement involved four electrodes, and the resistivity was measured using the four-electrode Wenner method, with two outer electrodes supplying current and two inner electrodes measuring the potential. As the measurement probes moved downward, the resistivity increased due to the resistivity contrast between the water and sediment. When all four electrodes were within the sediment, the measured resistivity reflected the true resistivity of the seabed sediment.

Sediment Resistivity: Measured between 0.09 and 1.64 m below the seabed, the resistivity varied from 0.05 to 0.19 Ω·m, with a change of 0.14 Ω·m. Resistivity generally decreased with depth, though two significant anomalies were observed. The first anomaly occurred at a depth of 0.83 m, where resistivity abruptly increased from 0.16 Ω·m to 0.19 Ω·m, a change of 0.03 Ω·m. This increase indicates a change in sediment properties, leading to a higher resistivity state. The second anomaly appeared at 1.58 m, where resistivity abruptly decreased from 0.16 Ω·m to 0.05 Ω·m, a decrease of 0.11 Ω·m. This drop, approaching the resistivity of seawater, suggests the potential presence of a high-salinity fluid migration pathway in this area.

4.2.2. Temporal Variation in Seabed Resistivity

The seabed resistivity measurements were conducted from 9 November 2019 to 14 December 2019, for a total of 36 days.

The resistivity of seawater showed a consistent overall trend (Figure 8). For example, at a depth of 1.05 m above the seabed, the seawater resistivity increased from 0.08 Ω·m to 0.10 Ω·m, a change of 0.02 Ω·m, with a slight upward trend over time. The seawater resistivity can be divided into three distinct periods:

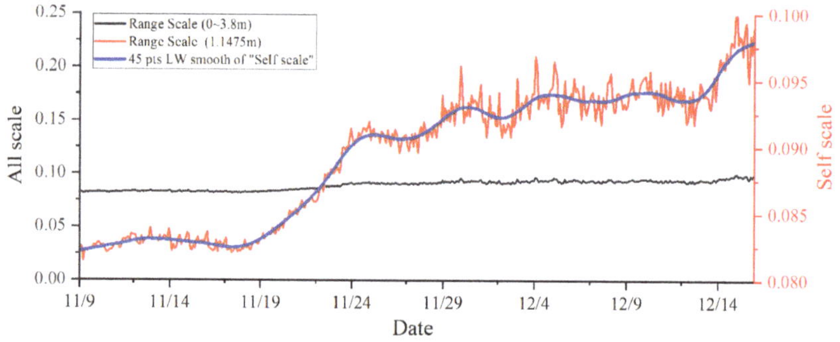

Figure 8. Resistivity curve in seawater over time ("45 pts LW" means each of the 45 points was subjected to smoothing fitting using "the Locally Weighted Scatterplot Smoothing" method).

Period 1 (9 November–19 November): The resistivity remained relatively constant at around 0.08 Ω·m.

Period 2 (19 November–24 November): The resistivity increased linearly from 0.083 Ω·m to 0.091 Ω·m, a 10% increase.

Period 3 (24 November–16 December): The resistivity showed a fluctuating increase, rising from 0.091 Ω·m to 0.11 Ω·m.

The seabed boundary layer resistivity was measured over the same period as the seawater (Figure 9). During this time, the resistivity of the boundary layer increased from 0.12 Ω·m to 0.21 Ω·m, a change of 0.09 Ω·m. The temporal variation in seabed boundary layer resistivity can be divided into four periods:

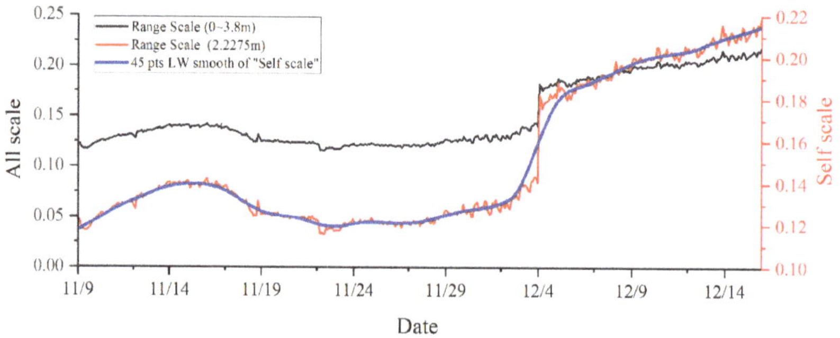

Figure 9. The curve of seafloor boundary layer resistivity over time ("45 pts LW smooth" refers to Figure 8).

Period 1 (9 November–22 November): The resistivity first increased, and then decreased, rising from 0.12 Ω·m to 0.14 Ω·m, before dropping to 0.11 Ω·m.

Period 2 (22 November–4 December): The resistivity gradually increased from 0.11 Ω·m to 0.14 Ω·m.

Period 3 (4 December 11:24–13:24): A sharp increase in resistivity occurred, from 0.14 Ω·m to 0.18 Ω·m.

Period 4 (4 December 13:24–16 December): The resistivity continued to increase gradually, from 0.18 Ω·m to 0.21 Ω·m.

The temporal variation in the seabed boundary layer resistivity reflects the dynamic process of sediment erosion, deposition, and resuspension at the seabed.

Seabed sediment resistivity at different depths showed varying trends over time. At a depth of 1.04 m below the seabed, the sediment resistivity first decreased and then increased, divided into two periods (Figure 10):

Period 1 (9 November–13 November): The resistivity decreased from 0.17 Ω·m to 0.16 Ω·m.

Period 2 (13 November–16 December): The resistivity steadily increased from 0.16 Ω·m to 0.17 Ω·m.

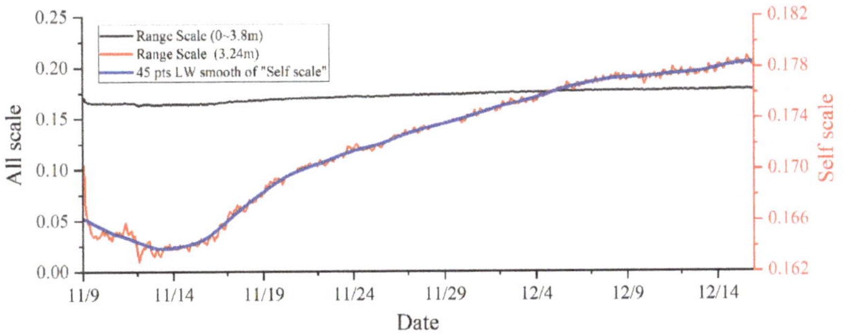

Figure 10. The variation in sediment resistivity with time at the depth of 1.04 m below the seabed ("45 pts LW smooth" refers to Figure 8). At a depth of 1.38 m below the seabed, the resistivity initially increased, then decreased, with the following three periods (Figure 11).

Period 1 (9 November–22 November): The resistivity increased steadily with minor fluctuations, from 0.09 Ω·m to 0.10 Ω·m.

Period 2 (22 November–27 November): The resistivity increased sharply from 0.10 Ω·m to 0.15 Ω·m.

Period 3 (27 November–16 December): The resistivity gradually decreased, from 0.15 Ω·m to 0.11 Ω·m.

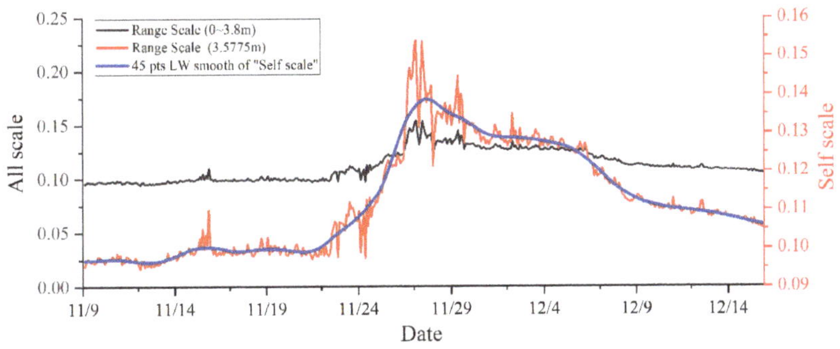

Figure 11. The variation in sediment resistivity with time at the depth of 1.38 m below the seabed ("45 pts LW smooth" refers to Figure 8).

4.3. Seabed Sound Velocity Observation Results

The vertical spatial distribution of the seabed sound velocity indicates that sound velocity increases gradually with depth (Table 1), exhibiting some fluctuations (Figure 12). For example, on November 9, the sound velocity in seawater was 1529 m/s. At a depth of 0.3 m below the seabed, the sediment sound velocity increased to 1610 m/s. At a depth of 0.6 m, the sound velocity slightly decreased to 1598 m/s. As the depth increased to 0.8 m below the seabed, the sediment sound velocity rose to 1713 m/s. At a depth of 1.3 m, the sound velocity slightly decreased to 1710 m/s but then increased again to 1786 m/s at 1.6 m.

Table 1. Summary of six channels of sound velocity of seafloor sediments.

Receiving Channel	Depth (m)	Min (m/s)	Max (m/s)	Average (m/s)	RMSE	Standard Deviation
CH1	0	1516	1571	1540	74	9
CH2	0.3	1558	1774	1569	321	18
CH3	0.6	1563	1885	1580	521	23
CH4	1.0	1557	1864	1810	594	24
CH5	1.3	1565	1823	1767	442	21
CH6	1.6	1561	1917	1845	5818	76

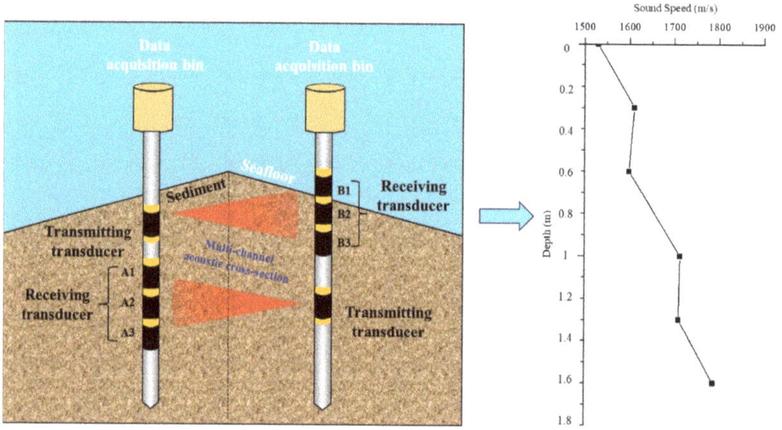

Figure 12. Illustration of seabed sound velocity measurements and vertical spatial distribution of sound velocity.

Temporal Variation in Seabed Sound Velocity at Different Depths

Seabed sound velocity at different depths exhibited varying temporal changes (Figure 13).

In seawater above the seabed: The sound velocity fluctuated slightly over time, with a variation range from 1516 to 1571 m/s. The average sound velocity was 1540 m/s, with a variance of 74 and a standard deviation of 9.

At 0.3 m below the seabed: The sediment sound velocity decreased slightly over time by about 50 m/s. The variation range was from 1558 to 1774 m/s, with an average of 1569 m/s, a variance of 321, and a standard deviation of 18.

At 0.6 m below the seabed: The sediment sound velocity fluctuated slightly, with a variation range from 1563 to 1885 m/s. The average sound velocity was 1580 m/s, with a variance of 521 and a standard deviation of 23.

At 1.0 m below the seabed: The sediment sound velocity showed greater fluctuations in the initial period, becoming more stable in the later period. The variation range was from 1557 to 1864 m/s, with an average of 1810 m/s, a variance of 594, and a standard deviation of 24.

At 1.3 m below the seabed: The sediment sound velocity remained relatively stable in the initial period, but fluctuated more in the later period, with an increase of approximately 90 m/s, representing a 5% increase. The variation range was from 1565 to 1823 m/s, with an average of 1767 m/s, a variance of 442, and a standard deviation of 21.

At 1.6 m below the seabed: The sediment sound velocity was stable in the early period, and then increased by about 100 m/s in the later period, followed by fluctuations. The variation range was from 1561 to 1917 m/s, with an average of 1845 m/s, a variance of 5818, and a standard deviation of 76.

On 12 November, at noon, changes in sediment sound velocity were observed in all but the 0.6 m depth channel. Specifically, the following were observed:

At 0.3 m below the seabed, the sediment sound velocity decreased.

At the seabed, and at depths of 1, 1.3, and 1.6 m below the seabed, the sound velocity increased to varying degrees.

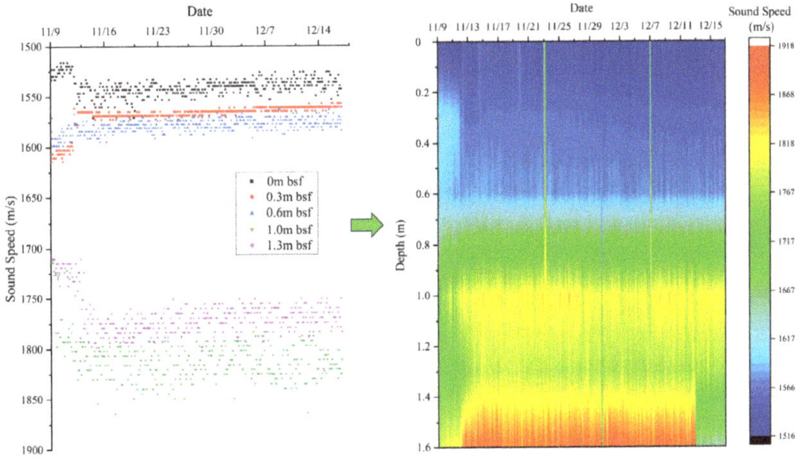

Figure 13. Spatial and temporal distribution of submarine sound velocity with cloud and scatter graphs.

4.4. Seabed Pore Pressure Observation Results

Sediment pore pressure sensors were installed at depths of 0.6 m and 1.6 m below the seabed to measure the excess pore water pressure (Figure 14).

At a depth of 0.6 m below the seabed, the excess pore pressure in the sediment remained relatively stable, with minimal variation.

At a depth of 1.6 m below the seabed, the excess pore pressure in the sediment initially decreased and then increased. The changes can be divided into four distinct stages:

Stage 1 (15–21 November): The excess pore pressure slightly increased, from 5.32 kPa to 5.68 kPa, an increase of 7%.

Stage 2 (21–25 November): The excess pore pressure rapidly decreased, dropping from 5.68 kPa to 2.48 kPa, a decrease of 56%.

Stage 3 (25 November–6 December): The excess pore pressure gradually decreased, from 2.48 kPa to 1.64 kPa, a decrease of 34%.

Stage 4 (6–18 December): The excess pore pressure gradually increased again, rising from 1.64 kPa to 2.79 kPa, an increase of 70%.

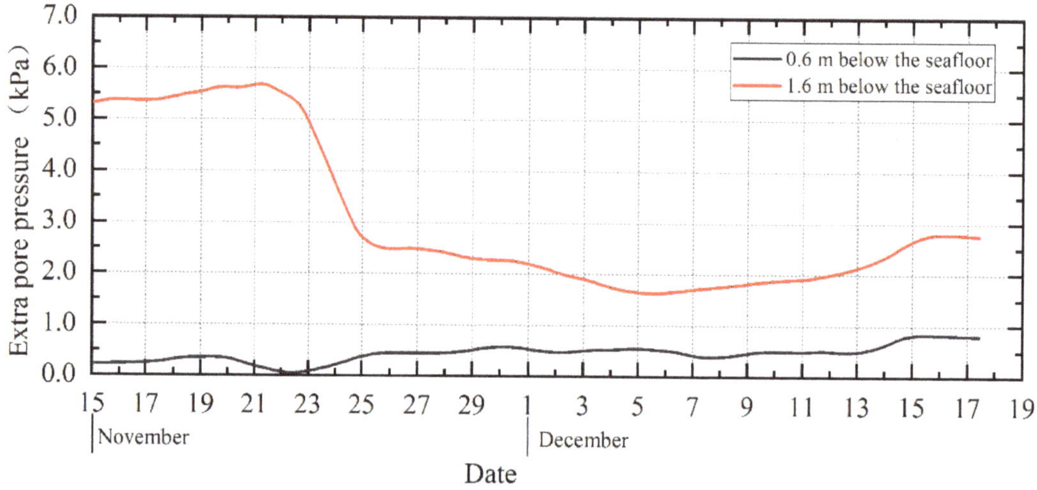

Figure 14. Variation curve of excess pore water pressure with time.

5. Analysis of Dynamic Changes in Seabed Sediments off the Coast of Qingdao

5.1. Inversion Model of Acoustic and Electrical Parameters and Physical and Mechanical Properties of Sediments

We measured many submarine sediment samples. With data from the literature and the Keras module in Python 3, we constructed a single-input (resistivity) and multi-output (density, water content, and porosity) model with a six-layer dense neural network. The training was carried out 180 times with the best prediction robustness, and five-fold model validation was adopted. Compared with the traditional empirical model, the prediction accuracy of this model improved by 33%, and the accuracy of the prediction range increased by 50%. The correlations between resistivity and the physical properties of the seabed manifested as follows: the resistivity gradually increases with the increase in density; the resistivity gradually decreases with the increase in porosity; and the resistivity gradually decreases with the increase in water content (Figure 15).

The sound velocity of the sediment was the most correlated with the compression coefficient, followed by the internal friction angle and cohesion (Figure 16). The sound velocity of sediment and the compression coefficient showed an exponential functional relationship. The sound velocity of the sediment is a parabolic function with the cohesion and internal friction angle.

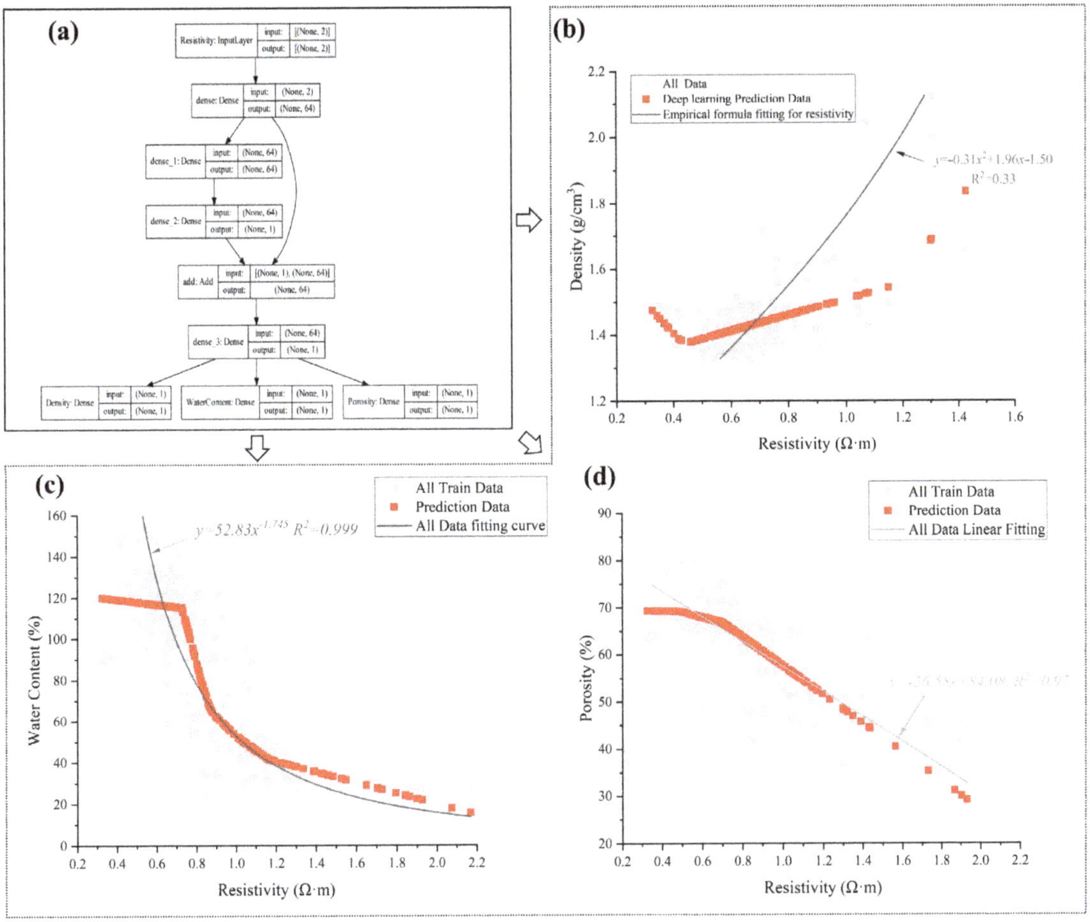

Figure 15. Construction algorithm and prediction model of seafloor sediment resistivity and density, water content and porosity ((**a**) is the framework of the deep learning neural network model based on the resistivity inversion of the seafloor sediment density, water content, and porosity; (**b**–**d**) show the relationship between resistivity and density/water content/porosity based on an empirical equation and deep learning model (modified from [29])).

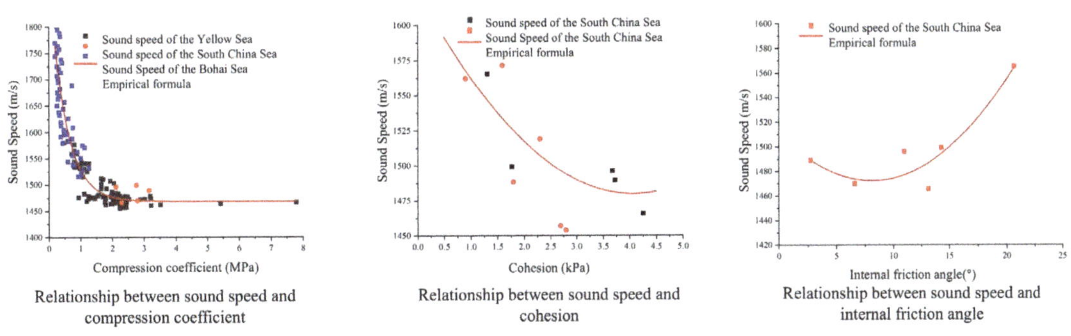

Figure 16. The relationship between the sound speed, compression coefficient, cohesion, and internal friction angle (Data source: Yellow Sea [30], South China Sea [25,29] Bohai Sea [31]).

5.2. Dynamic Variation in Seabed Physical Properties in the Qingdao Offshore Area

Based on the relationship between seabed sediment resistivity and physical properties previously established by the author [29], the dynamic variations in sediment density, water content, and porosity in the Qingdao offshore seabed have been inferred from resistivity measurements (Figure 17).

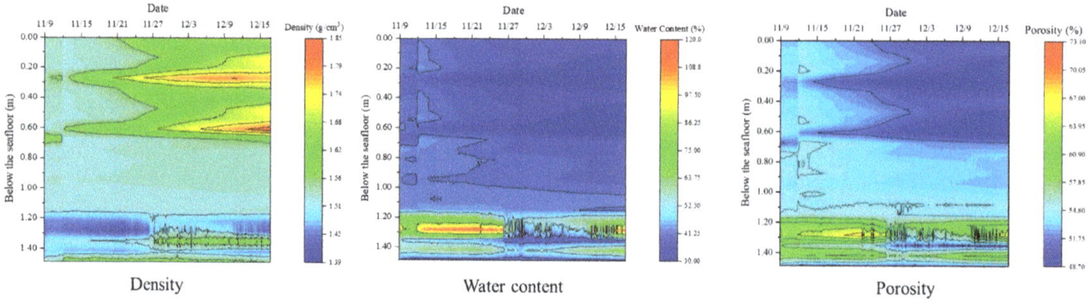

Figure 17. Variation in sediment density, water content, and porosity of Qingdao offshore sediment with time based on deep learning algorithm.

Sediment Density: The variation in seabed sediment density ranges from 1.39 to 1.85 g/cm^3.

In the depth range of 0 to 0.7 m below the seabed, the density gradually increases over time. Significant increases in sediment density are observed at depths of 0.3 m and 0.6 m.

In the range of 0.7 to 1.15 m, the density shows minimal variation over time.

Between 1.15 and 1.51 m, the density slightly increases over time, but the values remain relatively low.

Water Content: The variation in water content ranges from 30% to 120%.

In the depth range of 0 to 1.2 m below the seabed, the water content decreases gradually over time.

Between 1.2 and 1.5 m, a significant peak in water content is observed, with a trend of first decreasing and then increasing as time progresses. This notable increase in water content suggests the presence of fluid migration pathways, such as water fractures, in this region.

Porosity: The variation in porosity ranges from 48% to 73%.

In the depth range of 0 to 0.7 m below the seabed, porosity gradually decreases over time.

Between 0.7 and 1.2 m, porosity slightly increases over time, though the change is smaller compared to the variation observed in the upper 0 to 0.7 m.

In the range of 1.2 to 1.5 m, porosity initially increases, then decreases, and increases again over time. The changes in porosity may be influenced by tidal variations in seawater.

5.3. Dynamic Changes in the Mechanical Properties of Subsea Sediments in the Qingdao Offshore Area

Based on the experimental data on the relationship between the velocity of sound in seabed sediments and their mechanical properties, we established empirical formulas that link sound velocity to parameters such as compressibility, shear strength, cohesion, and internal friction angle. With the results from the sound velocity measurements in the Qingdao offshore area, the spatiotemporal variation patterns of the mechanical properties of the seabed sediments were derived (Figure 18).

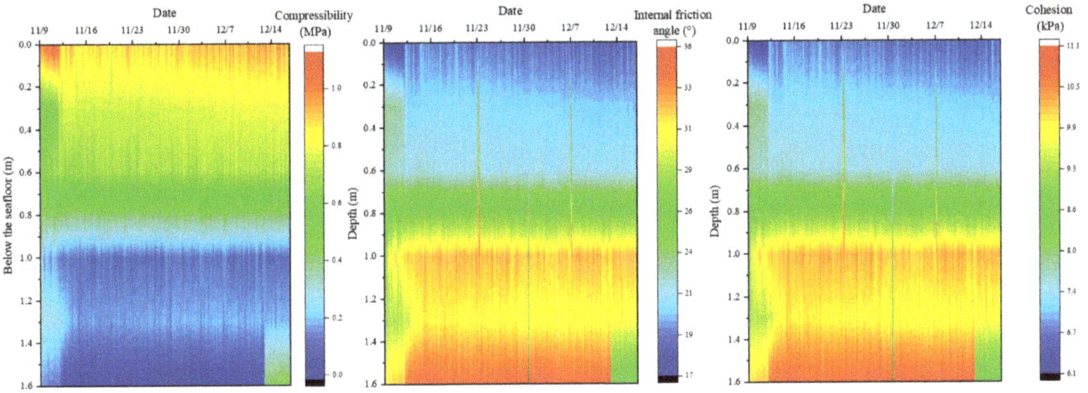

Figure 18. Time variation in compressive coefficient of seafloor sediment internal friction angle cohesion.

The range of compressibility in seabed sediments varies between 0.01 and 0.8 MPa. Vertically, the compressibility decreases with depth. Over time, the compressibility at different depths exhibits different trends. At a depth of 0.2 m below the seabed, the compressibility first increases rapidly, and then increases slowly. At 0.5 m in depth, the compressibility increases gradually over time. At depths of 0.9 m and 1.2 m, the compressibility increases rapidly at first, then stabilizes. At 1.5 m depth, the compressibility shows a trend of decrease, stabilization, increase, and stabilization. These dynamic changes in sediment compressibility reflect the evolving compressibility characteristics of the seabed sediments.

The range of cohesion in seabed sediments is between 7.05 and 11.13 kPa. Vertically, cohesion increases with depth, although it experiences an abnormal decrease at a depth of 1.2 m. Over time, at a depth of 0.2 m below the seabed, the cohesion first decreases sharply, then decreases slowly. At 0.5 m depth, cohesion decreases slowly over time. At depths of 0.9 m and 1.2 m, the cohesion initially increases rapidly, then stabilizes. At 1.5 m depth, the cohesion follows an increasing–stable–decreasing trend over time.

The internal friction angle of the seabed sediments varies between 20° and 36°. Vertically, the internal friction angle increases with depth but experiences an abnormal decrease at a depth of 1.2 m. Over time, at 0.2 m below the seabed, the internal friction angle decreases rapidly at first, and then decreases slowly. At 0.5 m in depth, the internal friction angle gradually decreases. At depths of 0.9 m and 1.2 m, the internal friction angle increases rapidly at first, and then remains constant. At 1.5 m in depth, the internal friction angle follows a pattern of increase, stabilization, decrease, and stabilization.

5.4. Analysis of the Dynamic Variation Process of Seabed Sediments in the Qingdao Offshore Area

During the observation period in the Qingdao offshore area, two significant wave events with wave heights greater than 1 m occurred. The first event took place from 11 November to 13 November, lasting 2 days, with a maximum wave height of 1 m. The second event occurred from 16 November to 20 November, lasting 4 days, with a maximum wave height of 2.2 m (Figure 19).

During the first wave event, from 11 November to 13 November, the maximum wave height was 1 m. During this period, the resistivity in the water slightly increased, as did the resistivity in the seabed boundary layer. The water resistivity showed a good linear relationship with the suspended sediment concentration [32], indicating that this wave event caused a significant increase in the concentration of suspended sediments. At a depth of 0.6 m below the seabed, resistivity decreased significantly, suggesting a reduction in sediment density [33] and an increase in porosity [34]. In this layer, sediment sound velocity decreased markedly, approaching that of seawater, and the shear strength of the sediments also significantly dropped [35].

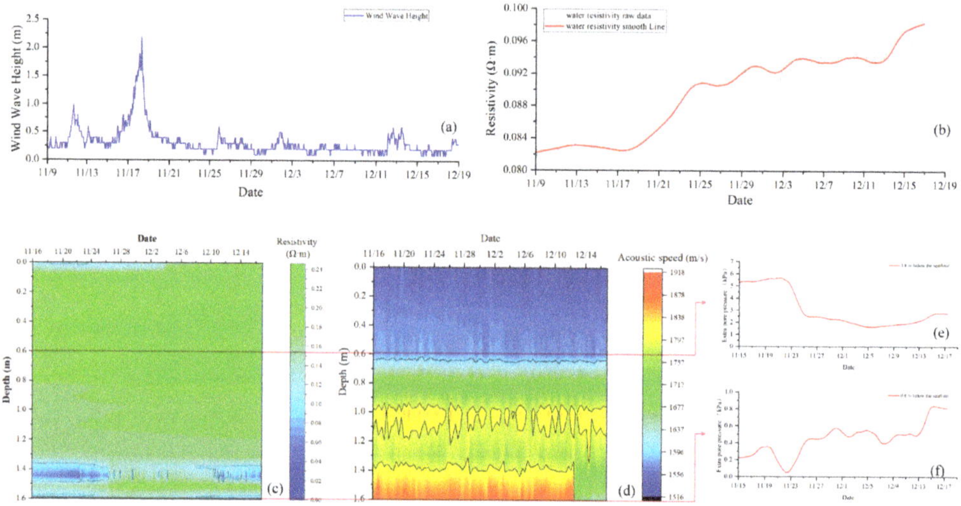

Figure 19. Dynamic change process of Qingdao offshore engineering geological environment: (**a**) high waves, data source: Wheat Island Observatory; (**b**) water resistivity; (**c**) sediment resistivity cloud; (**d**) sediment sound velocity cloud; (**e**) 0.6 m pore water pressure below the seabed; (**f**) 1.6 m pore water pressure below the seabed (the red arrows indicate the same depth). First Wave Event (11–13 November).

In the vertical structure, the resistivity of the overlying layer increased notably, indicating that this layer acted as an impermeable barrier [36], characterized by low porosity and high density [37]. Under wave pressure, pore water was mobilized upwards [38]. Due to the presence of this impermeable barrier, pore water accumulated in the layer, causing a sharp decrease in sediment resistivity [24]. At the same time, since the sound velocity in seawater was lower than that in sediment, the sound velocity in the sediment also decreased significantly.

At depths between 1.0 m and 1.8 m below the seabed (with pore pressure represented at 1.6 m depth), the resistivity increased significantly. The sediment porosity decreased, and the density increased [39]. The sediment sound velocity increased by 5%, compressibility decreased by 39%, cohesion increased by 7%, and the internal friction angle increased by 11% (Figure 20), resulting in a significant increase in shear strength [40]. In the vertical structure, the resistivity in this layer was low, approaching that of seawater, indicating that pore water was likely enriched in this layer. Under wave action, the excess pore water pressure decreased by 50%, releasing pore water from the sediment, which led to an increase in resistivity, a decrease in porosity, and an increase in density [41].

The second wave event occurred from 16 November to 20 November, with a maximum wave height of 2.2 m, peaking on 18 November. As the wave height gradually increased, water resistivity slightly decreased. This was likely due to the dilution of offshore seawater with low suspended particulate concentrations, which led to a decrease in water resistivity near the coast [42]. Concurrently, sediment deposition increased due to gravity, causing a slight increase in the resistivity of the seabed boundary layer.

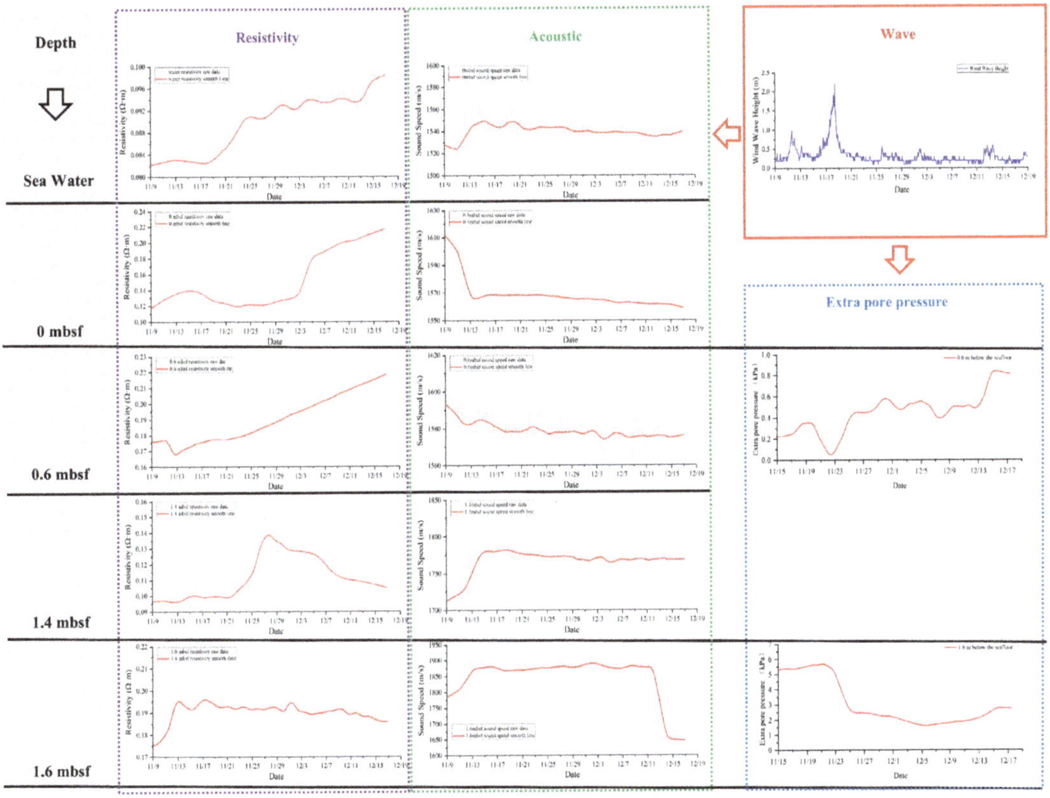

Figure 20. Dynamic changes in sediment resistivity, sound velocity, and excess pore water pressure at different depths in shallow seas. Second Wave Event (16–20 November).

Once the wave height began to decrease, significant changes in the sediment's engineering geological properties were observed [43]. From 18 November to 25 November, the water resistivity increased rapidly by 10%, indicating an increase in the concentration of suspended sediments. This increase was likely due to the resuspension of sediments from the seabed surface. Rijn [44] noted that under marine hydrodynamic forces, sediment particles undergo resuspension when the upward turbulent diffusion force balances the downward gravitational force.

From 16 November to 22 November, the resistivity of the seabed boundary layer continued to decrease, which confirmed the erosion of the seabed by the waves. From 22 November to 27 November, at a depth of 1.4 m below the seabed, sediment resistivity increased sharply. There are two possible reasons for this sharp increase.

In the vertical structure, the resistivity of this sediment layer was initially low, close to seawater resistivity, suggesting that this layer may have been enriched with pore water [29]. Under wave action, the pore water in this layer was likely released, leading to an increase in resistivity, a decrease in porosity, and an increase in density [45].

The wave action could have induced the vertical migration of pore water within the seabed sediments, which led to the upward migration of fine-grained sediments [46]. Due to the presence of an impermeable barrier in this layer, fine sediments accumulated, resulting in reduced porosity and increased compaction, which caused a rise in sediment resistivity. This finding closely matches the laboratory results of Jinguuji [33], who pointed out that permeability variations across different layers affect the dissipation of excess pore water pressure. Low-permeability layers may act as barriers, preventing the vertical flow of pore water. In the presence of a fine-grained low-permeability layer within the seabed sediments, such layers can trap water from the underlying coarser sand, leading to lower resistivity and density.

5.5. Evolution Model of Pore Water Accumulation and Release in Seabed Sediments Under Wave Loading

As early as 1985, Clukey et al. [47] proposed that under wave cyclic loading, pore water pressure within sediments is prone to accumulation and dissipation, exhibiting unique dynamic response characteristics. Studies have shown that after storm events, liquefaction occurs in the surface layer of the seabed, forming a muddy layer up to one-quarter of the thickness of the liquefied layer [48]. Under extreme conditions, wave loading can induce liquefaction in shallow sediments within a depth range of 4.2 m beneath the seabed [49]. Zhang and Suhayda [50] found that under large wave conditions, the shear strength of the shallow seabed sediments can decrease by up to 77.7% within 24 h. Furthermore, in such conditions, the liquefaction of the shallow seabed sediments can lead to a significant increase in the concentration of suspended sediments in the underlying water [51]. The resuspension process of seabed sediments is largely dependent on the dynamic response to wave loading.

In 2016, Zhang et al. [52] pointed out that under wave cyclic loading, the strength of the seabed sediments rapidly decreases or even dissipates, leading to a sharp decline in shear strength and critical shear stress. Wen [20] studied the dynamic changes in the seabed boundary layer of the Yellow River Delta, proposing four evolutionary models: calm sea conditions, seabed consolidation with uniform vertical sediment concentration; small waves, acceleration of seabed consolidation with the boundary layer extending upward; medium waves, accumulation of excess pore water pressure in the seabed, and further upward extension of the boundary layer; and large waves, liquefaction of the seabed, with the boundary layer extending inward and increasing suspended sediment concentration in the water. This study also observed an increase in the background concentration of suspended sediments in the seabed boundary layer after a large wave event with a wave height of 2 m.

Building upon the above studies and the dynamic changes observed in the Qingdao offshore seabed, this study summarizes the dynamic evolution model of seabed sediment properties under wave loading in the Qingdao offshore area.

In the initial stage, under wave action, residual excess pore water pressure in the seabed gradually increases, causing pore water to accumulate beneath the impermeable layer. When the wave height decreases, the excess pore water pressure in the seabed becomes greater than that in the overlying sediments. At this point, pore water gradually moves upward, carrying fine particles from the sediments into the water, which increases the concentration of suspended sediments in the near-bottom seawater. As pore water is released from the sediments, excess pore water pressure decreases, and the volume of pore water gradually diminishes. When the excess pore water pressure beneath the seabed becomes smaller than that in the overlying sediments, pore water stops migrating upward, and fine sediment particles no longer move into the water, leading to a decrease in suspended sediment concentration in the near-bottom seawater. As wave action continues, residual excess pore water pressure in the seabed gradually increases again, returning to the initial stage, and the cycle repeats (Figure 21).

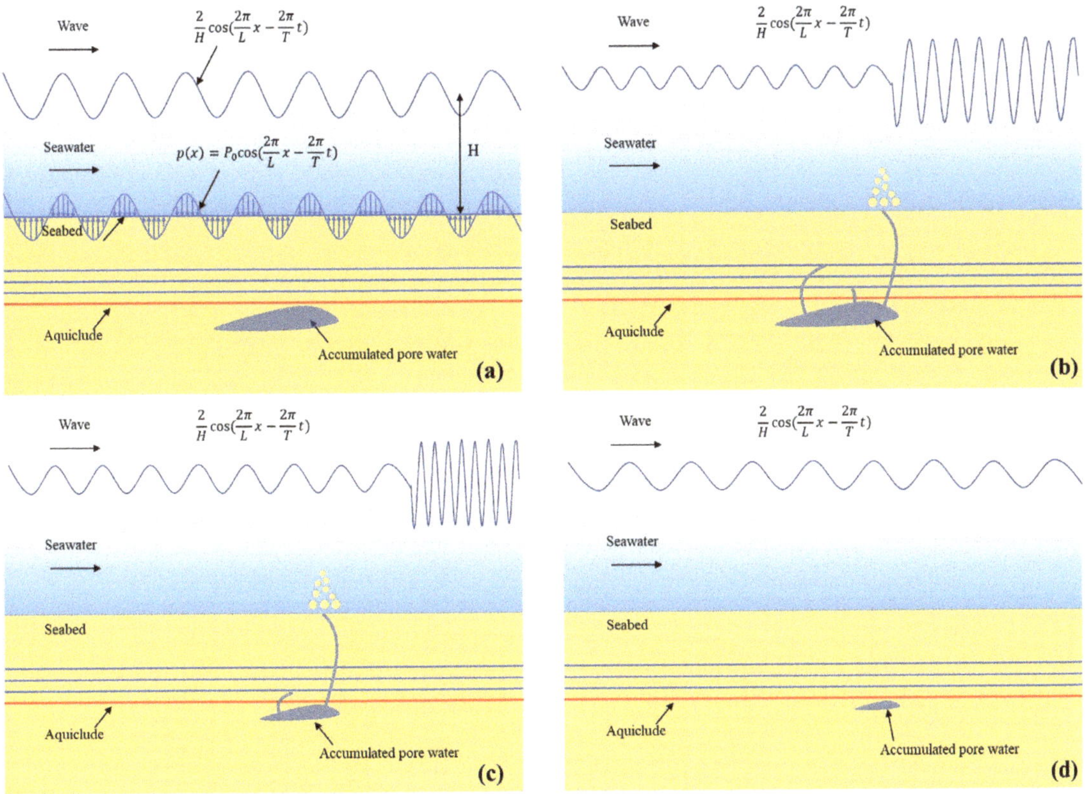

Figure 21. Dynamic variation patterns of pore water in sediments under different wave loads ((**a**) is the peak period of high waves when pore water accumulates [53]; (**b**) is the early stage of the recession period of high waves when pore water is released rapidly; (**c**) is the later stage of the recession period of high waves when pore water is released slowly; (**d**) is the calm period of waves when pore water accumulates partially).

Based on the observational data of acoustics, electrical resistivity, and pore pressure over one month, this study establishes a physical property model for electrical resistivity inversion and a mechanical property model for acoustic inversion. It reveals the dynamic changes in the engineering geological properties of offshore seabed sediments in the Qingdao offshore area under the action of wave loads and discovers the accumulation and release of seabed pore water caused by waves. The innovations in this paper are mainly reflected in two aspects. First, this study obtained in situ long-term observational data of the Qingdao offshore seabed for as long as one month, including acoustic data, electrical resistivity data, and pore water pressure data. Second, for the first time, it comprehensively analyzed the influence mechanism of wave loads on the cumulative release of seabed pore water and the changing process of the physical and mechanical properties of sediments by combining acoustics, electrical resistivity, and pore water pressure.

The limitations of this paper may be reflected in two aspects. First, since there were no wave and tide gauge in the observation equipment and the data from the nearby tide gauge station were adopted, there was a slight delay between the wave data and the real-time observation data. In the future, we will deploy integrated wave gauges for direct pressure gradient computation. Second, due to the multiplicity of solutions in electrical resistivity and acoustic inversions, the influencing factors are not unique. However, it is precisely because acoustics and electrical resistivity can be continuously measured that it becomes possible to continuously monitor the dynamic changes in the physical and mechanical properties of seabed sediments, breaking through the singleness of traditional destructive measurement methods such as sampling measurements and cone penetration testing.

Based on the in situ monitoring of electrical resistivity, acoustics, and pore pressure, combined with the physical property model of sediment electrical resistivity inversion and the mechanical property model of sediment acoustic inversion, the continuous long-term monitoring of the physical and mechanical properties of seabed sediments can be achieved. Meanwhile, the physical properties, mechanical properties, and pore pressure of sediments can be used as inputs for the Biot consolidation model, providing the possibility for further rigorous theoretical analysis. This research has taken a step forward from previous studies on continuous single pore pressure measurements, realizing the dynamic long-term monitoring of the physical and mechanical properties of seabed sediments. This research provides new methods and new perspectives for comprehensively studying the dynamic changing process of the engineering geological properties of seabed sediments.

6. Conclusions

This study, based on an independently developed in situ monitoring system for subsea engineering geological environments, revealed the dynamic response mechanisms of sediment physical–mechanical properties under wave loading through a 36-day continuous observation in the offshore area of Qingdao. Key findings include the following: (1) Wave events with heights $\geq$1 m (maximum 2.2 m) triggered the cyclic accumulation and release of pore water pressure at 1.6 mbsf, leading to fluctuations in shear strength (cohesion increased 7%, internal friction angle increased 11%, and resistivity anomalies). (2) A deep learning model integrating multi-source data (resistivity, sound velocity, pore pressure) demonstrated significant spatiotemporal correlations between sediment density (1.39–1.85 g/cm^3), porosity (48–73%), and a wave energy input within 0–1.5 mbsf. (3) An evolutionary model of "wave-induced pore water migration-rebalance" was proposed, elucidating the coupling mechanism of fine-particle resuspension (seawater resistivity increased 10%) and low-permeability layer blockage effects, which establishes a dynamic theoretical framework for assessing the long-term stability of subsea pipelines and foundations in monsoon-dominated coastal regions.

Author Contributions: Conceptualization, Z.S. (Zhiwen Sun); Methodology, Y.L.; Software, Z.F.; Validation, K.L.; Formal analysis, Z.S. (Zhongqiang Sun); Investigation, X.S.; Resources, L.X.; Project administration, N.W.; Funding acquisition, Y.J. All authors have read and agreed to the published version of the manuscript.

Funding: This research was supported by the China Postdoctoral Science Foundation, 2024M761565; Shandong Postdoctoral Innovative Talent Support Program, SDBX2023071; Open foundation project of the Key Laboratory of Polar Geology and Marine Mineral Resources (China University of Geosciences, Beijing), Ministry of Education, PGMR-2023-303; Taishan Scholar Programs; and the Laoshan Laboratory, No. LSKJ202203506.

Data Availability Statement: The raw data supporting the conclusions of this article will be made available by the authors on request.

Conflicts of Interest: The authors declare no conflict of interest.

References

1. Shanmugam, G. Modern Internal Waves and Internal Tides along Oceanic Pycnoclines: Challenges and Implications for Ancient Deep-Marine Baroclinic Sands. *AAPG Bull.* **2013**, *97*, 799–843. [CrossRef]
2. Wang, L.; Hu, Z.; Zhang, H.; Li, L.; Yan, L. Simulation Experimental Study on the Acoustic Characteristics of Shallow Gas Hydrate Formations in the Polar Cold Sea. *China Offshore Oil Gas* **2022**, *34*, 218–224.
3. Fan, Z.; Jia, Y.; Chu, F.; Zhu, X.; Zhu, N.; Li, B.; Quan, Y. Effects of Migration and Diffusion of Suspended Sediments on the Seabed Environment during Exploitation of Deep-Sea Polymetallic Nodules. *Water* **2022**, *14*, 2073. [CrossRef]
4. Jamil, M.; Siddiqui, N.A.; Umar, M.; Usman, M.; Ahmed, N.; Rahman, A.H.A.; Zaidi, F.K. Aseismic and Seismic Impact on Development of Soft-Sediment Deformation Structures in Deep-Marine Sand-Shaly Crocker Fan in Sabah, NW Borneo. *J. King Saud Univ. - Sci.* **2021**, *33*, 101522. [CrossRef]
5. Jia, Y.; Chen, T.; Li, P.; Li, Z.; Hu, C.; Liu, X.; Shan, H. Research Progress on In-Situ Monitoring Technologies of Marine Geological Disasters. *J. Geol. Hazards Prev. China* **2022**, *33*, 1–14. [CrossRef]
6. Busson, J.; Teles, V.; Mulder, T.; Joseph, P.; Guy, N.; Bouziat, A.; Danquigny, C.; Poli, E.; Borgomano, J. Submarine Landslides on a Carbonate Platform Slope: Forward Numerical Modelling of Mechanical Stratigraphy and Scenarios of Failure Precondition. *Landslides* **2021**, *18*, 595–618. [CrossRef]
7. Zhao, Y.; Kong, L.; Liu, L.; Hu, G.; Ji, Y.; Bu, Q.; Bai, C.; Zhao, J.; Li, J.; Liu, J.; et al. Mechanical Behaviors of Natural Gas Hydrate-Bearing Clayey-Silty Sediments: Experiments and Constitutive Modeling. *Ocean Eng.* **2024**, *294*, 116791. [CrossRef]
8. Bennett, R.H. Pore-water Pressure Measurements: Mississippi Delta Submarine Sediments. *Mar. Geotechnol.* **1977**, *2*, 177–189. [CrossRef]
9. Richards, A.F.; Øten, K.; Keller, G.H.; Lai, J.Y. Differential Piezometer Probe for an in Situ Measurement of Sea-Floor. *Géotechnique* **1975**, *25*, 229–238. [CrossRef]
10. Bennett, R.H.; Li, H.; Valent, P.J.; Lipkin, J.; Esrig, M.I. In-Situ Undrained Shear Strengths and Permeabilities Derived from Piezometer Measurements. *Strength Test. Mar. Sediments Lab.-Situ Meas.* **1985**. [CrossRef]
11. Prior, D.B.; Suhayda, J.N.; Lu, N.-Z.; Bornhold, B.D.; Keller, G.H.; Wiseman, W.J.; Wright, L.D.; Yang, Z.-S. Storm Wave Reactivation of a Submarine Landslide. *Nature* **1989**, *341*, 47–50. [CrossRef]
12. Fan, Z.; Zhu, X.; Xu, H.; Sun, Z.; Zhang, H.; Bi, X.; Hu, C.; Lu, D.; Sun, Z.; Li, K.; et al. A New Method for Long-Term in Situ Monitoring of Seabed Interface Evolution: A Self-Potential Probe. *Ocean Eng.* **2023**, *280*, 114917. [CrossRef]
13. Liu, T. Evaluation Model of Seabed Stability Under Wave Action and Its Engineering Applications. Master's Thesis, Ocean University of China, Qingdao, China, 2005.
14. Sun, Z.; Jia, Y.; Quan, Y.; Guo, X.; Liu, T.; Meng, Q.; Sun, Z.; Li, K.; Fan, Z.; Chen, T.; et al. Research and Development and Application of an In-Situ Long-Term Monitoring System for Complex Deep-Sea Engineering Geology. *Earth Sci. Front.* **2022**, *29*, 216–228. [CrossRef]
15. Sultan, N.; Cattaneo, A.; Sibuet, J.-C.; Schneider, J.-L.; Aftershocks, S. Deep Sea in Situ Excess Pore Pressure and Sediment Deformation off NW Sumatra and Its Relation with the December 26, 2004 Great Sumatra-Andaman Earthquake. *Int. J. Earth Sci.* **2009**, *98*, 823–837. [CrossRef]
16. Yang, Z.; Zhu, Y.; Liu, T.; Sun, Z.; Ling, X.; Yang, J. Pumping Effect of Wave-Induced Pore Pressure on the Development of Fluid Mud Layer. *Ocean Eng.* **2019**, *189*, 106391. [CrossRef]
17. Ma, X.; Fan, F.; Yan, J. In-situ Monitoring Platform for Marine Sedimentary Dynamic Processes and Its Applications. *Mar. Geol. Quat. Geol.* **2011**, *31*, 179–185. [CrossRef]
18. Zhang, S. The Influence of Wave-Induced Liquefaction of Silt Seabed on the Erosion and Resuspension of Sediments. Ph.D. Thesis, Ocean University of China, Qingdao, China, 2017.
19. Feng, C.; Liu, H.; Liu, J.; Jia, Y.; Hou, F.; Xue, L.; Quan, Y. A Deep Learning Prediction Method for the Contribution Rate of Wave-Induced Transient Liquefaction of Seabed to Resuspension. *J. Eng. Geol.* **2021**, *29*, 1788–1795. [CrossRef]
20. Wen, M. Study on the Dynamic Change Process of the Seabed Boundary Layer under Wave Action - Taking the Yellow River Subaqueous Delta as an Example. Ph.D. Thesis, Ocean University of China, Qingdao, China, 2019.
21. Wei, X.; Ding, Z.; Wu, J.; Liu, B. Research on the High Precision Resistivity Probe with Four Point-Electrodes for Marine Sediments. *J. Electron. Meas. Instrum.* **2013**, *27*, 810–816. [CrossRef]
22. Kermabon, A.; Gehin, C.; Blavier, P. A Deep-Sea Electrical Resistivity Probe for Measuring Porosity and Density of Unconsolidated Sediments. *Geophysics* **1969**, *34*, 554–571. [CrossRef]
23. Bellmunt, F.; Gabàs, A.; Macau, A.; Benjumea, B.; Vilà, M.; Figueras, S. Sediment Characterization in Deltas Using Electrical Resistivity Tomography: The Ebro Delta Case. *J. Appl. Geophys.* **2022**, *196*, 104520. [CrossRef]
24. Wu, J.; Guo, X.; Sun, X.; Sun, H. Flume Experiment Evaluation of Resistivity Probes as a New Tool for Monitoring Gas Migration in Multilayered Sediments. *Appl. Ocean Res.* **2020**, *105*, 102415. [CrossRef]
25. Chen, C. Experimental Study on the Relations Between Acoustic Velocity with Physical and Mechanical Properties of Seabed Sediments Based on the Triaxial Test. Master's Thesis, Guangdong University of Technology, Guangzhou, China, 2015.

26. Zou, D.; Wu, B.; Lu, B.; Zhang, W.; Chen, F. Studies on Clustering Analysis of Acoustic and Physical-Mechanical Properties of Seabed Sediments. *J. Trop. Oceanogr.* **2008**, *27*, 12–17.
27. Yang, G. Research on the Combination of Multiple Probes for Measuring the Mechanical Properties of Subsurface Sediments. Master's Thesis, Ocean University of China, Qingdao, China, 2023.
28. Sun, Z.; Jia, Y.; Shan, H.; Fan, Z.; Song, X.; Xue, L.; Li, K. Monitoring and Early Warning Technology of Hydrate-Induced Submarine Disasters. *IOP Conf. Ser. Earth Environ. Sci.* **2020**, *570*, 062030. [CrossRef]
29. Sun, Z.; Fan, Z.; Zhu, C.; Li, K.; Sun, Z.; Song, X.; Xue, L.; Liu, H.; Jia, Y. Study on the Relationship between Resistivity and the Physical Properties of Seafloor Sediments Based on the Deep Neural Learning Algorithm. *J. Mar. Sci. Eng.* **2023**, *11*, 937. [CrossRef]
30. Cheng, J. Study on Acoustic Physical Characteristics of Sediments in the Middle of the South Yellow Sea. Master's Thesis, First Institute of Oceanography, State Oceanic Administration, Qingdao, China, 2011.
31. Wang, F.; Tao, C.; Lin, X.; Dong, L. Statistical Characteristics of Submarine Sound Velocity in Jinzhou Bay Based on High-Resolution Shallow Profile and Borehole Information Comparison. *J. Oceanogr.* **2020**, *42*, 112–122.
32. Dai, X.; Shan, H.; Cui, W.; Jia, Y. Experimental Study on the Relationship between Suspended Sediment Content and Conductivity and Its Influencing Factors. *Acta Oceanol. Sin. Chin. Ed.* **2011**, *33*, 88–94.
33. Jinguuji, M. Visualization Technique for Liquefaction Process in Chamber Experiments by Using Electrical Resistivity Monitoring. *Soil Dyn. Earthq. Eng.* **2007**, *27*, 191–199. [CrossRef]
34. Archie, G.E. The Electrical Resistivity Log as an Aid in Determining Some Reservoir Characteristics. *Trans. AIME* **1942**, *146*, 54–62. [CrossRef]
35. Buckingham, M.J. Wave Speed and Attenuation Profiles in a Stratified Marine Sediment: Geo-Acoustic Modeling of Seabed Layering Using the Viscous Grain Shearing Theory. *J. Acoust. Soc. Am.* **2020**, *148*, 962–974. [CrossRef]
36. Lévesque, Y.; Walter, J.; Chesnaux, R.; Dugas, S.; Noel, D. Electrical Resistivity of Saturated and Unsaturated Sediments in Northeastern Canada. *Environ. Earth Sci.* **2023**, *82*, 303. [CrossRef]
37. Tian, H.; Liu, L.; Zhu, L.; Ge, X.; Ding, P.; Cai, J. Relationship between Normalized Permeability and Resistivity Index in Hydrate-Bearing Sediments: Fractal Model and Numerical Simulation. *Geophys. J. Int.* **2023**, *234*, 684–698. [CrossRef]
38. Tian, Z.; Jia, Y.; Zhu, J.; Chen, T.; Wang, H.; Ji, C.; Liu, C.; Lu, L.; He, M. Microseismic Observations Reveal That Internal Waves Intensify the Release of Methane from the Seabed. *Sci. China Earth Sci.* **2024**, *54*, 3237–3254. [CrossRef]
39. Sun, Z.; Miao, Y.; Fan, Z.; Song, X.; Xue, L.; Zhu, N.; Zhu, X.; Li, X.; Qiao, Y.; Sun, Z.; et al. Relationship between Resistivity and Physical Properties of Seafloor Sediments in the Hydrate Production Test Area of the South China Sea. *Mar. Georesources Geotechnol.* **2024**, *42*, 1131–1145. [CrossRef]
40. Zhang, Z. Study on the Correlation Between the Mechanical Properties and Shear Wave Characteristics of Shallow Sediments in the Pilot Production Area of Natural Gas Hydrates in the South China Sea. Master's Thesis, Ocean University of China, Qingdao, China, 2020.
41. Chen, T.; Jia, Y.; Liu, T.; Liu, X.; Shan, H.; Sun, Z. Review and Prospect of In-Situ Long-Term Observation Technologies for Pore Pressure of Seabed Sediments. *Earth Sci. Front.* **2022**, *29*, 229–245. [CrossRef]
42. Zhang, S.; Zhao, Z.; Li, G.; Wu, J.; Wang, Y.-G.; Nielsen, P.; Jeng, D.-S.; Qiao, L.; Wang, C.; Li, S. Estimation of Sediment Transport Parameters From Measured Suspended Concentration Time Series Under Waves and Currents With a New Conceptual Model. *Water Resour. Res.* **2024**, *60*, e2023WR034933. [CrossRef]
43. Tian, Z.; Xiang, J.; Huang, J.; Zhang, S. Suspension and Transportation of Sediments in Submarine Canyon Induced by Internal Solitary Waves. *Phys. Fluids* **2024**, *36*, 022112. [CrossRef]
44. van Rijn, L.C. *Principles of Sediment Transport in Rivers, Estuaries, and Coastal Seas*; Aqua Publications: Amsterdam, The Netherlands, 1993; ISBN 978-90-800356-2-1.
45. Xu, J.; Dong, J.; Zhang, S.; Sun, H.; Li, G.; Niu, J.; Li, A.; Dong, P. Pore-Water Pressure Response of a Silty Seabed to Random Wave Action: Importance of Low-Frequency Waves. *Coast. Eng.* **2022**, *178*, 104214. [CrossRef]
46. Zhao, Z.; Zhang, S.; Wu, J.; Qiao, L.; Li, G.; Li, H.; Li, S. Analysis of Fine-Grained Sediment Dynamics from Field Observations with a Vector Autoregressive Model. *J. Hydrol.* **2024**, *644*, 132100. [CrossRef]
47. Clukey, E.; Kulhawy, F.; Liu, P.-F. Response of Silts to Wave Loads: Experimental Study. In *Strength Testing of Marine Sediments: Laboratory and In-Situ Measurements*; Chaney, R., Demars, K., Eds.; ASTM International: West Conshohocken, PA, USA, 1985; Volume STP883, ISBN 978-0-8031-0431-0.
48. Liu, X.; Jia, Y.; Zheng, J.; Wen, M.; Shan, H. An Experimental Investigation of Wave-Induced Sediment Responses in a Natural Silty Seabed: New Insights into Seabed Stratification. *Sedimentology* **2017**, *64*, 508–529. [CrossRef]
49. Sun, Y.; Dong, L.; Song, Y. Analysis of the Characteristics and Causes of the Disturbed Soil Layers of Silt in the Yellow River Subaqueous Delta. *Rock Soil Mech.* **2008**, 1494–1499. [CrossRef]
50. Zhang, Q.; Suhayda, J.N. Behavior of Marine Sediment under Storm Wave-loading. *Mar. Georesources Geotechnol.* **1994**, *12*, 259–269. [CrossRef]

51. Guo, L. Research on the Development and Application of an In-Situ Integrated Observation System for the Seabed Boundary Layer. Ph.D. Thesis, Ocean University of China, Qingdao, China, 2016.
52. Zhang, S.; Jia, Y.; Liu, X.; Guo, L.; Shan, H. Characteristics and Mechanisms of the Dynamic Change Process of Sediments in the Modern Yellow River Delta. *Mar. Geol. Quat. Geol.* **2016**, *36*, 33–44. [CrossRef]
53. Ishihara, K.; Towhata, I. Sand Response to Cyclic Rotation of Principal Stress Directions as Induced by Wave Loads. *Soils Found.* **1983**, *23*, 11–26. [CrossRef] [PubMed]

MDPI AG
Grosspeteranlage 5
4052 Basel
Switzerland
Tel.: +41 61 683 77 34

Journal of Marine Science and Engineering Editorial Office
E-mail: jmse@mdpi.com
www.mdpi.com/journal/jmse

www.ingramcontent.com/pod-product-compliance
Lightning Source LLC
LaVergne TN
LVHW070539071126
851858LV00011BA/229